SONOCHEMISTRY

An Emerging Green Technology

SONOCHEMISTRY

An Emerging Green Technology

Edited by

Suresh C. Ameta, PhD

Rakshit Ameta, PhD

Garima Ameta, PhD

AAP | APPLE ACADEMIC PRESS

Apple Academic Press Inc.
3333 Mistwell Crescent
Oakville, ON L6L 0A2 Canada

Apple Academic Press Inc.
9 Spinnaker Way
Waretown, NJ 08758 USA

International Standard Book Number-13: 978-1-77188-629-1 (Hardcover)
International Standard Book Number-13: 978-1-315-10274-0 (eBook)

Library and Archives Canada Cataloguing in Publication

Sonochemistry : an emerging green technology / edited by
Suresh C. Ameta, PhD, Rakshit Ameta, PhD, Garima Ameta, PhD.

Includes bibliographical references and index.

Issued in print and electronic formats.

ISBN 978-1-77188-629-1 (hardcover).--ISBN 978-1-315-10274-0 (PDF)

1. Sonochemistry. I. Ameta, Suresh C., editor II. Ameta, Rakshit, editor III. Ameta, Garima, editor

QD801.S66 2018 541'.3 C2018-900505-X C2018-900506-8

Library of Congress Cataloging-in-Publication Data

Names: Ameta, Suresh C., editor. | Ameta, Rakshit, editor. | Ameta, Garima, editor.

Title: Sonochemistry : an emerging green technology / editors, Suresh C. Ameta, PhD,
Rakshit Ameta, PhD, Garima Ameta, PhD.

Description: Toronto : Apple Academic Press, 2018. | Includes bibliographical references and index.

Identifiers: LCCN 2018002163 (print) | LCCN 2018004391 (ebook) | ISBN 9781315102740
(ebook) | ISBN 9781771886291 (hardcover : alk. paper)

Subjects: LCSH: Sonochemistry. | Chemistry, Physical and theoretical. | Ultrasonic waves--
Physiological effect. | Ultrasonic waves--Industrial applications.

Classification: LCC QD801 (ebook) | LCC QD801 .S65 2018 (print) | DDC 541/.3--dc23

LC record available at https://lccn.loc.gov/2018002163

Apple Academic Press also publishes its books in a variety of electronic formats. Some content that appears in print may not be available in electronic format. For information about Apple Academic Press products, visit our website at **www.appleacademicpress.com** and the CRC Press website at **www.crcpress.com**

CONTENTS

LIST OF CONTRIBUTORS

Chetna Ameta
Department of Chemistry, M. L. Sukhadia University, Udaipur, India
E-mail:chetna.ameta@yahoo.com

Garima Ameta
Department of Chemistry, M. L. Sukhadia University, Udaipur, India
E-mail:garima_ameta@yahoo.co.in

Rakshit Ameta
Department of Chemistry, J.R.N. Rajasthan Vidyapeeth (Deemed to be University), Udaipur, India
E-mail: rakshit_ameta@yahoo.in

Satish K. Ameta
Department of Environmental Science, Mewar University, Chittorgarh, India
E-mail: skameta2@gmail.com

Suresh C. Ameta
Department of Chemistry, PAHER University, Udaipur, India
E-mail: ameta_sc@yahoo.com

Surbhi Benjamin
Department of Chemistry, PAHER University, Udaipur, India
E-mail: surbhi.singh1@yahoo.com

Anil Kumar Chohadia
Department of Chemistry, M. P.Govt PG College, Chitttorgarh, India
E-mail:anilchohadia@yahoo.co.in

Chetna Gomber
Department of Chemistry, JIET Group of Institutions, Jodhpur, India
E-mail:chirag11gomber@gmail.com

Abhilasha Jain
Department of Chemistry, St. Xavier's College, Mumbai, India
E-mail: jainabhilasha5@gmail.com

Sangeeta Kalal
Department of Chemistry, M. L. Sukhadia University, Udaipur, India
E-mail:sangeeta.vardar@yahoo.in

Gunjan Kashyap
Department of Chemistry, M. L. Sukhadia University, Udaipur, India
E-mail:gunjankashyap0202@yahoo.in

Neelam Kunwar
Department of Chemistry, PAHER University, Udaipur, India,
E-mail: neelamkunwar13@yahoo.com

Kiran Meghwal
Department of Chemistry, M. L. Sukhadia University, Udaipur, India
E-mail:meghwal.kiran1506@gmail.com

Arpita Paliwal
Department of Chemistry, Sangam University, Bhilwara, India
E-mail:paliwalarpita7@gmail.com

Arpita Pandey
Department of Chemistry, Sangam University, Bhilwara
E-mail:pandeyarpita88@gmail.com

Arpit Kumar Pathak
Department of Chemistry, M. L. Sukhadia University, Udaipur, India
E-mail:arpitpathak2009@gmail.com

P. B. Punjabi
Department of Chemistry, M. L. Sukhadia University, Udaipur, India,
E-mail: pb_punjabi@yahoo.com

Sanyogita Sharma
Department of Chemistry, PAHER University, Udaipur, India,
E-mail: sanyogitasharma22@gmail.com

Gupta Sharoni
Department of Chemistry, M. L. Sukhadia University, Udaipur, India
E-mail:sharoni290490@gmail.com

Neetu Shorgar
Department of Chemistry, PAHER University, Udaipur, India,
E-mail:nshorgar@gmail.com

Meenakshi Singh Solanki
Department of Chemistry, B. N. University, Udaipur, India,
E-mail : meenakshisingh001989@gmail.com

Dipti Soni
Department of Chemistry, PAHER University, Udaipur, India
E-mail: soni_mbm@rediffmail.com

Yasmin
Department of Chemistry, Techno India NJR Institute of Technology, Udaipur, India
E-mail: ali_yasmin2002@yahoo.com

PREFACE

Nature needs protection from the fast growing chemical pollution. The primary challenge for chemists is to make chemical processes more environmentally benign and sustainable. World has witnessed a tremendous outburst in modifying chemical processes to make them sustainable for making our environment clean and green. One such environmental friendly technique is the use of ultrasound.

Sonochemistry deals with the effect of ultrasonic waves on chemical systems. It has green value because of nonhazardous acoustic radiation and therefore, it is duly recognized as a part of Green Chemistry by synthetic chemists as well as environmentalists. There is no direct interaction of ultrasound with molecular species, but the observed chemical and physical effects of ultrasound are due to the cavitational collapse, which produces drastic conditions of temperature and pressure locally. It induces the formation of various chemical species, which cannot be easily attained under conventional conditions. Sometimes, these species are responsible for driving towards an unusual reactivity in molecular entities.

Exposure to ultrasonic radiation and the resultant sonochemical and/or sonophysical effects have established this technique for driving a particular chemical reaction more efficiently and that too with high yields and selectivity. Sonochemistry utilizes less hazardous starting materials, reagents and solvents. In this process, product selectivity and product yields are increased; in addition, energy consumption is also reduced. This book provides the complete development of sonochemistry starting from introduction, basic concepts of sonochemistry, different types of sonochemical reactions, instrumentation, use of ultrasound in driving particular chemical reactions and its applications in various fields such as polymer synthesis, decontamination of water and waste water, preparation of nanomaterials, food technology, pharmaceutical sciences and so forth.

Apart from this, some fields are also discussed in brief, which do not fall in the actual arena of sonochemistry, but utilize ultrasounds of different frequencies. These are food products and their processing, anaerobic digestion of waste, medical applications such as ultrasonography, sonodynamic therapy, drug delivery and so on. Sonochemistry will be successfully used on industrial scale in pharmaceutical drugs, polymers, nanomaterials, food technology, material science, biogas production, etc. in years to come and will be an established as green chemical technology of the future.

ABOUT THE EDITORS

Suresh C. Ameta, PhD, is currently Dean of Faculty of Science at PAHER University, Udaipur. He has served as professor and head of the department of chemistry at North Gujarat University Patan (1994) and at M. L. Sukhadia University, Udaipur (2002–2005), and as head of the department of polymer science (2005–2008). He also served as Dean of Postgraduate Studies (2004–2008). Prof. Ameta has held the position of President, Indian Chemical Society, Kotkata, and is now a life-long advisor. He was awarded a number of prestigious awards during his career, such as national prizes twice for writing chemistry books in Hindi; the Prof. M. N. Desai Award (2004), the Prof. W. U. Malik Award (2008), the National Teacher Award (2011), the Prof. G. V. Bakore Award (2007), Life Time Achievement Awards by Indian Chemical Society (2011) as well as Indian Council of Chemist (2015), etc. He has successfully guided 81 PhD students and having more than 350 research publications to his credit in journals of national and international repute. He is also the author of many undergraduate- and postgraduate-level books. He has published three books with Apple Academic Press, *Microwave-Assisted Organic Synthesis, Green Chemistry, and Chemical Applications of Symmetry and Group Theory,* and two with Taylor and Francis: *Solar Energy Conversion and Storage* and *Photocatalysis.* He has also written chapters in books published by several international publishers. Prof. Ameta has delivered lectures and chaired sessions at national conferences and is a reviewer of number of international journals. In addition, he has completed five major research projects from different funding agencies, such as DST, UGC, CSIR and Ministry of Energy, Government of India.

Rakshit Ameta, PhD, is Associate Professor of Chemistry, J.R.N. Rajasthan Vidyapeeth (Deemed to be University), Udaipur, India. He has several years of experience in teaching and research in chemistry as well as industrial chemistry and polymer science. He is presently guiding six research students for their PhD theses, and eight students have already obtained their PhDs under his supervision in green chemistry.

Dr. Rakshit Ameta has received various awards and recognition in his career, including being awarded first position and the gold medal for his MSc and receiving the Fateh Singh Award from the Maharana Mewar Foundation, Udaipur, for his meritorious performance. He has served at M. L. Sukhadia University, Udaipur; the University of Kota, Kota, and PAHER University, Udaipur. He has over 100 research publications to his credit in journals of national and international repute. He holds one patent, and two more are under way. Dr. Rakshit has organized several national conferences as Organizing Secretary at the University of Kota and PAHER University. He has delivered invited lectures and has chaired sessions in conferences held by the Indian Chemical Society and the Indian Council of Chemists. Dr. Rakshit was elected as council member of the Indian Chemical Society, Kolkata (2011–2013) and council member (2012–2014) and zonal secretary (2016–2018), Indian Council of Chemists, Agra as well as associate editor, Physical Chemistry Section (2014–2016) of Indian Chemical Society. Dr. Rakshit has also worked as Scientist-in-Charge in the Industrial and Applied Chemistry Section of Indian Chemical Society (2014–2016). He has written five degree-level books and has contributed chapters to books published by several international publishers. He has published three previous books with Apple Academic Press: *Green Chemistry* (2014), *Microwave-Assisted Organic Synthesis* (2015), *Chemical Applications of Symmetry and Group Theory* (2016) and two books *Solar Energy Conversion and Storage* (2016) and *Photocatalysis* (2017) with Taylor and Francis. His research areas focus on wastewater treatment, photochemistry, green chemistry, microwave-assisted reactions, environmental chemistry, nanochemistry, solar cells and bioactive and conducting polymers.

Garima Ameta, PhD, is working as a post-doctoral fellow at M. L. Sukhadia University, Udaipur, India. She has 13 publications including chapters in the fields related to sonochemistry. She has received Young Scientist Award (2016) for best presentation in a National Conference. Her research interests are in the field of use of ultrasound in wastewater treatment and synthesis of ionic liquids.

ABOUT THE BOOK

Normally, heat and light are thought as energy sources to drive a particular chemical reaction, but now, ultrasound can also be used as a promising energy source for this purpose. Collapse of a bubble generates a wide range of high temperatures and pressures and, therefore, use of ultrasound has a considerable potential in chemical and allied sciences. This book presents a complete picture of ultrasound-assisted reactions and technologies such as organic synthesis, polymer synthesis and degradation, nanomaterials, waste water treatment, food ingredients, and products, pharmaceutical, bioenergy applications, and so forth. Ultrasound-assisted reactions are green and economically viable alternatives to conventional techniques. This book is aimed to throw light on the diversified applications of ultrasound and its significant role as a green chemical pathway.

CHAPTER 1

INTRODUCTION

SURESH C. AMETA[1]

[1]*Department of Chemistry, PAHER University, Udaipur, India*
E-mail: ameta_sc@yahoo.com

CONTENTS

1.1 INTRODUCTION

The concept of the present-day generation of ultrasound was established and it dates back to the early 1880s, with the discovery of the piezoelectric effect by the Curies (1880a, 1880b). The first commercial application of ultrasonics came into existence in the form of echo-sounder (Chilowsky and Langévin, 1917). This original echo-sounder eventually became underwater sound navigation and ranging (SONAR) for submarine detection during World War 2. The transducer used was simply a mosaic of thin quartz crystals glued between two steel plates (the composite having a resonant frequency of about 50 kHz), which was mounted in a housing so that it can work under water (submersion). The echo-sounder simply sent a pulse of ultrasound from the keel of a boat to the bottom of the sea wherein it was reflected back to a detector available on the keel.

Wood and Loomis (1927) were the first to note the advantages of using ultrasound in chemistry. They observed that chemical reactions were not a consequence of the heat but a consequence of sonic energy. There was a very slow growth in the field of sonochemistry even though it was known for the last five decades. Significantly, an interest in ultrasonics was renewed only in the 1980s. Initially, sonochemistry was used only for chemical effects induced by ultrasound but now this field has been extended to other applications too. Currently, the term "sonochemistry" is used to describe the applications of ultrasound in diverse fields such as food processing, stabilization of oil emulsions, particle size reduction, filtering systems for suspended particles, homogenization, atomization, environmental protection, and so forth.

Ultrasound finds its application in animal communications (e.g. bat navigation and dog whistles). It is also used in the medical field for fetal imaging, in underwater range finding (SONAR), or in the nondestructive testing of metals for flaws, etc. A few decades back, a chemist would probably not consider sound as the type of energy that could be used for the enhancement of a chemical reaction and the use of ultrasound in chemistry was something of a curiosity. One would probably turn towards heat, pressure, light, or the use of any catalysts, etc., instead of choosing ultrasound to increase chemical reactivity. It is perhaps surprising that for several years sound was not considered as a potential source of increasing chemical reactivity. The only exception to this being the green-fingered chemists, who in the privacy of their own laboratories, talk, sing or even shout at their sonochemical reactions. In fact, sound is transmitted through a medium as a pressure wave and only this act of transmission must cause some excitation in the medium in the form of enhanced molecular motion. However, in order to produce real effects, the sound energy must be transmitted within the liquid itself. This is attributed due to the fact that transfer of sound energy from the air into a liquid is not an effective process.

Sonochemistry is still considered a strange word by most of the chemists till date. The term sonochemistry comprises two terms: the prefix sono means sound and the suffix chemistry has its own meaning. The combination of these two terms could lead to an understanding that this is a kind of chemistry in the presence of sound. Thus, sonochemistry is the use of sound waves of high frequency and power to input energy into a liquid reaction mixture. When high-frequency sound waves are passed through a liquid medium, compression and rarefaction zones are formed. The liquid molecules are pulled apart, when the power exceeds a certain minimum value in the rarefaction zone, thereby creating voids, bubbles, or cavitations filled with gas or liquid vapours. These bubbles suddenly collapse after several acoustic cycles and release a large temperature (nearly 5000°C) and very high pressure (1000 bar or 14,000 psi into the liquids). During this process, the mechanical energy of ultrasound is transformed and transferred through bubbles into renewable energy through a chemical reaction. Thus, cavitational phenomenon is the essence of sonochemistry and sonochemical effects.

These cavitations can create extreme conditions in liquids of both the types: physical and chemical. Similar phenomena can be also observed in liquids containing solid, when these are exposed to ultrasound. However, in this case, cavitation occurs near an extended solid surface. Near this surface, the cavity collapse is nonspherical and drives high-speed jets of liquid to the surface. These jets and associated shock waves can damage the highly heated surface. Liquid power suspensions produce high-velocity interparticle collisions. Such interparticle collisions can change the surface morphology, composition and reactivity.

As these bubbles are very small and are found to collapse rapidly, these can act as microreactors offering an opportunity of enhancing rates of reactions and also permitting some novel reactions to proceed in an absolutely controlled safe manner.

1.2 HOMOGENEOUS SONOCHEMISTRY

Chemical reactions in homogeneous systems proceed via radical intermediates. It means that sonication is able to affect reactions proceeding via radical formations and it does not affect ionic reactions. The bubbles or cavities act as microreactors in the case of volatile molecules as these molecules enter the microbubbles. High temperature and pressure is generated during cavitation which eventually breaks the chemical bonds. The short-lived chemical species are returned to the bulk liquid at room temperature and they can react with other species. Compounds of low volatility do not enter bubbles and therefore, these will be directly exposed to these extreme conditions of temperature and pressure. Thus, they experience a high-energy environment, which is due to the pressure changes. These pressure changes are associated with the propagation of the acoustic wave or with bubble collapse (shock waves).

1.3 HETEROGENEOUS SONOCHEMISTRY

The chemical reactions in a heterogeneous system are basically through the mechanical effects of cavitations. These effects include surface cleaning, particle size reduction, and improved mass transfer. If cavitation occurs in a liquid near a solid surface, then there is a significant change in the dynamics of cavity collapse.

The cavity remains in a spherical shape in a homogeneous system during cavity collapse because its surroundings are uniform. Cavity collapse is mostly asymmetric in proximity to a solid boundary in heterogeneous systems with generation of high-speed jets of liquid. These jets hit the surface with tremendous force and this process can cause a lot of change at the point of impact and produce newly exposed highly reactive surfaces.

Enhancement of chemical reactions by ultrasound has been widely studied. It has found some beneficial applications in mixed-phase synthesis, materials chemistry, biomedical uses, etc. As cavitation can only occur in liquids, such chemical reactions are not observed in the ultrasonic irradiation of solids or solid–gas systems.

Most of the modern ultrasonic devices rely on transducers (energy converters), which are composed of piezoelectric materials. Such materials respond to the application of an electrical potential across opposite faces with a small change in dimension. This is the inverse of the piezoelectric effect. If the potential is alternated at high frequencies, the crystal converts the electrical energy to mechanical vibration (sound) energy. At sufficiently high alternating potential, high-frequency sound (ultrasound) will be generated.

A bath or ultrasonic horn (a high-power probe) is used to carry out sonochemical reactions. Various reactions are conducted using a simple ultrasonic cleaning bath and in that case, the amount of energy released during the reaction varies only between 1 and 5 W cm^{-2}, but the temperature control is normally poor. Large-scale sonochemical reactions can be performed using immersible ultrasonic probes that circumvent the transfer of energy through water and the reaction vessel. In such cases, applied energies can be higher in the order of several hundred times. Ordinary laboratory equipment uses frequency range between 20 and 40 kHz, but cavitation can be generated well above these frequencies. At present, much broader range of frequencies is used.

Ultrasound finds its application in manufacturing of some chalcogenides, coordination polymers, and organometallic compounds. It has wide applications in synthesizing various organic moieties including hetrocycles such as oxazoles, thiazoles, pyrazoles, benzimidazoles, triazoles and tetrazoles. These hetrocycles have proved their importance in pharmaceutical chemistry. The rates of various name reactions have also been found to be enhanced in the presence of ultrasound; some of these reactions worth mentioning are Knoevenagel condensation, Pinacol coupling reaction, Biginelli reaction, Suzuki reaction, Wittig reaction others.

Particles of nanosize have gained an increasing attention of all the scientists worldwide due to their unique properties. This is all due to their high surface-to-volume ratios. These may be also used as efficient catalysts in various reactions. Nanomaterials of different morphologies can be synthesized using ultrasonic irradiation. Variety of nanostructures can be accomplished using ultrasound such as nanoflowers, nanobelts, nanorings, nanonails, nanoribbons, nanosheets and so on. Even metal loading can be conveniently accomplished using this technique.

Now, polymers have almost replaced all of the traditional materials of construction, furniture, textile, etc., such as metals, wood, cotton, and therefore, the present era is perfectly named as Polymer Era. The society has to contribute a lot to this and is facing problems for this replacement, because polymers are not biodegradable and their disposal after use is becoming a challenge for all of us. Ultrasound comes to our rescue at this stage for the degradation of some polymers or at least convert these polymers to a biodegradable component so that this can be used in further applications. This source of energy is also used in the synthesis of some polymers.

Ultrasound plays a vital role in the treatment of waste water-containing phenols, dyes, chloro compounds, surfactants, thio compounds, nitrogen-containing compounds, etc. to decrease the organic load making aquatic life easier by increasing the sustainability of aquatic organism and also to increase the recyclability of water. Ultrasound irradiation is not only utilized in the treatment of sludge, sewage and solid wastes, can be also utilized in the conversion of such materials into some useful compounds such as biogas through anaerobic digestion.

Food technology also finds a number of applications of ultrasound, for example, in the extraction of some natural products from bioresources, homogenization, emulsification, food preservation and so on. Various other industries such as textile, pharma, sugar, are also benefited by ultrasound in the processes of dyeing, washing, finishing, cleaning, etc. Many industries that use boilers are affected by the problem of scaling, because these require the removal of scales, which is a difficult task, because it is hard enough. The use of ultrasound makes this scale relatively soft so that it can be easily removed, thus improving the life of boilers.

Ultrasound is commonly used in the form of ultrasonography and drug delivery. Therapeutic ultrasound is basically used for sonodynamic therapy. It has also been used for the treatment of cancer. Normally, high-intensity-focused ultrasound is medically used in killing tumor cells. Ultrasonic imaging is another aspect of sonochemistry in medical field. The use of sound waves leads to nondestructive testing of ailments.

The subject of sonochemistry has been excellently reviewed from time to time by various workers (Mason, 1999; Thompson and Doraiswamy, 1999; Ameta et al., 2001; Mason and Peters, 2002; Mason and Lorimer, 2002; Gedanken, 2004; Mousavi and Ghasemi, 2012; Xu et al., 2013; Grieser et al., 2015; Asgarzadehahmadi et al., 2016; Ashokkumar, 2016; Panda and Manickam, 2017; Chatel, 2017).

However, some of the applications required different frequencies, which do not fall under the range of sonochemical reactions, but a brief description of such applications has also been provided and discussed in this book. Currently, sonochemistry is considered a green chemical technology as it has many advantages such as reduction in the reaction time, increasing the product yields, eco-friendly in nature and so on.

REFERENCES

Ameta, S. C.; Punjabi, P. B.; Swarnkar, H.; Chhabra, N.; Jain, M. *J. Indian Chem. Soc.* **2001,** *78*, 627–633.

Asgharzadehahmadi, S.; Raman, A. A. A.; Parthasarathy, R.; Sajjadi, B. *Renewable Sustainable Energy Rev.* **2016,** *63*, 302–314.

Ashokkumar, M. Ed. *Handbook of Ultrasonics and Sonochemistry;* Springer: Singapore, 2016.

Chatel, G. *Sonochemisrty: New Opportunities for Green Chemistry;* World Scientific: Singapore, 2017.

Chilowsky, C.; Langevin, P. Production of Submarine Signals and the Location of Submarine Objects. U.S. Patent 1,471,547, 1917.

Curie, J.; Curie, P. *Bulletin de la Société Minérologique de France* **1880a,** *3*, 90–93.

Curie, J.; Curie, P. *Comptes Rendus* **1880b,** *91*, 294–295.

Gedanken, A. *Ultrason. Sonochem.* **2004,** *11*(2), 47–55.

Grieser, F.; Choi, P.-K.; Enomoto, N.; Harada, H.; Okitsu, K.; Yasui, K. Eds. *Sonochemistry and the Acoustic Bubble;* Elsevier: Amsterdam, 2015.
Mason, T. J. *Sonochemistry;* Oxford University Press: Oxford, 1999.
Mason, T. J.; Lorimer, J. P. *Applied Sonochemistry: The Uses of Power Ultrasound in Chemistry and Processing;* Wiley-VCH: Weinheim, 2002.
Mason, T. J.; Peters, D. *Practical Sonochemistry: Power Ultrasound Uses and Applications,* 2nd ed.; Woodhead Publishing: Sawston, 2002.
Mousavi, M. F.; Ghasemi, S. Eds. *Sonochemistry: A Suitable Method for Synthesis of Nano-Structured Materials*; Nova Science: New York, 2012.
Panda, D.; Manickam, S. *Ultrason. Sonochem.* **2017,** *36,* 481–496.
Thompson, L. H.; Doraiswamy, L. K. *Ind. Eng. Chem. Res.* **1999,** *38*(4), 1215–1249.
Wood, R. W.; Loomis, A. L. *Philos. Mag.* **1927,** *4,* 414–436.
Xu, H.; Zeiger, B. W.; Suslick, K. S. *Chem. Soc. Rev.* **2013,** *42,* 2555–2567.

CHAPTER 2

BASIC CONCEPTS

GARIMA AMETA[1]

[1]*Department of Chemistry, M. L. Sukhadia University, Udaipur, India*
E-mail: garima_ameta@yahoo.co.in

CONTENTS

2.1 INTRODUCTION

Chemistry is the interaction of energy with matter. Different forms of energy such as heat (thermo), light (photo), and so on can drive any chemical reaction in a particular direction. But in the last few decades, ultrasound has also emerged as a potential source of energy that can enhance chemical reactivity apart from its use in other applications, which are nonchemical such as material testing, medical diagnosis, navigation systems of bats, sound navigation and ranging (SONAR), and cleaning or drilling of teeth. We have found that the chemical reactivity of a chemical system is affected by ultrasound in low-frequency range passed through that system. The effect of ultrasound on chemical reactivity forms a basis of sonochemistry. The degradation of some compounds in the presence of ultrasound is known as sonolysis. Ultrasound can also be used to synthesize compounds in reduced reaction time, increase in yield or sometimes with diversion from the main route (chemical switching).

Scientifically, sound may be defined as the transmission of energy through the generation of acoustic pressure waves in the medium. As these pressure waves are mechanical in nature, a medium is required and the particles in the medium vibrate to transfer sound waves. The frequency of the wave determines its regime. The medium may be air, water, solid, or human soft tissue.

Sound is classified into four different categories depending on its frequency range, and these are given as:

- **Infrasound**—Sound waves having frequency less than 20 Hz
- **Audible sound**—Sound waves having frequency between 20 and 20 kHz
- **Ultrasound**—Sound waves with frequency more than 20 kHz to 10 GHz
- **Hypersound**—Sound waves with frequency higher than 10 GHz.

There are two types of ultrasonic waves: (i) bulk (fundamental) waves that can propagate inside an object, and (ii) guided waves that propagate near the surface or along the interface of an object (Brekhovskikh, 1980; Auld, 1990; Achenbach, 1990; Rose, 1999). Waves propagating completely inside an object are called bulk waves and these are independent of its boundary and shape. Bulk waves exist in two types in an isotropic medium. These are:

1. Longitudinal (dilatational, compression, or primary) waves and
2. Shear (distortional, transverse, or secondary) waves.

Distortion is effected on ultrasound irradiation and it depends on the condition of application, whether this force is applied normal or parallel to the surface at one end of the solid and whether it will produce compression or shear vibrations, respectively. Thus, two types of ultrasonic waves, that is longitudinal waves or transverse waves propagate through the solid and energy of the wave is also carried with it (Figs. 2.1 and 2.2).

Propagation of ultrasonic wave is commonly described in terms of the direction of motion of particles in relation to the direction, in which wave propagates. The longitudinal waves are defined as waves where the motion of particle is parallel to the direction of the wave propagation. On the other hand, shear waves are defined as waves where the motion of particle is perpendicular to the direction of the wave propagation. Both these waves can exist in solids because solids have rigidity to offer resistance to shear as well as compressive loads, unlike liquids and gases. But

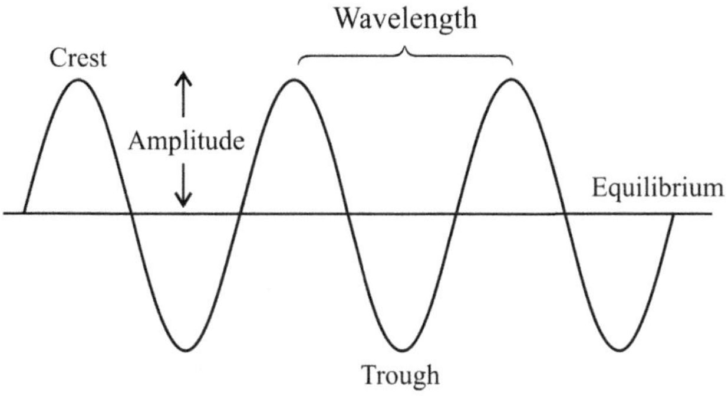

FIGURE 2.1 Longitudinal waves.

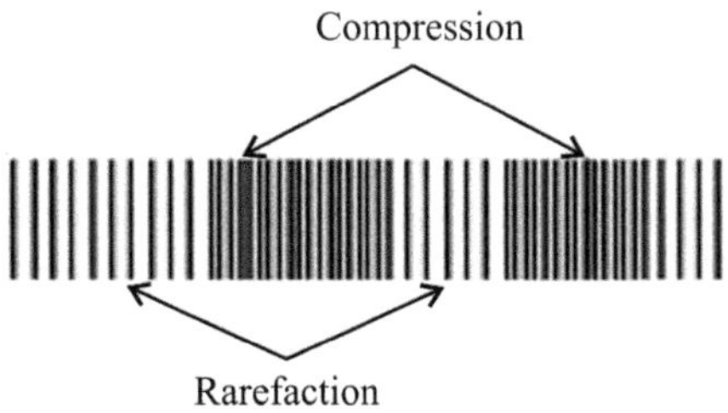

FIGURE 2.2 Transverse waves.

the shear waves cannot exist in liquid and gaseous medium as they offer no resistance to shear loads in such media.

Other type of waves includes the guided waves. These are propagated by considering the influences of the boundaries or the shape of an object. Such guided waves are of three types depending on the geometry of an object and these are:

1. Surface acoustic waves (SAWs)
1. Plate waves
2. Rod waves

Surface acoustic waves are defined as waves that propagate along a free surface with disturbance amplitude, which decays exponentially with depth into the object. Wave propagation characteristics of SAWs strongly depend on the properties of that material, surface structure, and nature at the interface of the object. Different kinds of SAWs such as Love, Scholte, Stoneley, and Rayleigh waves exist. When a SAW propagates along a boundary between a semi-infinite solid and air, the wave is normally called a Rayleigh wave. Here, particle motion is elliptical, and the effective penetration depth is of the order of one wavelength. Rayleigh wave is the most common and well-known SAW and therefore, any SAW is often called Rayleigh wave.

If ultrasonic wave propagates in a finite medium (like a plate), it is bounded within the medium and may resonate. Such waves in an object (with multilayered structure) of finite size are called plate waves, but these are called Lamb waves, if the object has a single layer. If the force is applied to the end of a slender rod, then an ultrasonic wave propagates axially along the rod. Wave propagations in structures such as thin rod and hollow cylinders have also been studied.

The propagation characteristics of guided waves depend strongly not only on the properties of a material but also on the thickness of the plate, the diameter of rod, and the frequency. The dependence of the wave velocity of guided waves on frequency is called frequency dispersion. This dispersion not only makes wave propagation behaviour much complicated but also provides unique material evaluations using these guided waves. Similar types of bulk and guided waves can exist for anisotropic materials but their behaviours become much more complicated than those for isotropic materials (Auld, 1990; Achenbach, 1990; Rose, 1999).

Two types of cavitation are distinguished. These are stable and transient. Stable cavitation usually takes place when microbubbles mainly contain a gas (e.g. air) and their mean life is relatively longer than a cycle of the ultrasound. As long as their resonance frequency is higher than that of the ultrasound during their growth, they are driven into pressure antinodes, where they may induce some chemical reactions. On the contrary, transient cavitation is a phenomenon of quite shorter duration. Here, a cavity is rapidly formed containing mainly vapor of the liquid and it vigorously collapses after a few cycles. Both these cavitations take place at ultrasound intensities of a few $W \cdot cm^{-2}$. Transient cavitation is quite often considered the most efficient way of driving different chemical reactions, but it is true with high ultrasound intensities, so that the formation of standing waves is avoided. At low ultrasound intensities, standing waves are there to achieve high bubble numbers and significant yields. In such a case, chemical reactions are induced through stable cavitation.

The cavitational bubbles increase in size in the positive cycle of the wave due to the positive acoustic pressure. Then these bubbles finally collapse, which leads to the formation of new nuclei for the next cavitation. Each one of the bubble would collapse and it acts as a hot spot. The process of collapse generates energy and, as a result, the temperature and pressure rises up to 5000 K and 1000 atm., respectively, and cooling rate as fast as 109 $K \cdot s^{-1}$. The process of formation, growth, and collapse of the cavitation bubbles is shown in Figures 2.3 and 2.4.

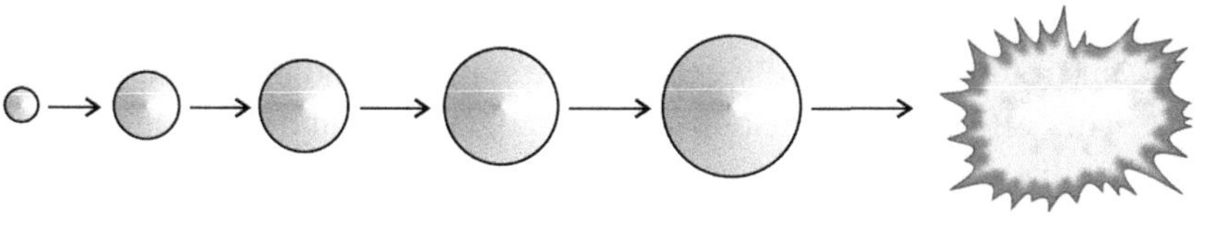

FIGURE 2.3 Growth and implosion of a bubble.

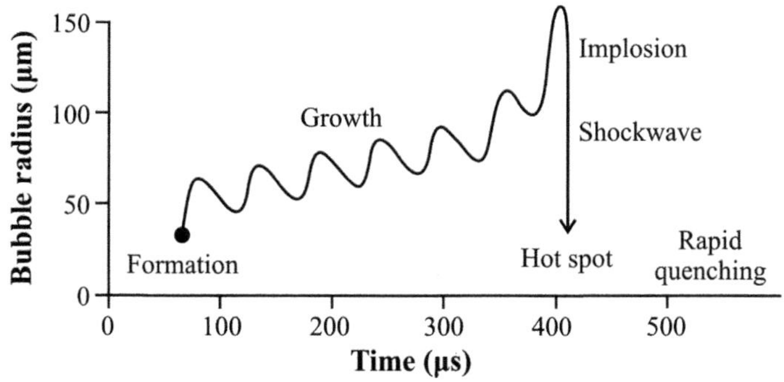

FIGURE 2.4 Formation, growth and collapse of a bubble.

Such collapsing bubbles create an unusual mechanism for high-energy chemical reactions due to abnormal increase in local temperatures and pressure.

This can be divided into three regions (Fig. 2.5):

- First one is the thermolytic center (hot spot). This is the core of the bubbles with localized hot temperature (5000 K) and high pressure (500 atm) during final collapse of cavitation. Bubble water molecules are pyrolyzed in this region forming ${}^{\bullet}OH$ and ${}^{\bullet}H$ in the gaseous phase. Then the substrate will either react with the •OH or it will undergo pyrolysis.

- Second one is an interfacial region between the cavitational bubble and bulk liquid. Here, similar reaction as hot spot occurs, but in the aqueous phase. However, some additional reactions may occur in this region, where •OH dimerize to form H_2O_2. Hydrophobic compounds are more concentrated in the interfacial region than in the bulk solution.
- Third one is the bulk region. In this region, the temperature remains almost similar to the room temperature because the process of cavitation is adiabatic in nature. The chemical reactions occurring in this region are basically between the substrate and the •OH or H_2O_2. The effects of ultrasound in homogeneous and heterogeneous media are clearly distinct. Sonochemical reactions are related to some new chemical species produced during cavitation in homogeneous media, whereas for the latter, enhancement of the reactions in heterogeneous media could also be related to mechanical effects induced in liquid system under sonication (Destaillats et al., 2003).

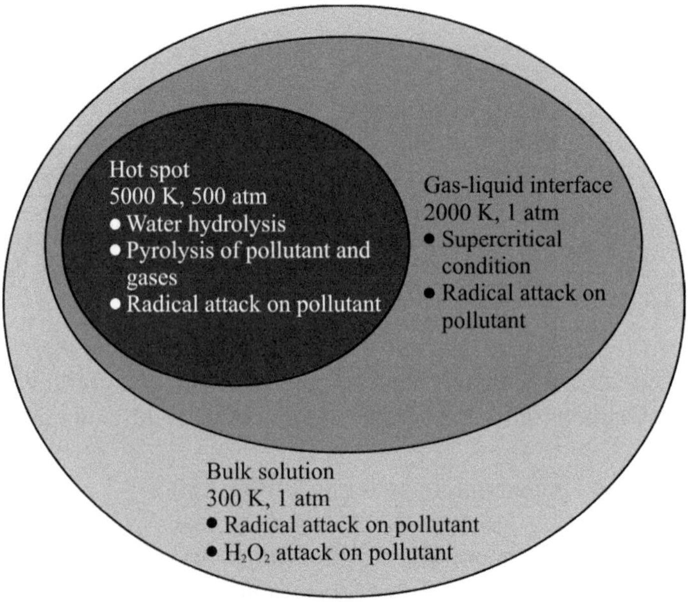

FIGURE 2.5 Reaction zone in cavitation process
Source: P. Chowdhury and Viraraghavan, 2009, with permission.

The study of sonochemistry is concerned with understanding the effect of ultrasound in forming acoustic cavitation in liquids, which is responsible for the initiation or enhancement of the chemical activity in the solution. The chemical effects of ultrasound are not due to a direct interaction of the ultrasonic sound wave with the molecules present in the solution. Sound waves propagating through a liquid at ultrasonic frequencies do these changes with a wavelength that is significantly longer than the bond length between atoms of the molecule. Therefore, the sound waves cannot affect vibrational energy of the bond, and therefore, it cannot increase the internal energy of a molecule directly (Suslick, 1990; Suslick and Flannigan, 2008). Sonochemistry arises from acoustic cavitation, that is, the formation, growth, and implosive collapse of bubbles in a liquid (Suslick and Kenneth, 1989).

The collapse of these bubbles is an almost adiabatic process and it results in the massive buildup of energy inside the bubble. This results in very high temperatures and pressures in a microscopic region of the sonicated liquid. Such extreme conditions of temperatures and pressures result in the chemical excitation of any matter that is either present inside the bubble, or in the close vicinity of the imploding bubble. A number of processes result from acoustic cavitation, an increase in chemical activity in the solution either due to the formation of primary and secondary radical reactions, and or through the formation of some new, relatively stable chemical species that can diffuse further into the solution to show chemical effects, etc. Hydrogen peroxide (H_2O_2) may be formed from the combination of two hydroxyl radicals generated in dissociation of water vapor inside the collapsing bubbles on exposed of water to ultrasound.

Wood and Loomis (1927) were the first to report the effect of sonic waves traveling through liquids. They have conducted experiments on the frequency of the energy that it took from sonic waves to penetrate the barrier of water. They concluded that sound does travel faster in water, but because of the density of water compared to our earth's atmosphere, it was very difficult to get the sonic waves into the water. They decided that sound can be dispersed into water by making loud noises into water by creating bubbles. They put sound into water by simply yelling. But another disadvantage of this was the ratio of the amount of time it took for the lower frequency waves to penetrate the bubble walls and access water around the bubble, and then the time from that point to a point on the other end of the body of water. But this exciting idea remained mostly ignored for a few decades till the 1980s, when some of inexpensive and reliable generators of high-intensity ultrasound were developed.

Acoustic cavitation usually occurs upon irradiation with high-intensity sound or ultrasound. Cavitation is the formation, growth, and implosive collapse of bubbles on exposure to ultrasound. It gave impetus to sonochemistry and sonoluminescence (Leighton, 1994). Bubble collapse in liquids produces large amounts of energy. It comes from the conversion of kinetic energy of the liquid motion into heating the contents of the bubble. The compression of the bubbles during cavitation is quite rapid than thermal transport, and it generates a short-lived localized hot spot. These bubbles have temperatures around 5000 K, pressures of roughly 1000 atm, and heating and cooling rates above 10^{10} $K \cdot s^{-1}$ (Suslick et al., 1986; Flint and Suslick, 1991). These cavitations can create extreme physical and chemical conditions in otherwise cold liquids.

Similar phenomena may occur with exposure to ultrasound within liquids containing solids. If this cavitation occurs near an extended solid surface, the cavity collapses in a nonspherical way and it drives high-speed jets of liquid to the surface (Leighton, 1994). These jets and associated shock waves can damage the highly heated surface. Suspensions of liquid powder produce high-velocity interparticle collisions, which can change the surface morphology, composition, and reactivity (Suslick and Docktyez, 1990).

Basically, sonochemical reactions can be classified into three classes. These are:

- Homogeneous sonochemistry of liquids
- Heterogeneous sonochemistry of liquid–liquid or solid–liquid systems, and, overlapping
- Sonocatalysis (Einhorn et al., 1989; Luche, 1996; Pestman et al., 1994)

Sonoluminescence is taken as a special case of homogeneous sonochemistry (Crum, 1994; Putterman, 1995). The enhancement of chemical reactions by ultrasound has found applications in mixed-phase synthesis, materials chemistry, and biomedical uses. Chemical reactions have not been studied in the solids or solid–gas systems in the presence of ultrasonic radiation as cavitation can occur only in liquids.

It has been observed that ultrasound greatly enhances chemical reactivity in the order of million-folds in a number of systems (Suslick and

Casadonte, 1987), acting as an effective catalyst by exciting the atomic and molecular modes of the system such as the vibrational, rotational, and translational modes. This is due to the fact that ultrasound breaks up the solid pieces from the energy released from the bubbles created by cavitation collapsing through them. This gives larger surface area to the solid reactant for the reaction to proceed, thus increasing the rate of reaction.

The intensity of sound decreases with distance on traveling through a medium. The amplitude of signal is reduced by the spreading of the wave in case of idealized materials, but in natural materials, further weakening is there from scattering and absorption of sound. Scattering is the process of reflection of the sound in different directions than its original direction of propagation whereas absorption is the conversion of the sound energy in other forms of energy. The combined effect of both these processes, scattering and absorption, is called attenuation. Attenuation is the decay rate of the wave as it propagates through some materials.

Attenuation can be used a tool for measurement to decide the factors decreasing the ultrasonic intensity. The change in the amplitude of a decaying plane wave can be expressed by the following expression:

$$A = A_0 e^{-\alpha z}$$

where A_0 is the original (unattenuated) amplitude of a propagating sound wave at some location, A is the reduced amplitude, when the wave has traveled a distance z from the starting point, α is the attenuation coefficient of the wave traveling in the direction of propagation, α is expressed in Nepers/length (Np/m), and a Neper is a dimensionless quantity. The term e is the exponential (or Napier's constant) and it is equal to approximately 2.71828.

It can be converted to other unit, that is decibels/length by dividing it by 0.1151, as decibel is a widely used unit.

The attenuation is generally proportional to the square of frequency of sound. The values of attenuation are normally given for a single frequency, or it is averaged over many frequencies. Actual value of the attenuation coefficient for a material depends on the process of manufacturing of that material.

Attenuation can be determined by evaluating the multiple backwall reflections seen in a typical A-scan display. The number of decibels between two adjacent signals is measured and this value is divided by the time interval between these two signals. This provides an attenuation coefficient in decibels per unit time U_t. This can be converted to Nepers/length as:

$$\alpha = \frac{0.1151}{v} U_t$$

where v is the velocity of sound in $m \cdot s^{-1}$ and U_t is in $decibels \cdot s^{-1}$.

2.2 CAVITATION

Collapse of a transient bubble is the main source of some chemical and mechanical effects of ultrasound. The collapsing bubble may act as a micro-reactor, where drastic conditions are created instantaneously (Suslick et al., 1986). This cavitation causes thermolysis of solute molecules to form highly reactive radicals such as hydroxyl radicals and others. This active species provides vigorous reaction conditions for the substrate in the reaction media. If some solid is also present in the solution, ultrasonic radiations causes solid disruption and, as a result, particle size of the solid is reduced. This results in the increase of the total solid surface giving higher rate of the reaction or greater yield of the products. No method other than ultrasonic irradiation provides such an effect and therefore, ultrasonication becomes a unique technique.

2.2.1 FACTORS AFFECTING CAVITATION

Cavitation under ultrasonic irradiation is a physical phenomenon, and different factors are affecting cavitation. Some of the major factors that affect cavitation are:

- Intensity of sonication
- Frequency of ultrasound
- Temperature
- Solvent
- Pressure and bubbled gas

2.2.1.1 INTENSITY OF SONICATION

Intensity of sonication is directly proportional to the amplitude of ultrasonic source and, therefore, any increase in amplitude of vibration will also increase the intensity of sonication. This will give an increased sonochemical effect. A minimum intensity of sonication is required for any reaction so that the cavitation threshold can be accomplished. It only means that it is always not necessary to use higher amplitudes. Higher amplitudes of vibration should be avoided, unless necessary as it will lead to deterioration of the source. It may also result in the agitation of liquid rather than cavitation and also a poor transmission of the ultrasound in liquid media.

2.2.1.2 FREQUENCY OF ULTRASOUND

The cavitation bubbles are easily produced at low ultrasonic frequencies, but at higher ultrasonic frequencies, it is very difficult to produce cavitation bubbles. Cavitation can be accomplished at higher frequency by increasing the intensity of applied ultrasound so that voids are formed by overcoming cohesive forces of liquid. Normally, more power is required to induce cavitation at higher frequencies, because the cycle of compression and rarefaction becomes too short that liquid molecules cannot be separated. As a consequence, void is not formed and no cavitation is there.

2.2.1.3 TEMPERATURE

It was observed that better cavitation is there at lower temperatures. The use of higher temperatures of the solvent assists in disrupting some strong solute–matrix interaction involving hydrogen bonding, dipole interaction, and Van der Waal forces. It is also accompanied by faster diffusion rates. The vapour pressure of the solvent increases with the increasing temperature and these solvent vapours fill the cavitation bubbles. Here, the effect of sonication is less intense, because the bubbles will collapse less violently and therefore, one has to make a compromise between cavitation and the temperature of the solvent to achieve better sonochemical reactions.

2.2.1.4 SOLVENT

The selection of the solvent should be made with utmost care. Normally water is used as a solvent, but sometimes less polar liquids (polar organic solvents or a mixture of solvents) are also used. Two physical properties of the solvent inhibit cavitation. These are surface tension and viscosity. These two natural cohesive forces of liquids are acting within a liquid (solvent) acting against the formation of voids to attain cavitation.

2.2.1.5 PRESSURE AND BUBBLED GAS

As the cavitation is produced by breaking the molecular forces, this is not supported by high external pressure. In such cases, more ultrasonic energy is required to induce cavitation. An enhanced sonochemical reaction will be achieved at a particular external pressure with a specific reference. Ultrasonication processes are favored by dissolved gas bubbles as they can act a nucleus for cavitation. The gas should be continuously bubbled into the solvent to obtain more cavitation. Generally inert monoatomic gases such as helium, argon, neon and so forth are used for this purpose.

Cavitation is the basic mechanism involved in sonochemistry. The high temperature and pressure generated during cavitation results in sonochemical reactions and sometimes, even sonochemical switching. Some reactions, which are difficult to be performed under normal conditions, can be carried out under ultrasound exposure. An enhancement in the rate of certain reactions is also observed along with improvement in yields, when performed using sound waves.

REFERENCES

Achenbach, J. D. *Wave Propagation in Elastic Solids;* Elsevier: Amsterdam, **1990**.

Auld, B. A. *Acoustic Fields and Waves in Solids*, 2nd ed.; Krieger Publishing: Florida, **1990**.

Brekhovskikh, L. M. *Waves in Layered Media*, 2nd ed.; Academic Press: New York, **1980**; Vol. 1 & 2.

Chowdhury, P: Viraraghavan, T. Sci. Total Environ. **2009**, 407, 2474-2492.

Crum, L. A. *Phys. Today* **1994,** *47*, 22–30.

Destaillats, H.; Hoffmann, M. R.; Wallace, H. C. Sonochemical degradation of Pollutants. In *Chemical Degradation Methods for Wastes and Pollutants: Environmental and Industrial Applications;* Tarr, M. A., Ed.; Marcel Dekker: New York, USA, **2003**.

Einhorn, C.; Einhorn, J.; Luche, J. L. *Synthesis* **1989,** 787–813.

Flint, E. B.; Suslick, K. S. *Science* **1991,** *253,* 1397–1399.
Leighton, T. G. *The Acoustic Bubble;* Academic Press: London, **1994**.
Luche, J. L. *Comptes Rendus Series IIB* **1996,** *323,* 203 and 307.
Pestman, J. M.; Engberts, J. B. F. N.; de Jong, F. *Recueil des Travaux Chimiques des PaysBas* **1994,** *113,* 533–542.
Putterman, S. J. *Sci. Am.* **1995,** *272,* 46–51.
Rose, J. L. *Ultrasonic Waves in Solid Media;* Cambridge University Press: Cambridge, **1999**.
Suslick, K. S. *Sci. Am.* **1989,** *260,* 80–86.
Suslick, K. S. *Science* **1990,** *247,* 1439–1445.
Suslick, K. S.; Casadonte, D. J. *J. Am. Chem. Soc.* **1987,** *109,* 3459–3461.
Suslick, K. S.; Doktycz, S. J. *Science* **1990a,** *247,* 1067–1069.
Suslick, K. S.; Doktycz, S. J. *Adv. Sonochem.* **1990b,** *1,* 197–230.
Suslick, K. S.; Flannigan, D. J. *Ann. Rev. Phys. Chem.* **2008,** *59,* 659–683.
Suslick, K. S.; Hammerton, D. A.; Cline, Jr. R. E. *J. Am. Chem. Soc.* **1986,** *108,* 5641–5642.
Wood, R. W.; Loomis, A. L. *Philos. Mag.* **1927,** *4,* 414–436

CHAPTER 3

INSTRUMENTATION

GARIMA AMETA[1]

[1]*Department of Chemistry, M. L. Sukhadia University, Udaipur, India*
E-mail: garima_ameta@yahoo.co.in

CONTENTS

3.1 INTRODUCTION

Richards and Loomis (1927) were the first to report chemical effects of ultrasound. The concept of the present-day generation of ultrasound was established and it dates back to the early 1880s, with the discovery of piezoelectric effect by the Curies (Jacques Curie and Pierre Curie) (1880, 1881). Crystalline materials showing this effect are known as piezoelectric materials. Ultrasonic devices consist of transducers (energy converters) which are composed of these piezoelectric materials. An inverse piezoelectric effect is used in transducers, that is a rapidly alternating potential is passed across the faces of piezoelectric crystal, which generates dimensional changes and thus, converts electrical energy into sound energy.

The first ultrasonic transducer was a whistle developed by Galton (1883), who was then investigating the threshold frequency of human hearing.

3.2 ULTRASONIC TRANSDUCERS

Ultrasonic sound can be produced by transducers which operate either by the piezoelectric effector themagnetostrictive effect. The magnetostrictive transducers can be used to produce high-intensity ultrasonic sound in the range of 20–40 kHz for ultrasonic cleaning and other mechanical applications.

Ultrasonic medical imaging typically uses relatively higher ultrasound frequencies in the range of 1–20 MHz. Such ultrasound is produced by applying the output of an electronic oscillator to a thin wafer of piezoelectric material.

Transducer is a device capable of converting one form of energy into another. Ultrasonic transducers are designed to convert either mechanical or electrical energy into high-frequency sound. These transducers are of three types:

- Gas-driven
- Liquid-driven
- Electromechanical

3.3 GAS-DRIVEN TRANSDUCERS

These are, quite simple, in the form of whistles with high-frequency output. The generation of ultrasound via whistles is more than a century old, when a whistle-generating sound of known frequencies was produced. Galton was able to determine that the normal limit of human hearing is around 18 kHz. This whistle was constructed from a brass tube with an internal diameter of about 2 mm. It is operated by passing a jet of gas through an orifice into a resonating cavity. On moving the plunger, the size of the cavity could be changed to alter the pitch or frequency of the emitted sound (Fig. 3.1).

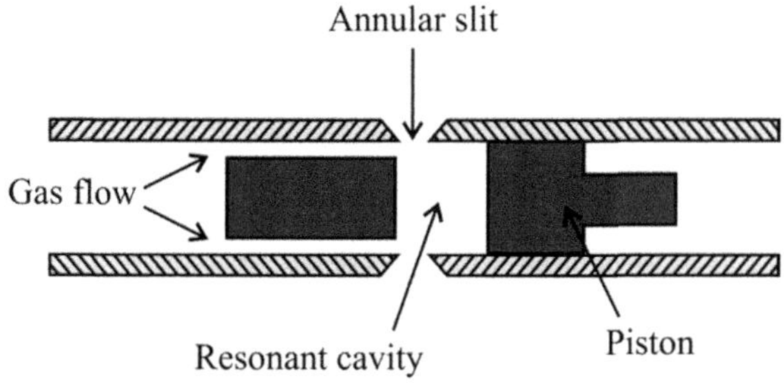

FIGURE 3.1 Galton's whistle.

When a solid object is passed rapidly back and forth across a jet of high-pressure gas, it interferes with the gas flow. As a result, a sound is produced having the same frequency at which the flow was disturbed. A siren can be designed in such a way that the nozzle of a gas jet impinges on the inner surface of a cylinder and as a result, there are a series of regularly spaced perforations. When the cylinder is rotated, the jet of gas emerging from the nozzle will rapidly alternate between facing a hole and the solid surface. The pitch of the sound generated by this device depends on the speed of rotation of the cylinder. Neither type of transducer has any significant chemical application since the efficient transfer of acoustic energy from a gas to a liquid is not possible. However, whistles are used for the atomization of liquids.

An atomized spray from a liquid is produced by forcing it at a high velocity through a small aperture. The disadvantage is blockage at the orifice of low-viscosity liquids.

A gas-driven atomizer comprises an air or a gas jet, which is forced into an orifice, where a shock wave is produced, on expansion (Fig. 3.2). As a result, an intense field of sonic energy is focused between the nozzle body

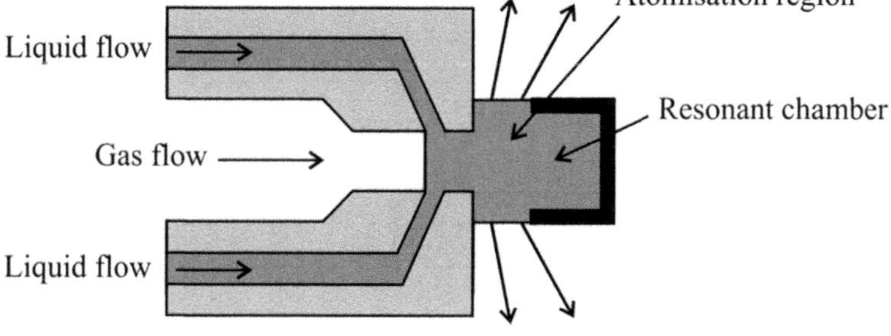

FIGURE 3.2 Gas-driven atomizer.

and the resonator gap. When any liquid is introduced into this region, it is vigorously sheared into droplets by the acoustic field. Very small droplets are produced having a low velocity.

3.4 LIQUID-DRIVEN TRANSDUCERS

This type of transducer is a liquid whistle. It generates cavitation via the motion of a liquid rather than a gas. The material is forced at high velocity by the homogenizer pump through an orifice (Fig. 3.3). It emerges as a jet which impacts upon a steel blade. There are two ways in which cavitational mixing can occur at this point and these are:

- Through the Venturi effect as the liquid expands rapidly into a larger volume on coming out of the orifice
- Via the blade which is caused to vibrate by the process material flowing over it

The activity of the blade can be optimized by controlling the relationship between orifice and blade. The required operating pressure and throughput are determined by using the orifices of different sizes and shapes. Velocity can also be changed to achieve the necessary particle size or the degree of dispersion. With no movement, only a pump is the moving part, none others and therefore, such a system is rugged and durable. When a mixture of immiscible liquids was forced through the orifice and across the blade, extremely efficient homogenization is produced by cavitational mixing.

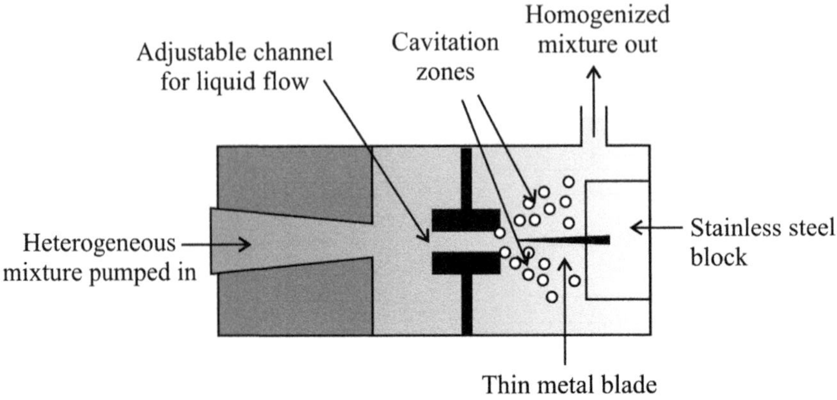

FIGURE 3.3 Liquid whistle.

3.5 ELECTROMECHANICAL TRANSDUCERS

Electromechanical transducers are of two types. These are either based on the piezoelectric or the magnetostrictive effect. The most commonly used transducers are piezoelectric transducers. These are generally used to power the bath- and probe-type sonicator systems. Electromechanical transducers are more commonly used as compared to mechanical transducers, in spite of the fact that these are relatively more expensive.

3.5.1 MAGNETOSTRICTIVE TRANSDUCERS

Magnetostriction can be explained as the corresponding change in length per unit length produced as a result of magnetization. Any material exhibiting this phenomenon is magnetostrictive in nature and this effect is known as magnetostrictive effect. The same effect can be reversed in the sense that, if an external force is applied on a magnetostrictive material, there will be a proportionate change in the magnetic state of the material. This property was first discovered by Joule (1842) by noticing the change in length of the material according to the change in magnetization. He called this phenomenon as Joule's effect. The reverse process is called Villari effect or magnetostrictive effect. This effect explains the change in magnetization of a material due to the force applied. Joule effect is commonly applied in magnetostrictive actuators whereas Villari effect is applied in magnetostrictive sensors. This process is highly applicable as a transducer as the magnetostriction property of a material does not degrade with time.

First is the case when no magnetic field is applied to the material under this condition, there will be no change in the length along with the magnetic induction produced. In the second case, the amount of the magnetic field (H) is increased or decreased to its saturation limits ($\pm H_s$). This causes an increase or decrease in the axial strain, respectively. The maximum strain saturation and magnetic induction are obtained at the point when the value of H_s is at its maximum. Beyond this limit, if the value of the field increases, it does not bring any change in the value of magnetization or field to the device. Thus, when the field value hits the saturation limit, the values of strain and magnetic induction will increase moving from the center outward.

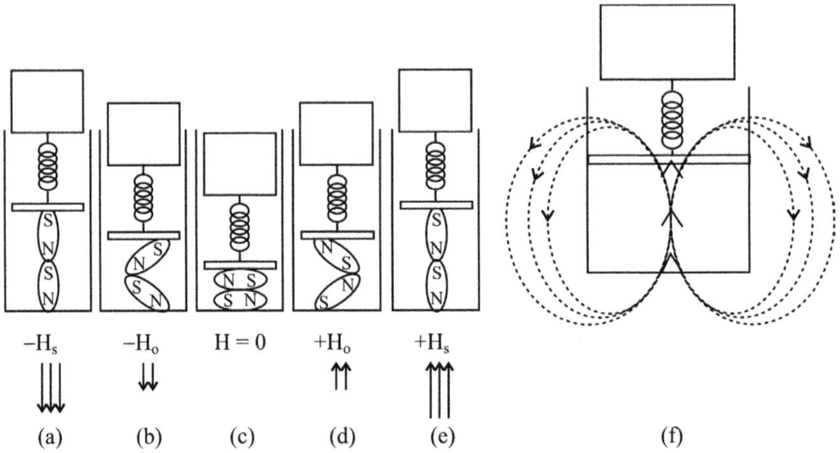

FIGURE 3.4 Magnetostrictive transducers (www.instrumentationtoday.com).

On the other hand, it the value of H_s is kept fixed and the magnitude of force on the magnetostrictive material is increased, then the compressive stress in the material will also increase on to the opposite side along with a reduction in the values of axial strain and axial magnetization.

There are no flux lines present due to null magnetization. Figures 3.4(b) and (d) have magnetic flux lines in a lesser magnitude, according to the alignment of the magnetic domains in the magnetostrictive driver. Figures 3.4(a) and 3(e) also have flux lines in the same design, but its flow will be in the opposite direction. The flux lines according to the applied field H_s and the placing of the magnetic domains are shown in Figure 3.4(f). These flux fields are measured using the principle of Hall effect or by calculating the voltage produced in a conductor kept in right angle to the flux produced. This value will be proportional to the input strain or force.

Such magnetostrictive materials transduce or convert magnetic energy to mechanical energy and vice versa. Bidirectional coupling between the magnetic and mechanical states of a magnetostrictive material provides a transduction capability, which is used for both actuation and sensing devices.

Magnetostrictive transducers consist of a large number of nickel (or some other magnetostrictive material) plates or laminations. These are arranged in parallel with one edge of each laminate attached to the bottom of a process tank or other surface undergoing vibration. A coil of wire is placed around this magnetostrictive material. A magnetic field is produced when a flow of electrical current is supplied through the coil of wire. This magnetic field causes the distortion in length of the magnetostrictive material, either it contracts or elongates; thus, a sound wave is introduced into the fluid. Of course, it is the oldest ultrasonic transducer technology, and was commonly used before the development of efficient and powerful piezoelectric transducers. Magetostrictive transducers have low electrical efficiency (50–60 %) as compared to the 95 % + electrical efficiency, offered by piezoelectric transducers. This efficiency difference is still retained and therefore, most of the ultrasonic equipment used are piezoelectric transducers.

With the discovery of giant magnetostrictive alloys, interest in magnetostrictive transducer technologies has been renewed. Ultrasonic magnetostrictive transducers have been developed for surgical tools, underwater sonar, chemical and material processing, magnetostrictive actuators, sensors, dampers, and so on. With more reliable and larger strain and force, magnetostrictive materials have become commercially available such as Terfenol-D Metglass. Wide range of applications for magnetostrictive devices can be found in ultrasonic cleaners, active vibration, or noise control systems, high-force linear motors, medical and industrial ultrasonics, positioners for adaptive optics, pumps, and sonar. Apart from these, magnetostrictive transducers find some applications in hearing aids, high-cycle accelerated fatigue test stands, razor blade sharpeners, mine detection, and seismic sources.

Magnetostrictive transducers were used to generate high-power ultrasound on an industrial scale. These are based on an effect, normally found in some materials. It was observed that a metal like nickel reduces in size, when it is placed in a magnetic field, but it returns back to its normal dimensions on removal of the field. This property is known as magnetostriction. When the magnetic field is applied as a series of short pulses to a magnetostrictive material, then it starts vibrating at the same frequency. Such transducers can be considered like a solenoid, where magnetostrictive material forms the core with copper wire winding.

The major advantages of these magnetostrictive systems are their durability, robustness, and very large driving forces, but these are also associated with some disadvantages. These are:

- There is an upper limit to the frequency range (~100 kHz), and above this range, the metal cannot respond fast enough to the magnetostrictive effect, and
- Its electrical efficiency is also less than 60 %, as there are significant losses in the form of heat.

Due to heat, normally magnetostrictive transducers have to be liquid cooled. On the contrary, piezoelectric transducers are more efficient and they operate over a relatively wider frequency range and therefore, these are considered more suited for sonochemical studies, particularly under laboratory conditions. Industries require heavy duty and continuous use of transducers and that too at high operating temperatures, the magnetostrictive transducers are gaining their importance in large-scale operations.

Presently, nickel-based alloys (which were used previously) have almost been replaced by more electrically efficient systems such as cobalt/iron combinations, aluminum/iron with a small amount of chromium. An alloy of the rare earths metals such as terbium and dysprosium with iron being used these days, which can be produced in different forms such as rods, laminates, tubes. A magnetostrictive transducer based on these materials can generate relatively more power than a conventional piezoelectric transducer.

3.5.2 PIEZOELECTRIC TRANSDUCERS

The term piezoelectricity means electricity resulting from pressure. It is derived from the Greek term piezō, which means to squeeze or press, and electron, which means amber, an ancient source of electric charge. The piezoelectricity effect was discovered by Curie brothers, Pierre and Jacques Curie (1880; 1881). They observed that a quartz sample becomes electrically charged, when subjected to compressive stress. Reverse is also true, when an electrical voltage is applied to the quartz crystal, it causes a deformation in the crystal.

When crystals of piezoelectric materials are deformed, compressed, twisted or distorted by applying mechanical stress upon them, an electrical charge will appear on their surfaces as a consequence. Here, mechanical energy is converted into electrical energy. This process of transformation of energy from one form to other is called the piezoelectric effect. This provides a convenient transducer effect between electrical and mechanical oscillations on the contrary, if an electrical oscillation is applied to these materials, then they will respond in such mechanical vibrations which produces ultrasound. This is the reverse piezoelectric effect.

3.5.2.1 PIEZOELECTRIC MATERIALS

There are many materials, both natural and man-made, that exhibit a range of piezoelectric effects. Some of the naturally occurring piezoelectric materials are:

- Berlinite (structurally identical to quartz)
- Quartz
- Cane sugar
- Rochelle salt
- Topaz
- Tourmaline
- Bone
- Silk
- Wood due to piezoelectric texture.

Examples of synthetic (man-made) piezoelectric materials including ceramic are:

- Titanates of lithium, barium, bismuth, and lead
- Langasite
- Gallium orthophosphate
- Niobates of lithium, sodium, and potassium
- Bismuth ferrite
- Lithium tantalate
- Lead zirconate titanate[$Pb(Zr_xTi_{1-x})O_3$], with $0 \leq x \leq 1$ more commonly known as PZT
- $Ba_2NaNb_5O_5$
- $Pb_2KNb_5O_{15}$

Piezoelectric ceramic materials are ionically bonded and these consist of atoms with positive and negative charges, called ions. These ions occupy positions in specific repeating units (called unit cells). If a unit cell is noncentro symmetric (means it lacks a center of symmetry), then the application of a stress produces a net movement of the positive and negative ions with respect to each other and this results in an electric dipole or polarization.

Polymers such as polyvinylidene fluoride and self-assembled diphenylalanine peptide nanotubes are also good examples of piezoelectric

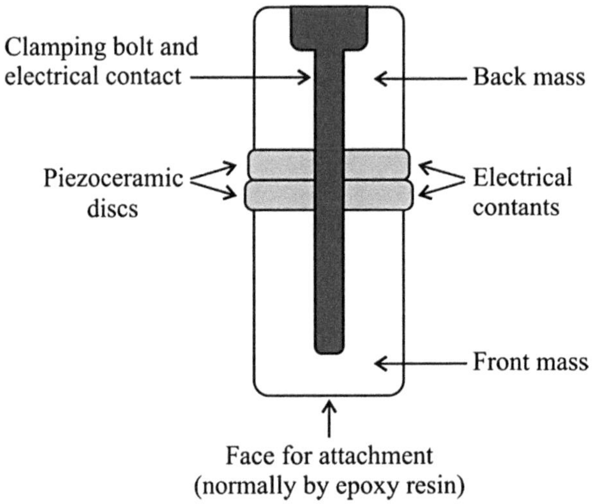

FIGURE 3.5 Piezoelectric transducer.

materials. Piezoelectric polymers exhibiting piezoelectric effect have some advantages like low acoustic impedance and softness.

Piezoelectric transducers are commonly used for the generation and detection of ultrasound (Fig. 3.5). It utilizes materials exhibiting the piezoelectric effect. Such materials show two effects. These are:

Direct Effect—When pressure is applied across the large surfaces of the section, then a charge is generated on each face, which is equal in size but have opposite signs. This polarity is reversed, if tension is applied across the surfaces.

The Inverse Effect—If a charge is applied to one face of the section and an equal but opposite charge is applied to the other face, then the whole section of crystal will either expand or contract and it depends upon the polarity of the applied charges. When rapidly reversing charges are applied to a piezoelectric material, then some fluctuations in dimensions will be produced. Such an effect can be harnessed to transmit ultrasonic vibrations from the crystal section through the medium.

Quartz was a piezoelectric material used earlier in devices such as underwater ranging equipment. It is not a good material for this purpose because it is somewhat fragile and difficult to machine. Instead, modern transducers are based on ceramics containing piezoelectric materials. As these materials cannot be obtained as large single crystals and, therefore, they are ground with binders and sintered under pressure at above 1000 °C so as to form a ceramic. When these are cooled above their ferroelectric transition temperature in a magnetic field, then the crystallites of the ceramic are aligned. These transducers can be produced in different shapes and sizes. At present, the most frequently employed piezoceramic materials contain PZT (P for plumbum, Z for zirconate and T for titanate).

The most common form of transducers is a disk with a central hole. Normally, two such piezoelectric disks are clamped between metal blocks in a power transducer. It serves both purposes: protecting the delicate crystalline material and also preventing it from overheating by acting as a heat sink. The resulting sandwich provides a durable unit with doubled mechanical effect. This unit is generally one half-wavelength long. (However, multiples of this can also be used.) The peak-to-peak amplitudes generated by such systems are of the order of 10–20 μ in general and they are electrically efficient. Piezoelectric devices cannot be used for long periods at high temperatures as the ceramic material will degrade under these conditions and therefore, under these conditions, these devices must be cooled.

Such piezoelectric transducers are highly efficient (>95 %) and can be used almost over a range of ultrasonic frequencies from 20 kHz to some MHz depending on the dimensions. These are good choice in medical scanning, which requires frequencies above 5 MHz.

The degree of polarization depends upon the stress and tensile or compressive stresses affecting the charge produced. Dipoles, present due to the noncentro symmetric structure, form regions having a same alignment of neighbouring dipoles.

Initially, these domains are randomly oriented and there is no overall polarization of the ceramic and therefore, no piezoelectric effect were exhibited. These domains are subjected to poling, when heat and strong DC field are applied causing the domains nearly aligned to the field to grow at the expense of those at differing alignments. When these are cooled to room temperature and DC field is removed, the domains are locked resulting in an overall alignment. As a result, the material shows piezoelectric behaviour.The piezoelectric materials used at present are based mostly on sintered materials. These materials are beneficial because they have high performance and are readily available in a large variety of shapes. They most often occur as a disc or ring with metal electrodes attached to their surfaces.

The thickness of the ceramic disc is e and it is subjected to a compressive force F. The voltage that appears at the electrode terminals is:

$$V = g_{33} \times e \times F / S$$

where g_{33} is the characteristic of the materials and S is the section of the disc.

If thickness of the ceramic disc e is subjected to a continuous electrical voltage V, the thickness of disc increases to $e + \Delta e$. The voltage and movement of the disc are related by the following expression:

$$\Delta e = d_{33}V$$

where $\Delta e = 0.5\ \mu$ per 1000 V of excitation and $d_{33} = 500\ 10 - 12$ m/V.

This effect is used particularly for thin ceramic multilayer stacks to produce stacks, which create movement of several tens of microns under a voltage of a few hundred volts up to 1000 V.

The relationships used to determine the thickness corresponding to a particular frequency are:

$$\text{Frequency} = \frac{\text{Velocity of ultrasound in crystal material}}{\text{Wavelength}}$$

$$\text{wavelength} = 2 \times \text{thickness } (t)$$

Therefore,

$$\text{Frequency } (f) = \frac{\text{Velocity } (v)}{2t}$$

or

$$t = \frac{v}{2f}$$

It means that thinner is the crystal, the higher will be the frequency.

A variety of ultrasonic transducers are manufactured depending on its different applications. A proper attention has to be paid in selecting a transducer based on the field of application. It is quite desirable to select transducers that have the desired frequency, bandwidth and focusing to optimize its capability for a particular application. Transducer is chosen so as to either enhance the sensitivity or resolution of the system.

3.6 SINGLE-CRYSTAL ULTRASONIC TRANSDUCER

A piezoelectric transducer consists of:

- Piezoelectric crystal
- Electrical connections
- Backing materials
- Front layers
- Casing and so forth

The typical construction of a single-crystal transducer is shown in Figure 3.6.

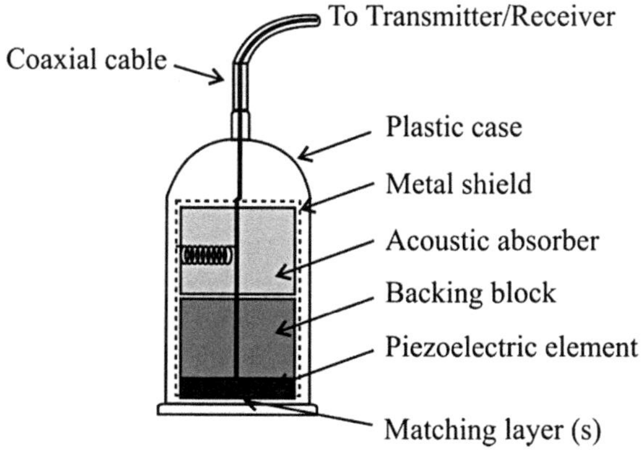

FIGURE 3.6 Components of the single-crystal transducer.

3.6.1 PIEZOELECTRIC CRYSTAL

A single-crystal ultrasonic transducer consists of a thin layer of piezoelectric material sandwiched between epoxy, a glue-like backing material, and a number of layers of facing. The active element is a piezoelectric- or single-crystal material. This crystal converts electrical energy into ultrasonic energy. It also then receives back the ultrasonic energy and converts it into electrical energy. The electrical energy pulse is generated from a device such as a flaw detector.

3.6.2 BACKING BLOCK

The backing material functions as a damping block. It is usually made of an epoxy-like material, and is glued to the inner surface of the crystal. It absorbs the reversed ultrasound waves that are transmitted to the back of the crystal. The material must have the same acoustic impedance as that of the crystal, in order to prevent an echo from a crystal/damping interface from returning the energy back to the crystal and creating reverberation noise. This damping block is also known as the mechanical pulse damper as it serves to limit the spatial pulse length by mechanically stopping the ringing of the crystal. This helps in optimizing axial resolution. Pulse damping also reduces the amplitude of ultrasound and, as a result, it reduces the sensitivity of the transducer. Excessive pulse damping gives a very wide frequency bandwidth. The backing is most commonly a highly attenuative and very dense material. It is used to control the vibration of the transducer crystal by absorbing the energy that is radiated from the back face of the piezoelectric element. When the acoustic impedance of the backing material matches with that of the piezoelectric crystal, it resulted in a highly damped transducer with high-quality resolution. By varying the backing material and in turn varying the difference in impedance between the backing and the piezoelectric crystal, a transducer is affected in a way that resolution may be relatively higher in signal amplitude or sensitivity.

3.6.3 INSULATION RING

It is also known as a sidewall acoustic insulator. This part of a transducer is made of the same material as that of the damping block. It serves primarily to absorb energy generated from the sides of the crystal.

3.6.4 TUNING COIL

The crystal is a capacitative device forming part of the pulser and receiver circuits. The tuning coil serves to offset the capacitative effect of the crystal by removing residual electrical charges. The tuning coil, by dusting off the excess electrical charges, improves the transmitting and receiving functions of the transducer.

3.6.5 ELECTRICAL SHIELD

The electrical shield is an isolation barrier that serves to eliminate unwanted, stray signals, also known as noise. It accomplishes this task by detecting, isolating, and sending the stray signals to ground. Noise is any unwanted vibration that interferes with the efficient production of a sonographic image. All electrical outlets that are isolated carry spurious vibrations and radiofrequency (RF) signals from respirators, elevators, typewriters, coffee grinders, etc., which are also connected to the same circuit. Shields built into the imaging system can eliminate most of this unwanted, low-level, electronic noise from being detected by the very sensitive receiver. Any vibratory or RF noise that is detected by the receiver will be displayed on the image and will degrade the overall image quality. During the operation of the imaging system outside the ultrasound department, external noise is most appreciable in a clinical environment, such as in the intensive or critical-care units, where multiple electronic and mechanical devices are operating in close proximity.

3.6.6 ELECTRIC CONNECTORS

These connectors serve to link the transducer electrically to the ultrasound instrument. It is through these connectors that the electrical impulse ringing the crystal is delivered, and the returning echoes are received. Generally, these are a pair of very thin wires attached to each crystal. Highly sophisticated ultrasound imaging probes may have more than 1000 crystals, each one of which is attached to a separate pair of electrical connectors. These wires are housed in the transducer cable that is attached to the probe port on the imaging system.

3.6.7 MATCHING LAYERS

The primary objective in designing transducers for diagnostic ultrasound imaging systems is to have highest sensitivity, penetration, optimal focal characteristics, and best possible resolution all at low acoustic power levels. It is made more difficult by the physical reality that the difference in acoustic impedance between the transducer crystal (PZT) and the surface of the patient's skin is significant. The magnitude of this difference in acoustic impedance effectively prohibits the adequate transmission of ultrasound energy into a patient's body so that an engineering solution has been devised to overcome this obstacle to imaging.

A layer of material is placed that possesses an intermediate acoustic impedance between the crystal surface and skin. Thus, a type of mechanical transformer has been created, which steps down the impedance change more gradually. Engineers have created a shoehorn that helps sound waves slip into the body more easily by placing multiple layers of "transforming" material in the probe face, thereby saving majority of the energies for imaging deeper inside. These layers of transforming material are called multiple matching layers.

The optimum thickness of this layer is one quarter of a wavelength. Since a broad spectrum of frequencies exists in any given ultrasound beam, the natural frequency is used to determine the appropriate wavelength. Basically, a phase reversal occurs by using one-quarter wavelength thickness, which increases signal performance and strengthens the wavefront entering the body. The quarter wavelength matching layer design provides increased sound transmission and reception.

3.6.8 ACOUSTIC COUPLANT

In addition to such matching layers, an acoustic gel is applied as a sonic couplant to the patient's skin. It also aids in reducing the large acoustic mismatch between the transducer and the patient. The transducer face is normally designed to have an impedance value midway between that of the crystal and that of the patient's skin, so that the gel is chemically engineered to have impedance value halfway between that of the transducer face and the skin. Both matching layers and acoustic couplant address the physical challenges of insonating the human body.

3.6.9 WEAR PLATE

The main purpose of the wear plate is to simply protect piezoelectric transducer element from the environment. Wear plates are selected to generally protect against wear and corrosion. In an immersion-type transducer, the wear plate also serves as an acoustic transformer between the piezoelectric transducer element and water, wedge, or delay line.

The front layer protects the piezoelectric element against external stresses and other influences of the environment. It also functions as an impedance matching layer with which the transfer of ultrasonic energy to the target medium is optimized whereas the backing material functions as a damping block to alter the resonance frequency of the piezoelectric element. In addition to this, the unwanted ultrasonic waves reflected from the back wall are also deleted. The electrical line is used to connect AC or DC voltage supplies that are operated at the resonant frequency of the piezoelectric element. Depending on the type of applications, other types of transducers are also available.

3.7 ULTRASONIC CLEANING BATH

The simple ultrasonic cleaning bath is the most widely available and low-cost source of ultrasonic irradiation in the chemical laboratory. It is possible to use the cleaning bath itself as a reaction vessel, but it is not used commonly, because of the problems associated with corrosion of the bath walls and containment of any evolved vapours and gases. Therefore, normal use involves the immersion of standard glass reaction vessels into the cleaning bath, which provides a fairly even distribution of energy into the reaction medium (Fig. 3.7). The reaction vessel does not need any special adaptation,

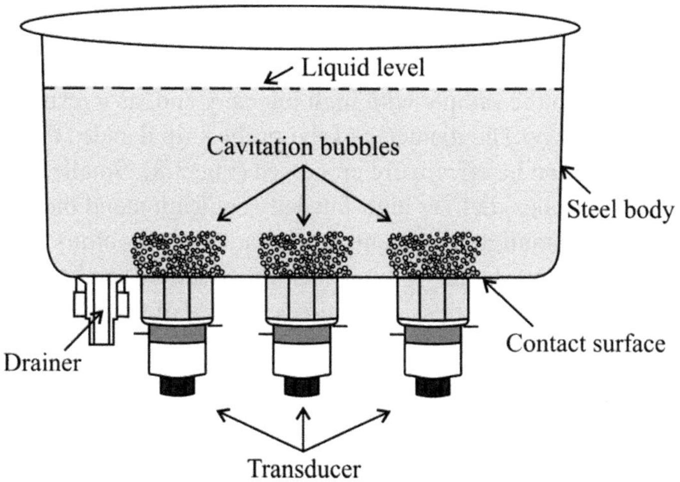

FIGURE 3.7 Ultrasonic cleaning bath.

since it can be placed into the bath. An inert atmosphere or pressure can be readily maintained during a sonochemical reaction. The amount of energy reaching the reaction through the vessel walls is low, normally between 1 and 5 $W \cdot cm^{-2}$. Temperature control in commercial cleaning baths is generally poor and so the system may require additional thermostatic control.

3.8 ULTRASONIC PROBE

This allows acoustic energy to be introduced directly into the system rather than relying on its transfer through the water of a tank and the reaction vessel walls as done in the case of ultrasonic cleaning bath. The power of such systems can be easily controlled and the maximum power ranges in the order of several hundreds of $W \cdot cm^{-2}$. Ultrasonic probe system is more expensive than the cleaning bath and it is slightly less convenient to use also as special seals will be needed, if the horn is to be used in reactions requiring reflux, inert atmospheres or pressures more or less than ambient conditions.

Ultrasonic probe is classified into two types:

1. Direct sonication
2. Indirect sonication

3.8.1 DIRECT SONICATION

It is the most common way to process a sample by inserting a probe directly into a sample vessel. In this case, energy is transmitted from the probe directly into the sample with high intensity and, as a result, sample is processed quickly. The diameter of the probe's tip decides the volume of the liquid that can be effectively processed (Fig. 3.8). Smaller tip diameters (microtip probes) deliver high-intensity sonication and the energy is focused within a small, but concentrated area. On the contrary, larger tip diameters can process larger volumes of liquids. Probes are offered with either replaceable or solid tips. These tips are made of titanium.

3.8.2 INDIRECT SONICATION

It eliminates the need for a probe to come in contact with sample. This technique is commonly described as a high-intensity ultrasonic bath. Here,

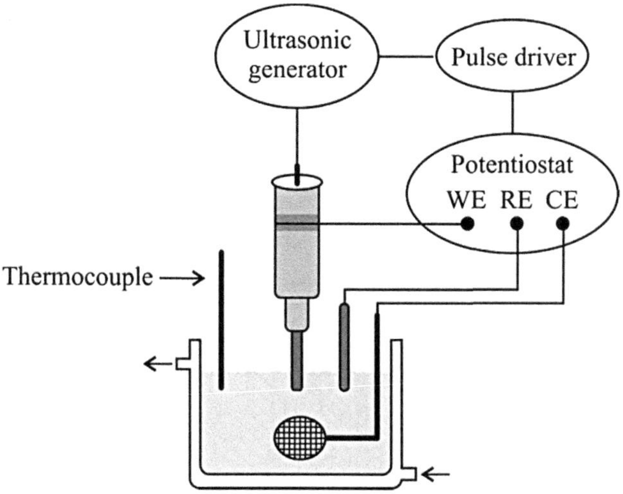

FIGURE 3.8 Probe system.

Note: WE is working electrode (platinum wire), RE is reference electrode (silver wire) and CE is counter electrode (platinum flag).

the ultrasonic energy is transmitted from the horn, through the water, and finally transmitted into a vessel or multiple sample tubes. Indirect sonication is most effective for small samples because most of the damages to the sample such as loss, over temperature, etc. are eliminated. Pathogenic or sterile samples are ideal for this method because aerosols and cross-contamination are prevented. The cup horns deliver indirect sonication and these are ideal for many applications, where the liquid sample is more in volume (Fig. 3.9).

3.9 FLOW CELL

There are basically two systems. These are (i) continuous stirred tank reactors (CSTRs) and (ii) continuous plug-flow reactors (CPFRs).

CSTRs are in principle equipped in the same way as batch systems while in ultrasonic plug flow reactors, the reaction mixture flows through one or more tubes. The setup is often designed as a loop in a mixed-flow system (mixed-flow reactor with recirculation), that is, the sonoreactor is placed in the loop linked to a conventional reactor containing the reaction mixture. It allows for a repeated flow and the coupling of several

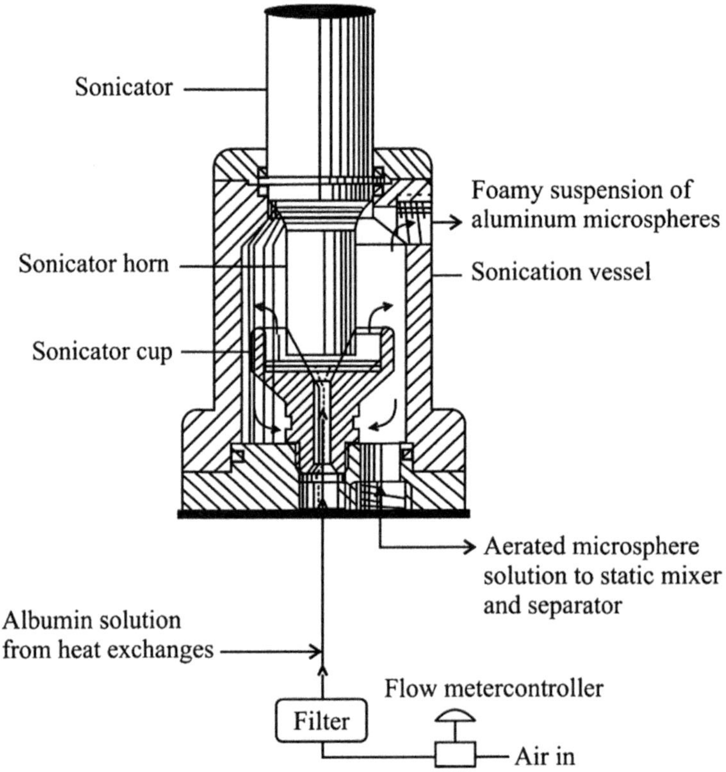

FIGURE 3.9 Cup horn.

modules, which may be used for pre-reactions or secondary processes. Several different types of plug-flow reactors are commercially available for their use in laboratory.

- Flow-cell attachment for the probe system
- Liquid whistle homogenizer
- Tube with transducers bonded around the outside so that the metal-forming tube becomes the ultrasonically vibrating source
- Bar arranged so that the vibrations are emitted radially from the cylindrical face rather than the end. This bar is placed concentrically in the center of a tube through which the medium flows.

Flow systems are best suited for processes requiring only a short sonication time since the residence duration is short in this case (Fig. 3.10).

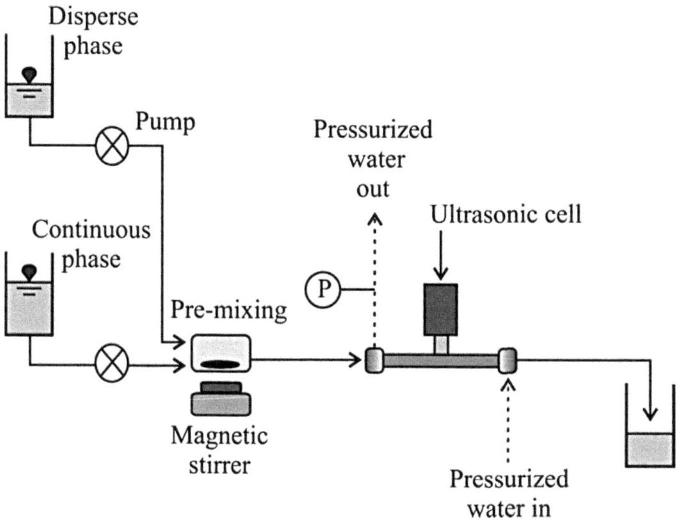

FIGURE 3.10 Flow cell.

They allow a high-processing throughput. It is known that the penetration depth of ultrasound decreases strongly with increasing frequency and, therefore, batch reactors are not suitable at high frequency. In such cases, a flow loop with or without recirculation can be used.

Immersible (submersible) transducers are designed for use in a heavy, industrial environment. These transducers are enclosed in a water-proof casing so that it can be immersed into a cleaning tank, used as part of an online cleaning process, or used to trouble shoot, where cleaning of hot spots is required. The hermetically sealed welded housings are manufactured from stainless steel. Plastic tanks are not suitable for ultrasonic applications because they dampen the ultrasonic waves and reduce the available ultrasonic cleaning power from anything between 60 and 90%.

Submersible cleaning systems are powered by ultrasonic generators that are housed in transducer packs, which are made of stainless steel. These packs contain a double stack of individual high-efficiency ultrasonic transducers. They are ordinarily mounted to the bottom or the side walls of an existing tank so that ultrasonic cleaning is better in the tank in which these transducers are submerged. The transducer packs have special mounting tabs with holes and these are so designed that they can be attached permanently/temporary to the tank. Submersible systems are available with single- or multiple-frequency ultrasonic systems.

3.10 TUBE REACTOR

It is a modern method of continuous or intermittent inline ultrasonic processing for any liquid material or food requiring ultrasonic treatment. Single or multiple transducers are connected to custom clamps. These clamps are designed in such a way that they fit into tubes or pipes of any size, constructed of aluminum, stainless steel, or titanium. Unique multi-frequency, multimode, and modulated (MMM) technology offers a highly efficient transfer of ultrasonic energy to this metal pipe or tube. This pipe or tube becomes a radiating element allowing internal or external material treatment. Such assemblies will turn nearly any suitable pipe or tube into a highly efficient ultrasonic reactor. Longer pipe sections may be used with multiple clamps powered by one or more MMM generators.

Ultrasonic transducers are manufactured for a variety of applications. These can be fabricated depending on the type of applications. However, a proper transducer should be carefully selected for a particular application. It is quite important to select transducers that have the desired frequency, bandwidth, and focus to optimize inspection capability for an application. Most often, the transducer is chosen to enhance the sensitivity or the resolution of system.

Various instruments have been designed to carry out different sonochemical reactions. If such reactions are to be performed on industrial scales, then these instruments need to be robust and more stable for a longer period of time. Another aspect, which could be improved for sonochemical instrumentation, is their ease of use and handiness. It will make sonochemistry a more developed branch of chemistry and gain a higher status as a green chemical technology.

REFERENCES

Curie, J.; Curie, P. *Comptes Rendus* **1880,** *91*, 294–295.

Curie, J.; Curie, P. *Comptes Rendus***1881,** *93*, 1137–1140.

Galton, F. *Inquiries into Human Faculty and Development;* McMillan: London, **1883**.

Joule, J. P. *Ann. Electr. Magn. Chem.* **1842,** *8*, 219–224.

Richards, W. T.; Loomis, A. L. *J. Am. Chem. Soc.* **1927,** *49*, 3086–3100.

www.instrumentationtoday.com.

CHAPTER 4

ORGANIC SYNTHESIS

CHETNA AMETA[1], ARPIT KUMAR PATHAK[2], P. B. PUNJABI[3]

[1]*Department of Chemistry, M. L. Sukhadia University, Udaipur, India*
E-mail: chetna.ameta@yahoo.com

[2]*Department of Chemistry, M. L. Sukhadia University, Udaipur, India*
E-mail: arpitpathak2009@gmail.com

[3]*Department of Chemistry, M. L. Sukhadia University, Udaipur, India*
E-mail: pb_punjabi@yahoo.com

CONTENTS

4.1 INTRODUCTION

The application of ultrasound in organic synthesis has gained considerable momentum in recent years and as a consequence, several organic transformations were effected using ultrasound. It has been observed that ultrasound reduces the requirement of higher reaction temperature, and higher reaction rates can be achieved even under ambient conditions.

Two types of effects are mediated by ultrasound. These are chemical and physical effects. Mainly physical rate acceleration plays a vital role, when the quantity of bubbles is low. A specific effect is the asymmetric collapse near a solid surface, forming microjets. Basically, this effect is the major reason for using ultrasound effectively in cleaning, but it is also responsible for rate enhancement in multiphasic reactions as surface cleaning and erosion lead to improved mass transport.

Water is a desirable solvent for synthesis because water is low cost, safe, and nontoxic. As the use of water as a solvent is limited to water-soluble compounds, it excludes most of the organic, lipophilic substances. A solution to this issue is to increase the solubility of organic compounds by using a surfactant to form micelles, thus increasing contacts with the reagent.

A number of name reactions have been carried out under sonochemical conditions, but only some of these name reactions have been discussed in this chapter. It does not mean that other reactions cannot be carried out with ultrasound.

4.2 NAME REACTIONS

There are several name reactions that have been carried out under sonochemical conditions, and we have restricted our discussion only to some of these name reactions in this chapter.

4.2.1 ALDOL CONDENSATION

An aldol condensation is a condensation reaction in organic chemistry in which an enol or an enolate ion reacts with a carbonyl compound to form a β-hydroxyaldehyde or β-hydroxyketone, followed by dehydration to give a conjugated enone. Aldol condensations are important in organic synthesis, because they provide a smooth way to form carbon–carbon bonds (Carey and Sundberg, 1993). These reactions are usually catalysed by strong acids or bases, and a variety of different Lewis acids have been evaluated in this reaction (Reeves, 1966). Unfortunately, the presence of a strong acid or base promotes the reverse reaction (Hathaway, 1987) and this leads to the self-condensation of the reacting materials to give the corresponding byproducts in low yields (Nakano et al., 1987).

Based on the importance of the methylene moiety found in many naturally occurring compounds and antibiotics and the use of aldols as precursors for the synthesis of bioactive compounds, Khaligh and Mihankhah (2013) proposed ultrasound-assisted aldol condensation reaction for a wide range of ketones with a variety of aromatic aldehydes using poly(*N*-vinylimidazole) as a solid base catalyst in a liquid–solid system. The catalyst can be recovered by simple filtration and reused at least 10 times without any significant reduction in its activity. The reaction is also amenable to the large scale, making the procedure potentially useful for industrial applications.

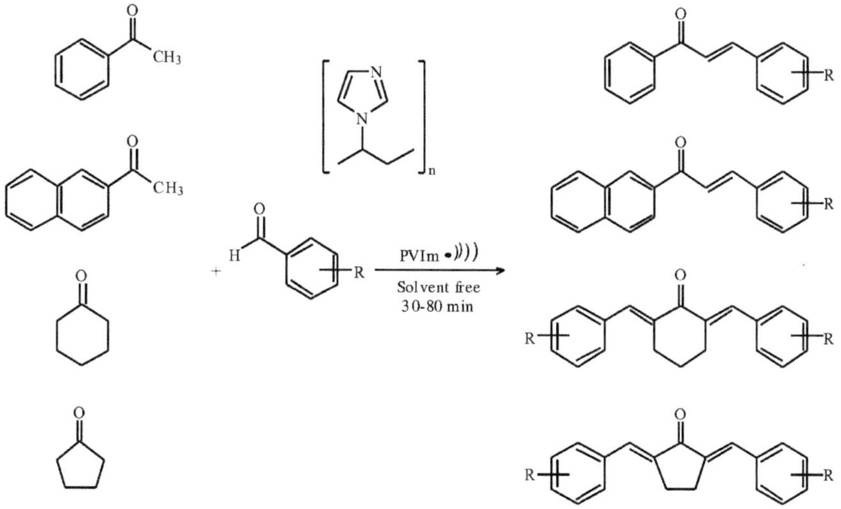

There are several advantages to the current methodology such as:

- Simple procedure as the use of poly(*N*-vinylimidazole) as a solid base catalyst does not require any inert or anhydrous conditions
- A green process because PVIm is environmentally benign
- Atom economical
- Avoids the occurrence of adverse-side reactions because PVIm can facilitate the aldol condensation under mild and neutral conditions and therefore eliminate any side-reactions resulting from strong basic conditions

- Catalyst can be conveniently recovered and reused without any significant impact on its activity

The current method represents an attractive alternative for the preparation of chalcones.

Cravotto et al. (2003) studied the aldol reaction under high-intensity ultrasound (HIU). HIU was employed to reinvestigate the aldol reaction in water. A number of aldols that would eliminate or form side products under conventional conditions were isolated in good yields within minutes under ultrasonics (Table 4.1).

According to Table 4.1, we can see that the results are highly reproducible because the sonochemical parameters were rigorously controlled.

TABLE 4.1 Aldol Condensation under Ultrasonics.

Product	Yield (%)	Product	Yield (%)
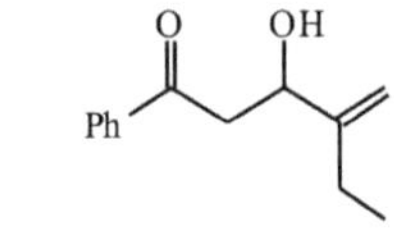	52	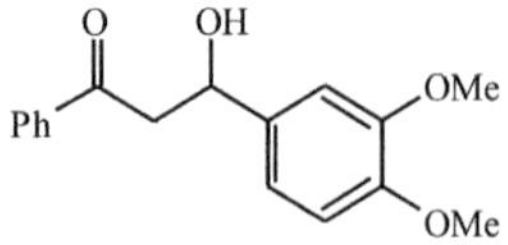	74
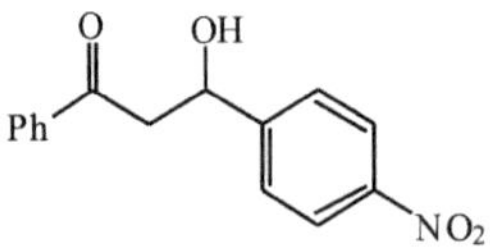	78	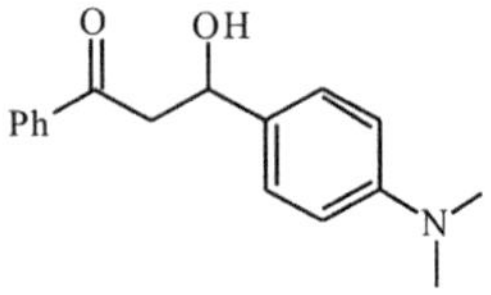	62

4.2.2 *BAYLIS–HILLMAN REACTION*

The Baylis–Hillman reaction is a carbon–carbon bond-forming reaction between the α-position of an activated alkene and an aldehyde, or generally a carbon electrophile. This reaction provides a densely functionalized product (e.g. functionalized allyl alcohol in the case of aldehyde as the electrophile) employing a nucleophilic catalyst, such as tertiary amine and phosphine (Baylis and Hillman, 1972). Roos and Rampersadh (1993) observed that the rate of reaction of 1,4-diazabicyclo [2. 2. 2] octane (DABCO) catalysed coupling of aldehydes with methyl acrylate increases by the application of ultrasound.

In Baylis–Hillman reaction, several aldehydes (aromatics as well as aliphatics) react with different α,β-unsaturated reactants. The effect of ultrasound radiation on the Baylis–Hillman reaction has been studied by Fernando et al. (2002).

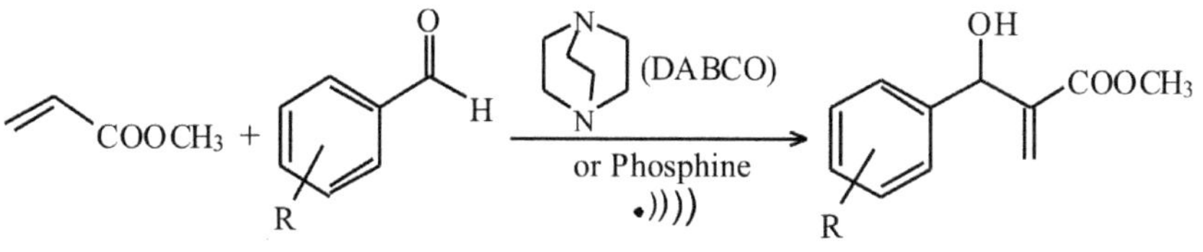

where R = H, 4–OCH_3, 3,4–OCH_2O, 4–Cl.

The utilization of ultrasound radiation in the reaction with several aldehydes (aromatics and aliphatics) and different α,β-unsaturated reactants has also been described by Coelho et al. (2002). The utilization of ultrasound sources augmented the reaction rate and the chemical yields for all aldehydes. The use of ultrasound with two different catalysts (tri-*n*-butylphosphine and DABCO) was also investigated. It was demonstrated that DABCO is relatively effective for catalyzing a Baylis–Hillman reaction under the influence of ultrasound than tri-*n*-butylphosphine.

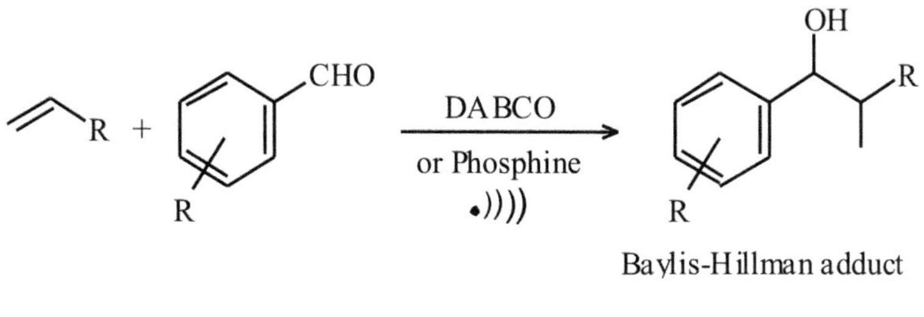

where R = CO_2CH_2, $COCH_2$, CN.

A convenient protocol was developed for the synthesis of the Baylis–Hillman adducts in the presence of weak lewis base and l-proline at room temperature under ultrasound irradiation (Mamaghani and Dastmard, 2009). This method provides products in good to high yields (65–90%) and that too in reasonable reaction times.

A catalyst-free one-pot four-component methodology for the synthesis of 2H-indazolo[2,1-b]phthalazine-triones under ultrasonic irradiation at room temperature has been described by Shekouhy and Hasaninejad (2012). They used 1-butyl-3-methylimidazolium bromide [BMIM]Br, as a neutral reaction medium. A broad range of structurally diverse aldehydes (aromatic aldehydes bearing electron-withdrawing and/or electron-releasing groups as well as heteroaromatic aldehydes) were used successfully, and the corresponding products were obtained in good to excellent yields without any byproduct.

Pourabdi et al. (2016) synthesized various derivatives of thiopyrano[4,3-b]pyran structure via a multicomponent procedure starting from tetrahydro-4H-thiopyran-4-one, aromatic aldehydes and malononitrile. This reaction takes place using catalytic quantities of $LiOH{\cdot}H_2O$ in one pot under ultrasonic conditions and in an ethanolic medium.

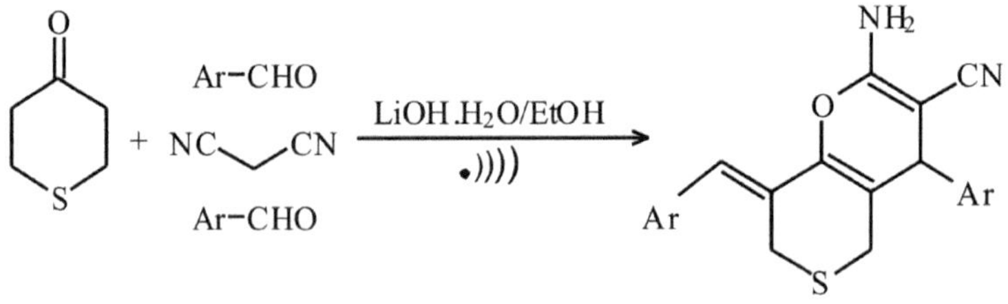

4.2.3 BENZOIN CONDENSATION

The benzoin condensation is a reaction between two aromatic aldehydes, particularly benzaldehyde. The reaction is catalysed by a nucleophile such as cyanide anion or an *N*-heterocyclic carbene. The reaction product is an aromatic acyloin with benzoin as the parent compound (Adams and Marvel, 1941).

Benzoin condensation is an important method for the formation of carbon–carbon bonds starting from aldehydes giving α-hydroxycarbonyl compounds, which are important building blocks for the synthesis of natural and pharmaceutical compounds (Iwamato et al., 2006). After the first report by Wohler and Liebig, who used cyanide as a precatalyst in 1832 (Wohler and Liebig, 1832), various other catalysts have been used to promote the benzoin condensation reaction effectively.

There is still a need to introduce novel methods to permit better selectivity under milder conditions and with easy work-up procedures. Ultrasonic irradiation has been considered as a clean and useful protocol in organic synthesis during the last few decades, as the procedure is more convenient compared with conventional methods.

Kinetics of the benzoin condensation of benzaldehyde in the presence of KCN as the catalyst in water and in ethanol–water binary solutions was investigated without sonication and under ultrasound at 22 kHz (Hagu et al., 2007). A statistically significant 20 % decrease in the rate was observed in water. The retardation effect of ultrasound gradually decreases up to 45 wt.% ethanol content. It is an evidence of ultrasonic retardation of reactions. Ultrasound can disturb solvation of the species in the solution. Sonication hinders the reaction whereas perturbation of the solvent stabilization of the reagents accelerates the reaction.

A new method for the benzoin condensation has been proposed by Estager et al. (2007) with imidazolium-based ionic liquids (ILs) as solvents/catalysts. The use of a 30-kHz ultrasonic irradiation as activation method provided a quantitative yield of benzoin in 1-octyl-3-methylimidazolium bromide without degrading the ionic medium, which can be reused at least three times with no loss of activity.

Safari et al. (2015) developed a rapid, highly efficient and mild green synthesis of benzoin using substituted benzaldehyde catalysed by KCN and imidazolium salts in EtOH/H_2O under ultrasonic activation.

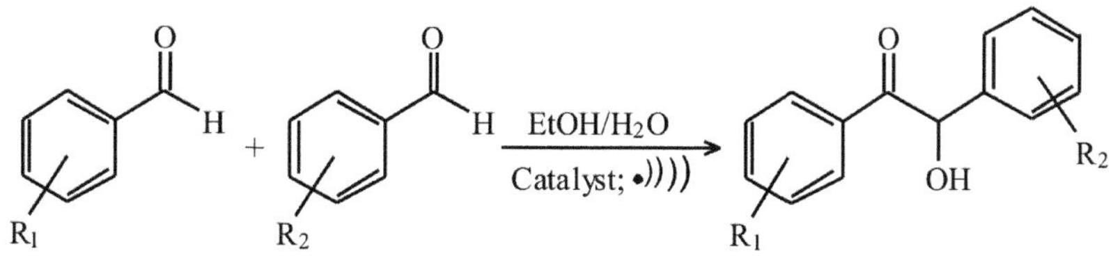

The products were obtained in good yields and high purity within short reaction times with *N,N*-dialkylimidazolium salts, which were found to be more effective precatalysts at room temperature for benzoin condensation in comparison to corresponding cyanide ion in heating method. This is a simple method affording benzoin derivatives at room temperature.

The imidazolium salts efficiently catalyse the benzoin condensation of aldehydes in the presence of a base. These tricationic and dicationic imidazolium salts show considerable catalytic potential, when compared with monocationic imidazolium salts and toxic cyanide anions. Safari et al. (2015) also made use of tricationic imidazolium salts in the acyloin condensation under ultrasound. This afforded benzoin derivatives at room temperature in short reaction times with high yield and purity. It can be used as a replacement for conventional thermal synthetic methodology, allowing rapid access for the synthesis of natural and pharmaceutical compounds.

4.2.4 BIGINELLI REACTION

Due to increasing environmental consciousness, a challenge exists for a sustainable clean procedure avoiding the use of harmful organic solvents, or it will be still better, if no solvent is used at all. Sonochemistry has offered such a versatile and environmentally benign condition for a large variety of syntheses and emerged as a new trend in organic chemistry. This has led to multistep synthetic strategies that can produce better yields; however, it lacks the simplicity of the original one-pot Biginelli protocol.

The Biginelli reaction is a multicomponent reaction that gives 3,4-dihydropyrimidin-2(1*H*)-ones from ethyl acetoacetate, an aryl aldehyde (such as benzaldehyde) and urea. This reaction was developed by Biginelli (1893). The reaction can be catalysed by Bronsted acids and/or by Lewis acids such as copper(II) trifluoroacetate hydrate and boron trifluoride (Kappe, 1993).

Li et al. (2003) reported that the condensation of aldehydes, β-keto esters and urea catalysed by NH_2SO_3H in ethanol results in the formation of dihydropyrimidinones with high yields under ultrasound irradiation.

Liu and Wang (2010) reported an efficient synthesis of novel 4-(2-phenyl-1,2,3-triazol-4-yl)-3,4-dihydropyrimidin-2(1H)-(thio)ones from 1,3-dicarbonyl compounds, 2-phenyl-1,2,3-triazole-4-carbaldehyde and urea or thiourea using samarium perchlorate $[Sm(ClO_4)_3]$ as a catalyst in water bath of the ultrasonic cleaner at 75–80 °C. The main advantages of this method are milder conditions, shorter reaction times, and higher yields than that of conventional methods.

Singhal et al. (2010) reported an efficient, simple, and environmentally clean synthesis of 3,4-dihydropyrimidinones in excellent yields in the presence of water without additional solvent/acid catalyst under conventional heating, and microwave irradiation/ultrasound. In this reaction, a small flask containing aldehyde (2 mmol), urea (2 mmol), β-dicarbonyl compound (2 mmol), and water (3–4 drops) was sonicated at 25 kHz.

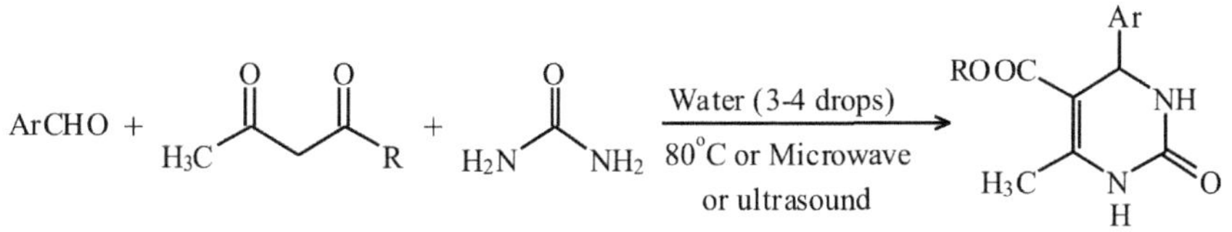

where Ar = Ph, 4–$CH_3C_6H_4$, $CH_3OC_6H_4$, $NO_2C_6H_4$, ClC_6H_4; 2–ClC_6H_4, Pyridyl, Furyl; $C_6H_5CH{=}CH$, C_6H_5; R = OEt, OMe, Me.

Mandhane et al. (2010) reported an environmentally benign aqueous Biginelli protocol for the synthesis of substituted 3,4-dihydropyrimidin-2(1*H*)-ones using thiamine hydrochloride as a catalyst under ultrasound. These reactions proceed efficiently in water in the absence of an organic solvent. Utilization of ultrasound irradiation, simple reaction conditions, isolation, and purification makes this manipulation very interesting from an economic and environmental perspective.

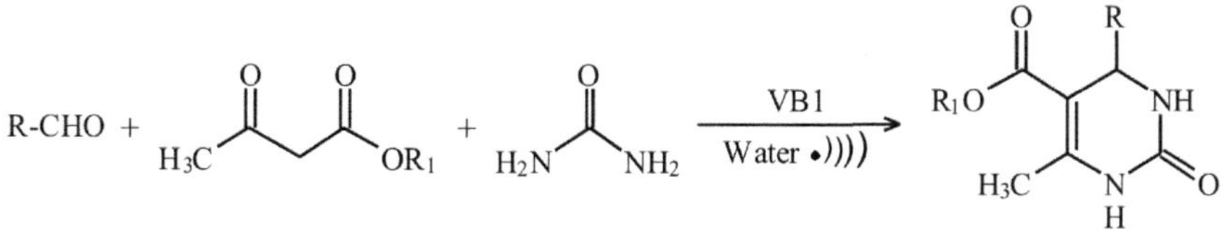

Biginelli reaction is carried out by simply heating a mixture of all the three components (aldehyde, urea or thiourea and ethyl acetoacetate) dissolved in ethanol with a catalytic amount of HCl at reflux temperature. This method has some drawbacks such as low yields (20–40%), particularly, when substituted aldehydes are used. It is also accompanied by the loss of acid-sensitive functional groups during this reaction. Kakaei et al. (2015) reported an ecofriendly method for the synthesis of dihydropyrimidinones under ultrasonic irradiation in the presence of holmium chloride hexahydrate as a catalyst under solvent-free conditions. The reactions were completed within 2–4 h with 87–96% yields.

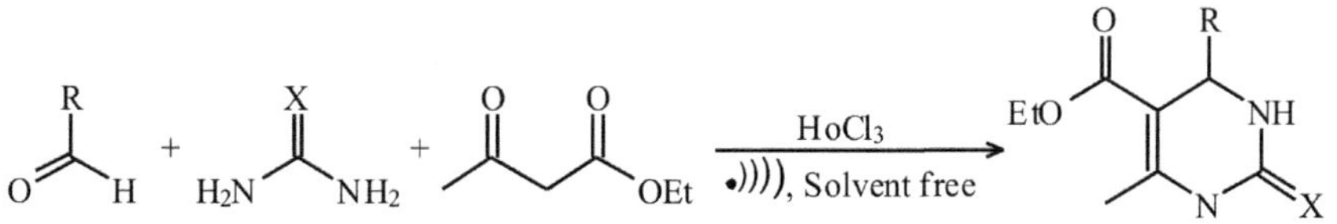

Dhawanpalli et al. (2015) described the Biginelli reaction for the synthesis of 3,4-dihydropyrimidin-2(1H)-one derivatives under solvent-free conditions (involving cyclocondensation) with ultrasound irradiation as the energy source. This synthesis was performed in the presence of polymer nanocomposite polyindole-Fe as the catalyst. The reactions were compared with the classical Biginelli reaction conditions. This method has the advantage of excellent yields (82–92 %).

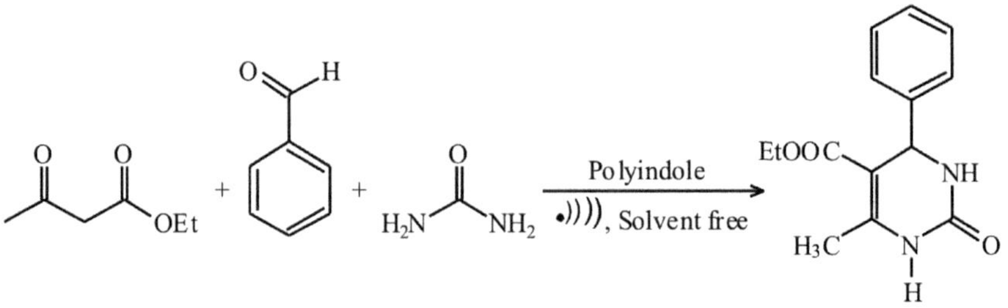

Girase et al. (2016) reported an efficient ultrasound promoted synthesis of 3,4-dihydropyrimidin-2(1H)-one/thione by using aluminium sulphate as a catalyst for aromatic aldehydes, 1,3-dicarbonyl compounds, and urea or thiourea in PEG as a reaction solvent. It was compared with the classical Biginelli reaction conditions. This new procedure is simple, rapid and high yielding. The catalyst also exhibited a remarkable reactivity and is reusable.

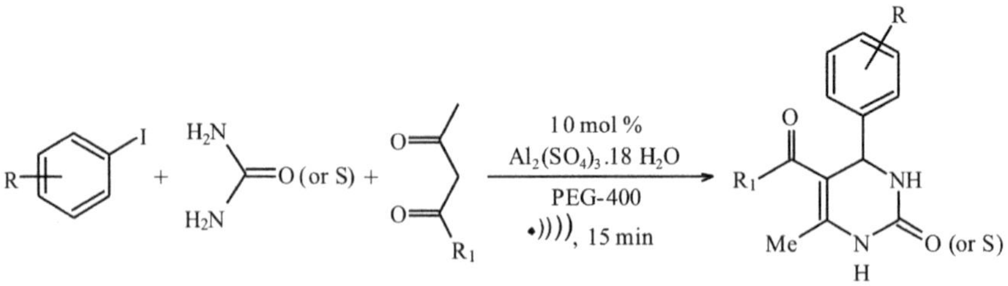

4.2.5 BLAISE REACTION

Blaise reaction is an organic reaction that forms a β-ketoester from the reaction of zinc metal with α-bromoester and a nitrile (Rao et al., 2008). The reaction was first reported by Edmond Blaise in 1901 (Blaise, 1901). The final intermediate is a metaloimine, which is then hydrolysed to give the desired β-ketoester (Cason et al., 1953).

Sonication is a very important tool for the formation of β-amino-α,β-unsaturated esters. These are produced by Blaise reaction of nitriles, zinc powder, zinc oxide, and ethyl bromoacetate in THF in the presence of ultrasound (Lee and Chang, 1997).

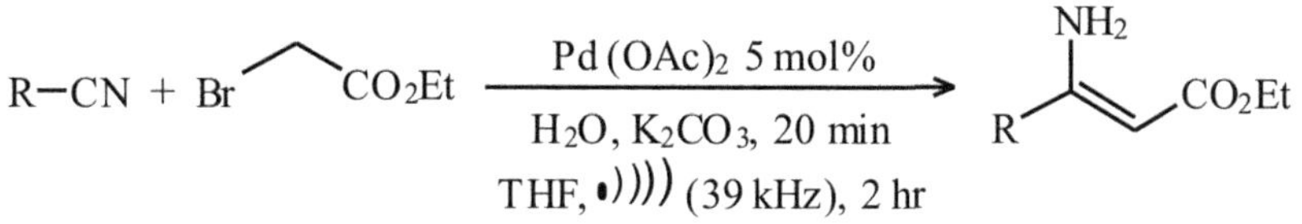

Where R = Alkyl, Ar, Bn.

2,3-*o*-Isopropylidene-d-glyceraldehyde was selected as the initial compound, which is a chiral material. Naturally occurring (s)-(+)-dihydrokavain was synthesized by a procedure that involves a sonochemical blaise reaction as the key step (Wang and Yue, 2005). The absolute configuration of (s)-(+)-dihydrokavain was demonstrated for the first time by total synthesis from a chiral source. Its opposite enantiomer, (r)-(–)-dihydrokavain, was also synthesized after the inversion of the chiral center in a Mitsunobu reaction.

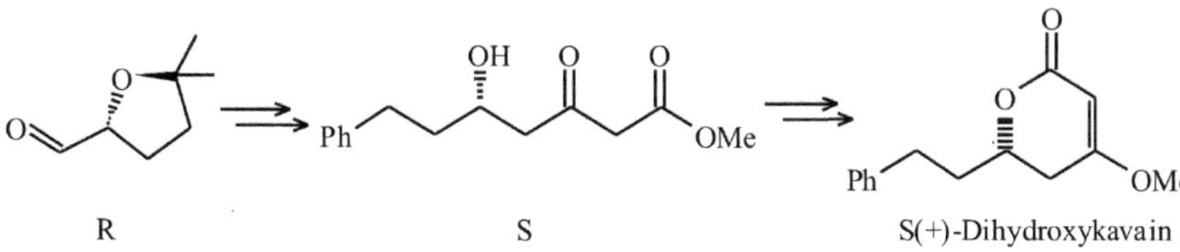

4.2.6 CANNIZZARO REACTION

Cannizzaro reaction was named after its discoverer Stanislao Cannizzaro. It is a chemical reaction that involves the base-induced disproportionation of an aldehyde lacking a hydrogen atom in the alpha position (Cannizzaro, 1853).

Ultrasound accelerates the Cannizzaro reaction of *p*-chlorobenzaldehyde under phase-transfer conditions. Three phase-transfer catalysts (PTCs), benzyltriethylammonium chloride (TEBA), aliquat, and 18-crown-6, were tested by Polackova et al. (1996), out of which benzyltriethylammonium chloride (TEBA) was found to be the most effective. Ferrocene carbaldehyde and *p*-dimethylaminobenzaldehyde gave 1,5-diaryl-1,4-pentadien-3-ones as the main product under similar conditions (TEBA as the catalyst).

Entezari and Shameli (2000) reported their observations on the Cannizzaro reaction under ultrasonic waves. The reaction of benzaldehyde with potassium hydroxide was chosen as the reference reaction. The kinetics of the reaction was followed by the amount of benzoic acid. They also discussed the effect of parameters such as temperature, PTC, aldehyde, frequency and other possible variables. As the PTC depends strongly on mass transfer between two phases, ultrasonic waves do have a greater efficiency of interface mixing than conventional agitation. The results showed that an ultrasonic wave of 20 kHz dramatically accelerates the reaction (Table 4.2).

Thus, there is a high potential for increasing the efficiency of this kind of heterogeneous reaction using high-power ultrasonic waves. By

TABLE 4.2 Cannizzaro Reaction with and without Sonication.

Aldehyde	Product yield (%)		
	Conventional	**900 kHz**	**20 kHz**
Benzaldehyde	33.3	36.2	62.9
p-Chlorobenzaldehyde	47.5	70.4	94.6

optimizing the conditions, it is possible to use ultrasonic waves instead of PTC. A combination of the PTC and ultrasonic waves could be more favorable for the reaction.

Abimannan and Rajendran (2016) synthesized 1-(isopentyloxy)-4-nitrobenzene under ultrasound-assisted liquid–liquid PTC. The reaction rate of 1-chloro-4-nitrobenzene with isoamyl alcohol catalysed by PTC combined with ultrasonic irradiation was investigated. Various process parameters (stirring speed, catalyst concentration, base concentration, different catalysts, and solvents, volume of isoamyl alcohol, inorganic salts and temperature) have been optimized. Apparent reaction rates were observed to obey the pseudo-first-order kinetics with respect to the 1-chloro-4-nitrobenzene. The reaction rate was found to increase with increase in temperature, catalyst amount, base concentration, and inorganic salts.

Chen et al. (2016) reported a novel three-component coupling reaction of substituted 2-aminobenzimidazoles, aromatic aldehydes, and 1,3-cyclohexadiones under ultrasonic irradiation to synthesize benzimidazo[2,1-b]quinazolin-1(1H)-ones.

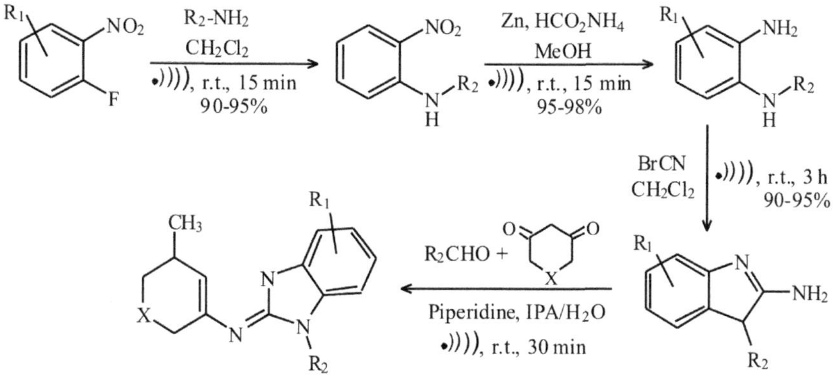

Ultrasound wave causes acoustic cavitations to trounce molecular attractive forces and to promote molecular mixing, which increases the intimate contact between different molecules to form highly reactive species. They studied a mild reaction condition to provide a simple access toward functionalized benzimidazo-quinazolinones by one-pot multicomponent reaction. Substituted 2-aminobenzimidazoles were synthesized via aromatic nucleophilic substitution of *o*-nitrofluoroarenes with various amines, followed by nitro reduction to furnish diamines. The diamines on cyclization with cyanogen bromide gave 2-aminobenzimidazoles. All these transformations were performed under ultrasound to give better yields in shorter reaction time compared to those under conventional heating conditions.

4.2.7 CLAISEN–SCHMIDT CONDENSATION

The α, β–unsaturated ketones are very common in natural product chemistry, which are usually synthesized via Claisen–Schmidt condensation of aromatic aldehydes with ketone using reagents (such as NaOH, KOH, $Ba(OH)_2$) as catalyst by heating method. However, there were always some problems due to long reaction time, or difficult workup. Fuentes et al. (1987) have reported that the sonochemical synthesis of chalcones catalysed by an activated $Ba(OH)_2$ catalyst can be carried out with very good yields. This sonochemical process took place at room temperature and with a lower catalyst weight and reaction time than the thermal process.

To test the selectivity of the reaction, Li et al. (1999) carried out the reaction in the ratio (furfural/cyclohexanone) of 2:1 and 1:1.1. Both the reactions provided the same product, α,α′-bisfurfurylidene cyclohexanone, and no α-furfurylidene cyclohexanone was obtained. Claisen–Schmidt condensation of furfural with acetophenones at room temperature results in 78–96 % yields of chalcones within 4–12 min under ultrasound. The synthesis of α,α′-bisfurfurylidene cycloalkanones was carried out in 68–96 % yields for Claisen–Schmidt condensation of furfural with cycloalkanones catalysed by KOH under ultrasound irradiation for 30–50 min in anhydrous ethanol at 20–50 °C.

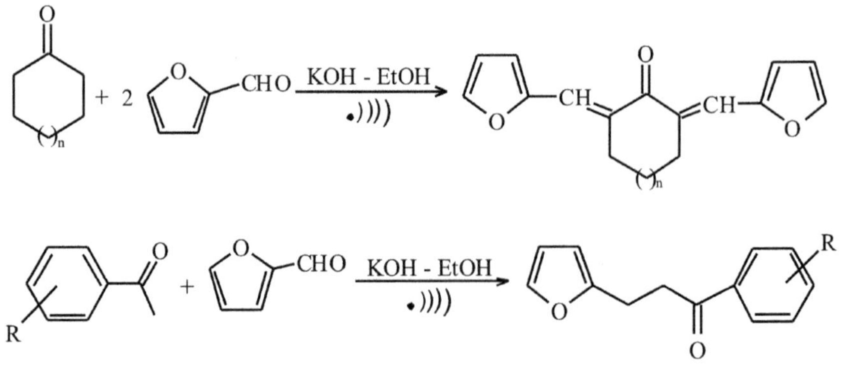

It involves condensation of an aromatic aldehyde with an aliphatic aldehyde or ketone in the presence of a base or an acid to form α,β-unsaturated aldehyde or ketone. Here, activated barium hydroxide is used as a catalyst. It is nontoxic, easy to handle, low cost, and high catalytic activity (Pinto et al., 2000).

Li et al. (2002) reported pulverized KOH-catalysed condensation of aromatic aldehydes with acetophenone to afford the corresponding chalcones under sonication. This synthesis of chalcones was carried out in good yields (52–97 %).

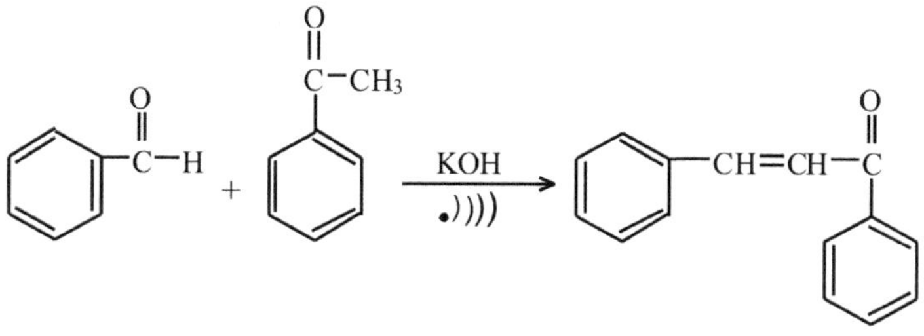

In the absence of ultrasound, the mixture of acetophenone and KOH has to be stirred at 26–30 °C for 5 h to produce 1,3-diphenylpropenone in 79 % yield, while the reaction was carried out with 80 % yield at room temperature for 25 min under ultrasonication. In the classical reaction catalysed by NaOH in 50 % aq. EtOH, 59 % yield was obtained, while the ultrasound procedure gave 91 % yield.

It is worth noting that the size of cycloalkanones and the electronic effect of substituents in benzene ring of aceophenones seems to have no effect on the yield of the products.

The condensation of cyclopentanone or cyclohexanone with a variety of aromatic aldehydes was improved using ultrasound. Catalysed by KF-Al_2O_3 in MeOH at 30–40 °C for 0.5–2.5 h, α,α′-bis(substituted benzylidene), cycloalkanones were prepared in good yields (Mahata et al., 2002).

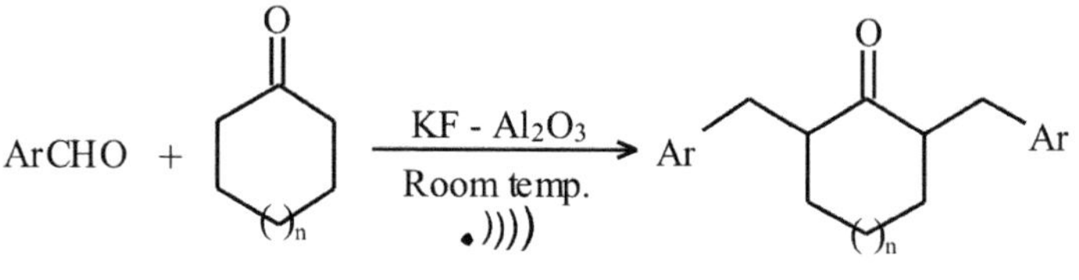

In the classical condensation of aromatic aldehydes with cycloalkanones catalysed by NaOH(aq.) in 95 % EtOH, products were obtained in moderate yields (40–50 %), whereas ultrasound method gave 80 % to 90 % yield. KF-Al_2O_3-catalysed condensation of cyclopentanone with furfural gave higher yield (96 %) than that of the procedure catalysed by KOH (85 %).

Meciarova et al. (2002) reported that Al_2O_3, $Ba(OH)_2$, and K_2CO_3 catalysed condensation of aromatic aldehydes with vinyl acetate yielded interesting substituted cinnamic aldehydes, but the results were not found to be reproducible and even with application of ultrasound, only 10–23 % yields of the product were isolated.

Calvino et al. (2005) synthesized chalcones under sonochemical irradiation by Claisen–Schmidt condensation between benzaldehyde and acetophenone. Two basic activated carbons (Na and Cs-Norit) have been used as catalysts. A significant enhancement in the yield was observed, when the carbon catalyst was activated under ultrasonic waves. This green method (combination of alkaline-doped carbon catalyst and ultrasound waves) has been applied for the synthesis of several chalcones with excellent activities and selectivity.

A comparative study under nonsonic activation showed that the yields are lower in silent conditions, which confirms that the sonication exerts a positive effect on the activity of the catalyst. Cs-doped carbon gives excellent activity for this type of condensation. Cs-Norit carbon catalyst can compete with the traditional NaOH/EtOH, when the reaction was carried out under ultrasounds.

1,5-Diarylpenta-2,4-dien-1-ones were also synthesized with ultrasound irradiation in the presence of activated barium hydroxide as the catalyst.

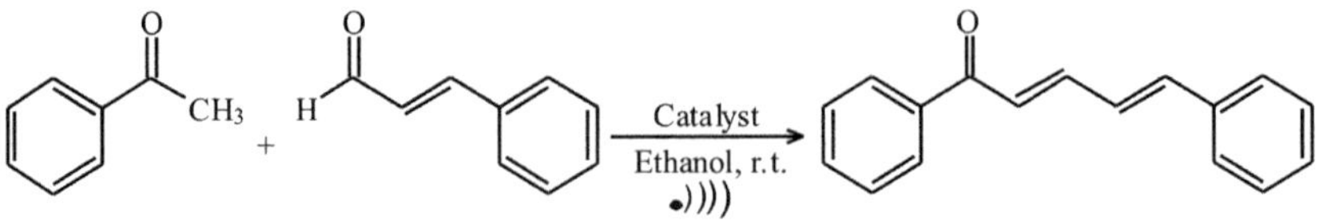

Activated barium hydroxide has also been used as a catalyst for the synthesis of 1,5-diarylpenta-2,4-dien-1-ones under ultrasound exposure (Zhang et al., 2007). This methodology offers several advantages over other conventional methods like simple procedure with excellent yields, shorter reaction time, and milder conditions. 1,5-Diarylpenta-2,4-dien-1-ones are important intermediates and raw materials, which are widely used as precursors to different drugs, nonlinear optical materials, and also for their biological activities. Conventional synthesis of 1,5-diarylpenta-2,4-dien-1-ones takes a longer time (20 h) and that too with poor yields (15–81 %) via Claisen–Schmidt condensation. The time of this condensation reaction was reduced to 180–420 min, in the presence of ultrasound (Batovska et al., 2007).

Ramesh et al. (2016) reported a methane sulphonic acid-catalysed efficient protocol for the synthesis of 2-hydroxy chalcones. Methane sulphonic acid was found to be an efficient catalyst for the synthesis of 2′-hydroxy chalcone via aldol condensation of 2-hydroxy acetophenone and substituted benzaldehyde. This catalyst also worked well with electron-donating as well as electron-withdrawing group present in the aromatic ring.

4.2.8 CURTIUS REARRANGEMENT

The Curtius rearrangement is the thermal decomposition of an acyl azide to an isocyanate with loss of nitrogen gas (Curtius, 1890 and 1894). Isocyanate then undergoes attack by a variety of nucleophiles, to yield a primary amine, carbamate, or urea derivative (Kaiser and Weinstock, 1988).

An efficient conversion of Nα-[(9–fluorenylmethyl)oxy]carbonyl (FMOC) amino acid azides to the corresponding isocyanates was carried out using ultrasound (Babukantharaju and Tantry, 2005). The Curtius rearrangement was observed using acid azides in toluene solution as well as solid powder at room temperature. All the synthesized isocyanates have been obtained as crystalline solids. Coupling of isocyanates with amino acid methyl ester hydrochloride salts in the presence of *N*-methylmorpholine (NMM) resulted in FMOC-protected dipeptidyl urea esters.

Hemantha et al. (2009) reported a facile one-pot procedure for the synthesizing urea-linked peptidomimetics and neoglycopeptides via Curtius rearrangement. They employed Deoxo-Fluor and $TMSN_3$ as efficient catalysts, which avoid the isolation of acyl azide and isocyanate intermediates. This reaction was carried out under ultrasound irradiation.

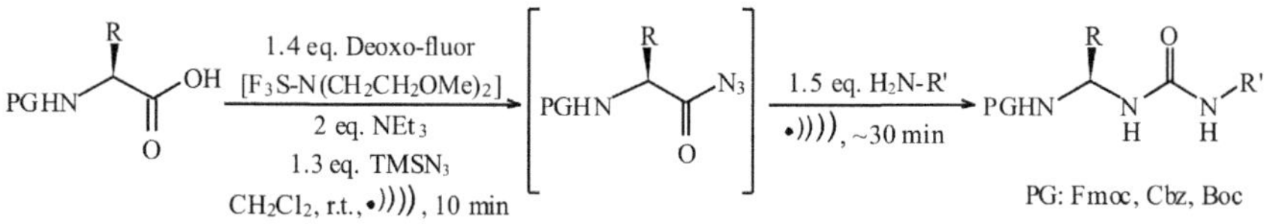

4.2.9 DIELS–ALDER REACTION

The Diels–Alder reaction is a reaction (specifically, a [4+2] cycloaddition) between a conjugated diene and a substituted alkene, to form a substituted cyclohexene system. It was first described by Diels and Alder (1928; 1929). Ultrasonic irradiation provides efficient promotion of the reaction between substituted buta-1,3-dienes with substituted 2,3-dimethoxycyclohexadiene-1,4-diones in a Diels–Alder cycloaddition, which affords a variety of bicyclo[4.4.0] fused-ring systems just in one step with high yields. Sonochemistry provides an increase in yield over a much shorter period than the conventional methodology for reactions performed in benzene, toluene, and methylene chloride. It provides a convenient route to naphthaquinols and lonapalene, a 5-lipooxygenase inhibitor in the treatment of psoriasis (Javed et al., 1995).

Ultrasound acts either by mechanical effects (as a very efficient stirring), by the generation of radicals acting as catalysts or via the in situ formation of molecules, which themselves act as catalysts. Ultrasound does not affect a cycloaddition reaction itself but it promotes in situ generation of a hydrogen halide acting as the catalyst. Thus, sonochemical effect is indirect in this case.

The yield of a silent (conventional) reaction is n% in a specific period of time, while the yield of this reaction is m% in the presence of ultrasound and m/n ratio is > 1, then this is considered as a sonochemical effect. Interestingly, this m/n ratio depends on n. It only means that if a silent reaction (especially a heterogeneous reaction) is performed without stirring or an inefficient stirring or if this reaction is carried out in the absence of a catalyst (which can be generated by ultrasound), it is expected that there will be some sonochemical effect.

Such a sonochemical effect has been observed on Diels–Alder reaction by Caulier and Reisse (1996). They demonstrated a clear-cut effect of ultrasound on a Diels–Alder reaction in various halogenated solvents.

Reactions were performed in strictly homogeneous solutions, and Caulier and Reisse observed such sonochemical effects in $CHCl_3$, CH_2Cl_2, and CH_2Br_2. The occurrence of these effects in halogenated solvents only suggests that solvent molecules take a predominant part in the sonochemical pathway. Additional experiments have been performed in dichloromethane to confirm this assumption. A possible explanation could be that direct sonication of halogenated molecules by a titanium horn leads to the formation of Lewis acids ($TiCl_4$, $TiBr_4$) known to catalyze this Diels–Alder reaction.

A sealed tube was immersed in a thermostated cleaning bath in order to avoid direct contact between the solution and the ultrasonic transducer in a control experiment. In these conditions, increases of rate and stereoselectivity were still observed, precluding the hypothesis of an artifact due to an in situ generation of a Lewis acid derived from the titanium horn. Then, a set of experiments was performed in the presence of radical scavenger such as 2,2-diphenyl-1-picrylhydrazyl.

A hetero Diels–Alder reaction between 1-dimethylamino-1-azadienes and electron-deficient dienophiles under ultrasound irradiation has been described by Villacampa et al. (1994). Other advantages of these sonicated reactions are lower reactions times, increased yields and in some cases, even a decrease in some side reactions.

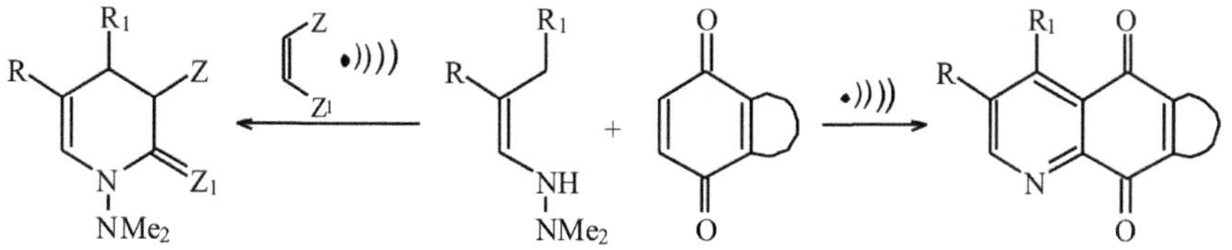

The $Yb(OTf)_3$ catalysed formal aza-Diels–Alder (or Povarov) reaction of cyclopentadiene and 1,3-cyclohexadiene with in situ generated *N*-arylimines under conventional/ultrasonic techniques (Pelit and Turgut, 2014).

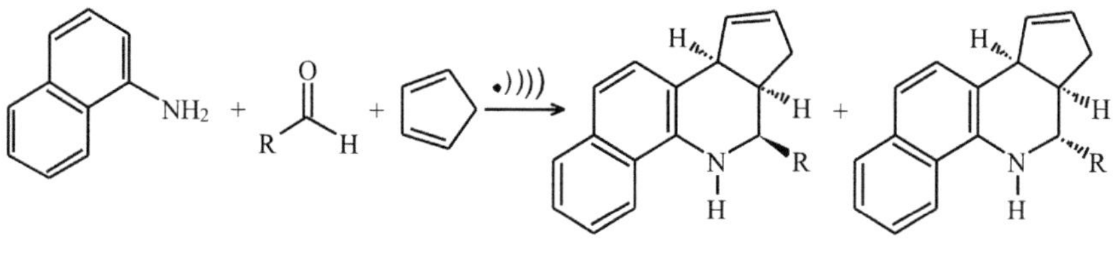

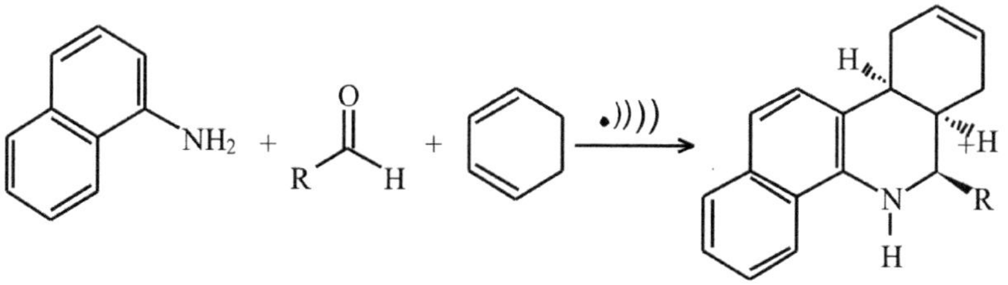

4.2.10 HECK REACTION

Heck reaction (also called the Mizoroki–Heck reaction) is the chemical reaction of an unsaturated halide (or triflate) with an alkene in the presence of a base and a palladium catalyst (or palladium nanomaterial-based catalyst) to form a substituted alkene (Mizoroki et al., 1971; Heck and Nolley, 1972; Drahl, 2010).

Heck reaction involves palladium-catalysed coupling of alkenes/alkynes with aryl and vinyl halides. Deshmukh et al. (2001) reported Heck reaction at an ambient temperature (30 °C) with enhanced reaction rate (1.5–3 h) via the formation of Pd–biscarbene complexes and stabilized clusters of zero-valent Pd nanoparticles in ILs under ultrasonic irradiation. Generally, such reactions are carried out in polar solvents such as dimethylformamide (DMF) and required longer reaction times (8–72 h) and that too at higher temperatures (80–140 °C).

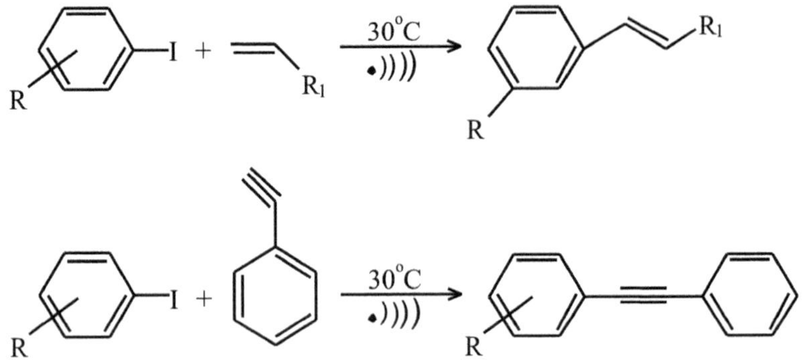

Where R = H, 4-OMe, 4-Cl; R_1 = COOMe, COOEt, Ph.

It was found that the formation of Pd–biscarbene complex and its subsequent sonolytic conversion to a highly stabilized cluster of zero-valent Pd nanoparticles increased the rate of reactions.

Heck reaction of iodobenzene with methyl acrylate was initially studied in the presence of $PdCl_2$ in water. A reaction mixture containing iodobenzene, methyl acrylate, $PdCl_2$, Na_2CO_3 (as base), and tetrabutylammonium bromide (TBAB) was stirred in 3 mL of water at 25 °C for 4.5 h. (*E*)-Methyl cinnamate was obtained with a very low yield of only about 10%. Observing the fact that ultrasonic irradiation may promote organic reactions, Zhang et al. (2006) carried out the Heck reaction in water under ultrasonic irradiation but without heating. They observed that this reaction can be carried out in aqueous solution at ambient temperature with good yields (86%). This was attributed to in situ formation of palladium nanoparticles, which can be recycled again.

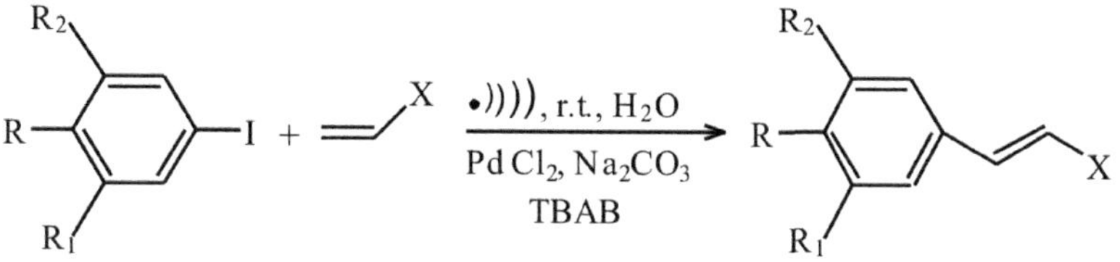

Where R = H, OMe, Cl, Br, OMOM, NHCOMe or NO_2; R_1 = I, CHO, or H; R_2 = I or H; X = COOMe, COOEt, COOH, CN or Ph.

The ultrasound-promoted Heck reaction was also carried out under mild and environmental-friendly conditions by An et al. (2011). It gave significantly high yield (96%) and products with high regioselectivities and stereoselectivities (*E*/*Z*>99:1). They used acrylamide and acrylic acid for the first time as substrates in the ultrasound-promoted Heck reaction. This reaction was conducted and completed within 20 min.

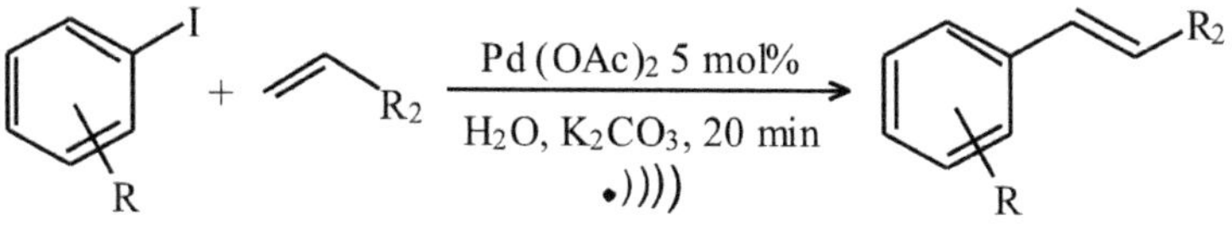

Where R_2 = $CONH_2$, COOMe, COOH.

Heck reaction of aryl bromide with styrene in *N,N*-DMF/water (solvent) has been studied using different catalysts under ultrasonic irradiation at room temperature by Said et al. (2011). The coupling reaction of aryl halides (10 mmol) and styrene (9 mmol) were carried out in DMF/water (4 mL/4 mL) with catalyst (1% mol), K_2CO_3 (15 mmol), and aliquat-336 (5 mmol) for 2 h at 95 °C. The reaction mixture was subjected to ultrasonic irradiation during 5 min at room temperature. The results show that ultrasonic irradiation increases the yield of Heck reaction and becomes faster with the amount of aliquat-336.

$$ArBr + CH_2{=}CH{-}C_6H_5 \xrightarrow[\bullet))))]{\text{Catalyst}} Ar{-}CH{=}CH{-}C_6H_5$$

Ghotbinejad et al. (2014) synthesized a novel and highly stable $[Pd(EDTA)]^{2-}$ salt as a catalyst, using a counter-cation of *N*-methylimidazolium bonded to 1,3,5-triazine-tethered superparamagnetic iron oxide nanoparticles. This complex catalysed the Mizoroki–Heck and Suzuki–Miyaura cross-coupling reactions quite efficiently. The cross-coupled products were produced under conventional heating and ultrasound irradiation at an extremely low-catalyst loading (as low as 0.032 mol% Pd). It was indicated that conventional synthesis took relatively a longer and gave only moderate yields, while in the presence of ultrasound irradiation, the reaction was very fast with high to excellent yields. It was found that catalyst can be quickly recovered by an external magnetic field and could be reused for several reaction cycles without any significant change in catalytic activity.

Mizoroki–Heck coupling afforded a number of 3-vinylindole derivatives in good to acceptable yields under ultrasound irradiation. Recently, it was revealed that Pd/C–PPh_3 catalyst system facilitated the C–C bond forming reaction between 3-iodo-1-methyl-1H-indole and various terminal alkenes under ultrasound irradiation (Bhavani et al., 2016).

4.2.11 HENRY REACTION

In this reaction, the addition of bromonitroalkanes to aldehydes afforded the corresponding nitroalkanols in good yields on using zinc in the presence of a catalytic amount of indium under sonication (Soengas and Silva, 2012) (Table 4.3).

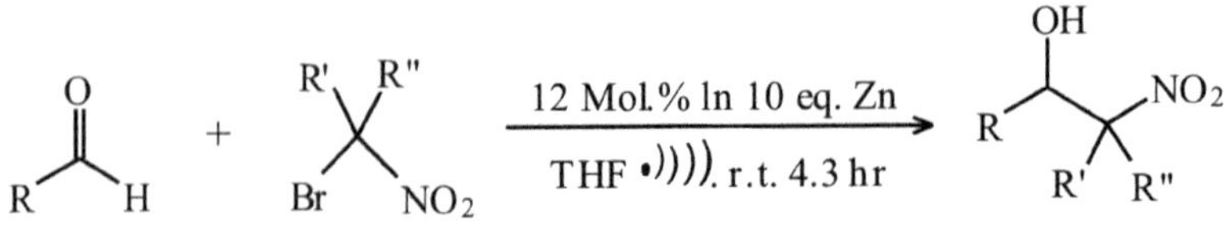

(Added after 20 min)

where R = Ar, Alkyl; R′ = Me, H; R″ = H, Me.

Indium-catalysed ultrasound-assisted Henry reaction is better than the classical base-catalysed reaction in terms of yields and selection of

TABLE 4.3 Henry Reaction Under Ultrasound.

Product	Yield (%)	Product	Yield (%)
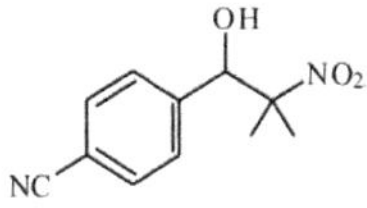	94	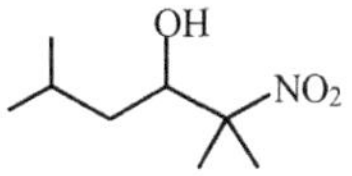	87
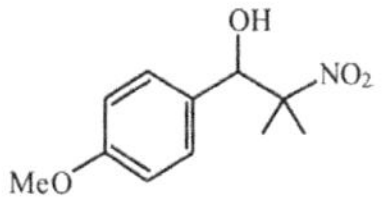	34	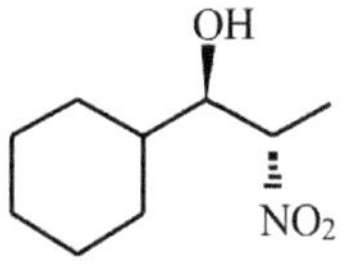 Anti : syn = 60 : 40	90

substrate. It avoids the use of larger amounts of the expensive indium powder as only catalytic amount is required here and it can also be scaled up easily.

Rodriguez and Dolors (2011) reported the Henry condensation reaction under microwave irradiation or ultrasound as the energy source. β-Nitroalcohol or nitroethylene derivatives can be obtained from aryl aldehydes and nitromethane with Henry condensation by the modification that microwave irradiation or ultrasound can be used as the energy source. Microwave irradiation allowed a novel one-pot synthesis of nitroolefins from aryl aldehydes using ammonium acetate as a catalyst without solvent. Different reaction conditions, such as base, solvent, and reaction time, were also studied by them. Only small amounts of β-hydroxyl nitro compounds were isolated, using microwave irradiation for less than 25 min, but on the contrary, the use of ultrasound increased the yield of the nitroalcohols.

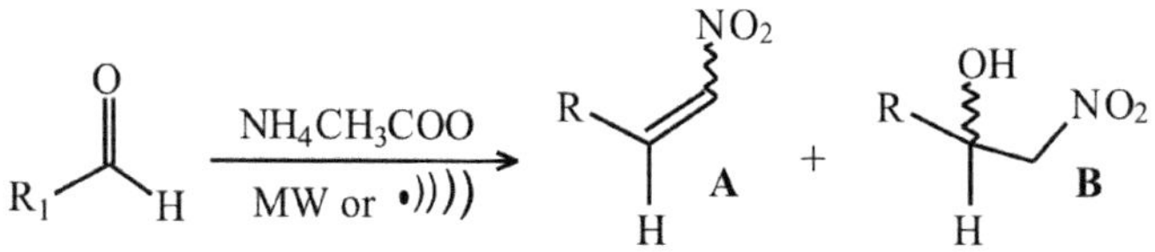

4.2.12 KNOEVENAGEL CONDENSATION

The Knoevenagel condensation reaction is an organic reaction named after Emil Knoevenagel. It is a modification of the aldol condensation

(Knoevenagel, 1898). A Knoevenagel condensation is a nucleophilic addition of an active hydrogen compound to a carbonyl group followed by a dehydration reaction, where a molecule of water is eliminated. The product is often an α,β-unsaturated ketone (a conjugated enone).

Li et al. (1999) reported an ultrasound-assisted and pyridine-catalysed Knoevenagel condensation of ethyl cyanoacetate with a variety of aromatic aldehydes afforded ethyl α-cyanocinnamates in a good to excellent yield (80–96 %), whereas the conventional methods gave relatively lower yields (47–92 %) (Popp, 1960).

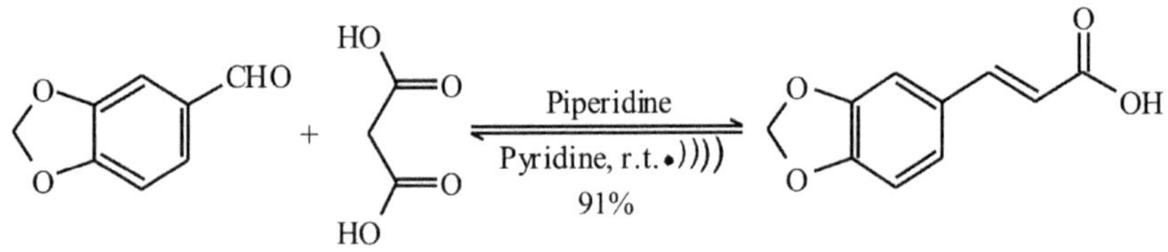

Ultrasound technique represented a better procedure in terms of milder conditions and easier work-up. The product was prepared in a moderate yield (58 %) in dry dioxane after withstanding overnight at room temperature, whereas the ultrasonic method without other solvent at 20 °C in only 3 h provided the same product in 94 % yield. Electron-donating groups such as Me_2N, H_2N, MeO, HO, and Me in the aromatic ring did not retard the condensation under these conditions.

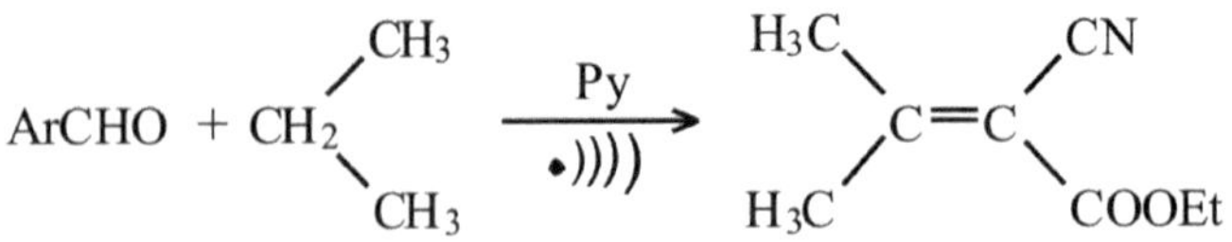

Li et al. (2001) studied ultrasound irradiation-accelerated condensation of ketones with ethyl cyanoacetate catalysed by NH_4OAc-AcOH, giving ethyl alkylidene α-cyanoacetates in 31–89 % yields. The synthesis of ethyl cycloalkylidene cyanoacetate was carried out in good yields (55–89 %) under ultrasound irradiation at 41–50 °C for 2.5–9 h, without any extra solvent than with that of conventional methods.

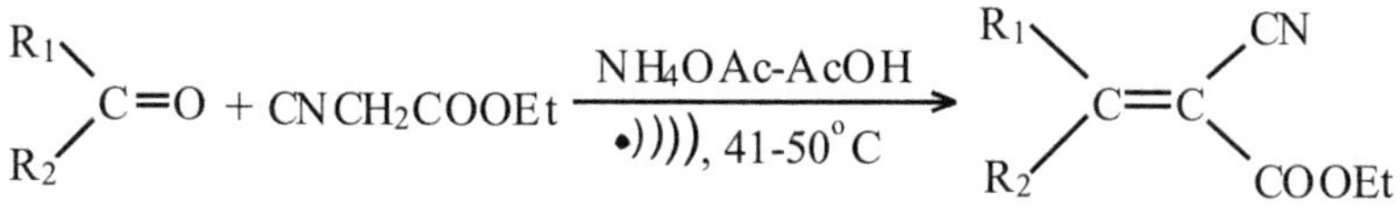

Application of ultrasound has been found to greatly assist the Knoevenagel condensation reaction of activated methylenes with aromatic aldehydes under mild conditions. The outcome of the ultrasound-promoted reaction depends upon the electronic nature of the aromatic aldehyde, the solvent employed, and the addition of acids, bases, or ammonium salts. Ultrasound efficiently promotes the condensation of aromatic aldehydes with active methylenes under mild conditions.

McNulty et al. (1998) reported that ultrasound promoted the condensation of nitromethane with a variety of aromatic aldehydes with electron-donating groups catalysed by NH_4OAc-HOAc to give nitroalkenes in good to excellent yields. Typical conditions to affect the condensation are heating HOAc solution of the aldehyde with appropriate nitromethane with NH_4OAc at 100 °C for a few hours.

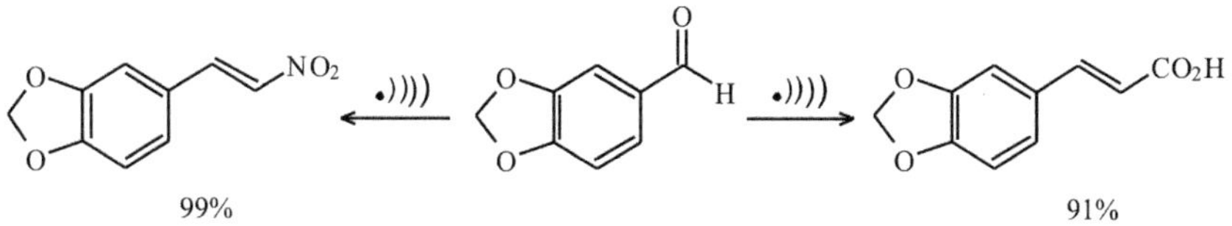

However, benzaldehyde reacted with same reagents at room temperature by the application of ultrasound to give nitroalkene in 99 % yield. It was observed that the ultrasound failed to promote the condensation without HOAc or NH_4OAc. However, on using diisopropylethylamine in place of HOAc, aromatic aldehydes with electron-deficient groups condensed with nitromethane to give nitroaldols in good yields under ultrasound.

Keipour et al. (2015) prepared potassium fluoride–clinoptilolite from commercially available and low-cost starting materials, which was used as an efficient and recyclable catalyst for Knoevenagel condensation of aromatic aldehydes with active methylene compounds under ultrasound irradiation. Short reaction times, high yields of products, ease of recovery, and catalyst reusability make this new method quite economic and efficient.

De-la-Torre et al. (2014) made a comparative study of the effects on three techniques in the reactions between heterocyclic aldehydes and 2-(benzo[d]thiazol-2-yl)acetonitrile, that is, stirring, ultrasound coupled to PTC conditions (US-PTC), and MW irradiation (MWI) under solvent and catalyst-free conditions. The effects of conditions on reaction parameters were evaluated and compared in terms of reaction time, yield, purity, and outcomes. The US-PTC method was found to be more efficient than the MWI and conventional methods. The reaction times were considerably shorter, with high yields (>90%), and good levels of purity.

Martín-Aranda et al. (1997) studied ultrasound-activated Knoevenagel condensation of malononitrile with carbonylic compounds catalysed by alkaline-doped saponites and β-unsaturated nitriles have been synthesized. Two alkaline promoted clays (Li^+- and Cs^+-exchanged saponites) have been employed as catalysts. The influence of the carbonylic compound (benzaldehyde or cyclohexanone) and the use of a solvent on the catalytic activity have been studied. A remarkable increase in the conversion values has been found, when the reaction is activated by ultrasound, as compared to thermal activation. In this green, solvent-free procedure, β-unsaturated nitriles have been produced in very high yields (97%), particularly, when the Cs^+-saponite was used as a catalyst.

Suresh and Sandhu (2013) reported ultrasound-assisted synthesis of 2,4-thiazolidinedione and rhodanine derivatives catalysed by task-specific ILs. Some 1,1,3,3-tetramethylguanidine-based task-specific ionic liquids (TSILs) were prepared and employed to catalysed solvent-free Knoevenagel condensation of 2,4-thiazolidinedione and rhodanine with a variety of aldehydes. Best results were obtained with 1,1,3,3-tetramethylguanidine lactate ([TMG][Lac]). The TSIL can be easily recovered and recycled, yielding products in excellent yields under ultrasonic environment without the formation of any side products or toxic waste.

4.2.13 LOSSEN REARRANGEMENT

Lossen rearrangement is the conversion of a hydroxamic acid to an isocyanate via the formation of an *o*-acyl, sulphonyl, or phosphoryl intermediate hydroxamic acid *O*-derivative and then conversion to its conjugate base (Lossen, 1872; Lossen, 1875).

Carboxylic acids are converted into hydroxamic acids in conventional Lossen rearrangement. Vasantha et al. (2010) used 1-propanephosphonic acid cyclic anhydride (T3P) to promote this reaction. They reported that the application of ultrasound accelerates this conversion, whereas T3P has also been employed to activate the hydroxamates, thus leading to isocyanates via Lossen rearrangement.

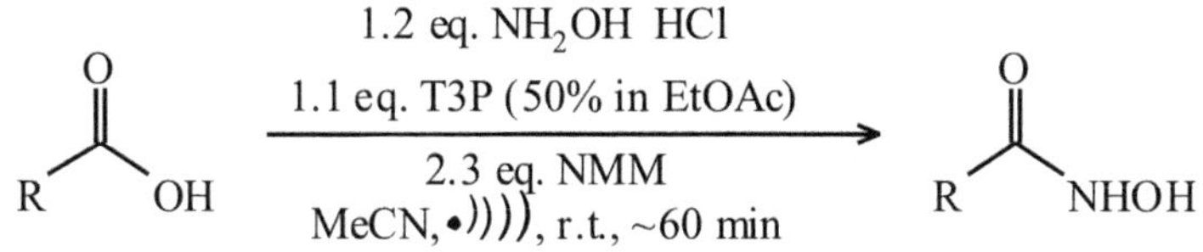

where T3P is 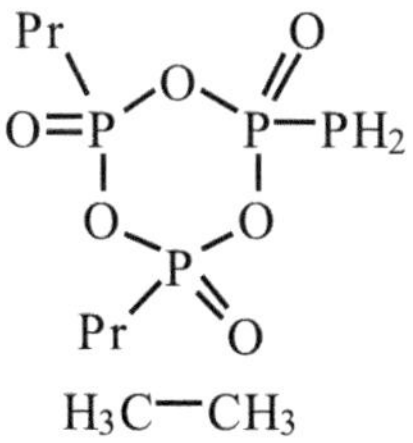

4.2.14 MANNICH REACTION

Mannich reaction involves amino alkylation of acidic proton placed next to a carbonyl functional group by formaldehyde and a primary or secondary amine or ammonia. The final product is a β-amino-carbonyl compound, which is also known as Mannich base. Reactions between aldimines and α-methylene carbonyls are also considered as Mannich reactions because these imines are formed between amines and aldehydes (Mannich and Krosche, 1912).

Zhang et al. (2009) reported the $Ga(OTf)_3$-catalysed three-component Mannich reaction of aromatic aldehydes, aromatic amines, and cycloketones in water promoted by ultrasound gave the corresponding *β*-amino cycloketones in good to excellent yields and good *anti* selectivities. Sulphamic acid (NH_2SO_3H, SA) was used as an efficient, inexpensive, nontoxic, and recyclable green catalyst for the ultrasound-assisted one-pot Mannich reaction of aldehydes with ketones and amines.

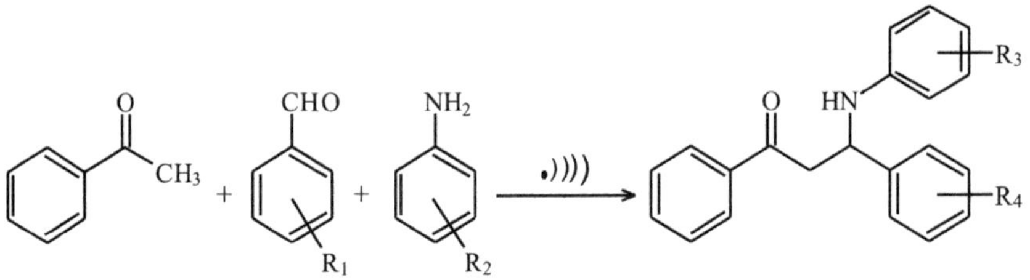

Li et al. (2011) reported that the synthesis of Mannich bases related to gramine via Mannich reaction of secondary amine, formaldehyde, and indole or *N*-methylindole can be carried out in 69–98% yields in acetic acid aqueous solution at 35 °C under ultrasound irradiation.

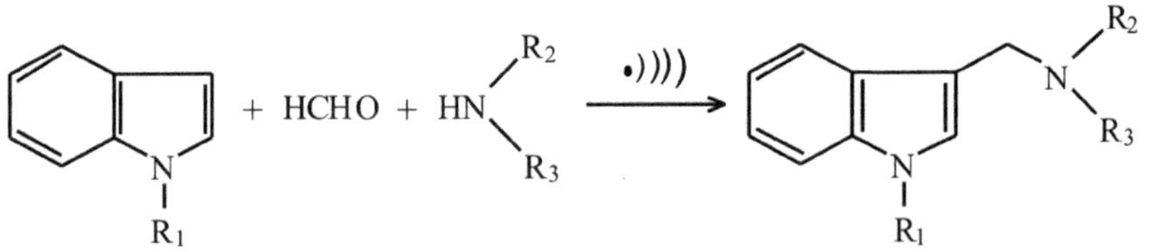

Magnetite nanoparticles (Fe_3O_4) (MNPs) were introduced as a heterogeneous novel catalyst for ultrasound-assisted stereoselective synthesis of β-aminocarbonyl using Mannich reaction. These MNPs with particle size <40 nm were synthesized via a simple chemical precipitation method. A mixture of benzaldehyde, aniline, cyclohexanone or acetophenone, and MNPs was sonicated in ethanol at room temperature for 30–75 min to afford β-aminocarbonyl compounds (Saadatjoo et al., 2012). The reaction product was eluted from the MNPs with hot ethanol, and the catalyst was removed by using a permanent magnet.

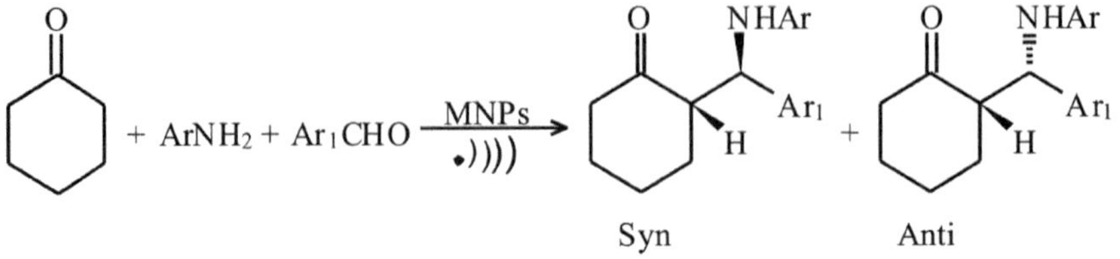

where $Ar_1=C_6H_5$, $2NO_2$-C_6H_4, 2-Naphtyl, 4-Br-C_6H_4; Ar = C_6H_5, 4-Br-C_6H_4.

The use of MNPs and ultrasonic-assisted method as an alternative to other synthesis methods involving Mannich reaction offers several advantages, such as high yield, reaction rate, short reaction times, low costs, and a recyclable catalyst with a very simplified procedure. As the catalyst can be magnetically separated, it improved the separation rate of product as the time-consuming column passing or filtration operation is avoided in this case.

A green, rapid, and highly efficient method was developed for the synthesis of *β*-aminoketone derivatives by Ozturkcan et al. (2012). The coupling reactions of cyclohexanone, various substituted aromatic amines, and aromatic or heteroaromatic aldehydes were successfully carried out under ultrasonic irradiation with bismuth(III) triflate in water at room temperature. It can be considered the first report of three-component Mannich-type reaction in water catalysed by bismuth(III) triflate under ultrasonic irradiation. The higher yields and anti-selectivity, versatility, convenient operation, low cost, and ecofriendly nature of the reaction support this as a practical and rapid method for the preparation of *β*-aminoketone derivatives.

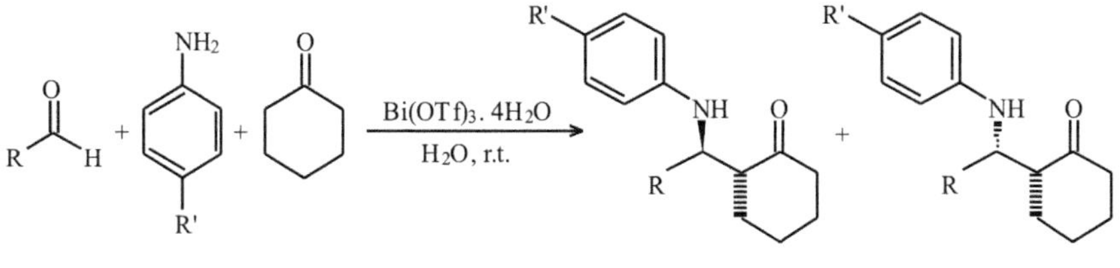

Direct-type three-component Mannich reaction of aromatic or heteroaromatic aldehydes with substituted aromatic amines and cyclohexanone proceeded smoothly and yielded products in moderate to good yields (38–79%). This three-component reaction using 4-bromobenzaldehyde was performed under the same conditions, but in the presence of ultrasonic irradiation at room temperature. It was found that this reaction was completed within 1 h and the product was obtained in reasonably good 89% yield. Some other aldehydes and amines were also tested for Mannich reactions with cyclohexanone in water under ultrasonic irradiation at room temperature. The three-component direct-type Mannich reactions proceeded very smoothly in a short time (1–2 h) in the presence of 5 mole% of $Bi(OTf)_3{\cdot}4H_2O$ in the presence of ultrasound irradiation with corresponding *β*-aminoketone with high yields.

Ghadami and Jafari (2015) carried out an efficient synthesis of Mannich bases by sonication in sodium dodecyl sulphate micellar media by Mannich reaction of aldehydes, aromatic amines, and acetophenone derivatives or cyclohexanone. The sonication of biphasic systems was performed at 37 kHz, 300 W, under neutral conditions. It was observed that the reaction can be carried out readily under milder conditions. The precipitated solid Mannich product was recovered by simple filtration without using any organic solvent.

TABLE 4.4 A Comparative study of Mannich Reaction.

Reaction conditions	Time (h)	Yield (%)
Conventional	12–24	<2
With 0.2 mol% acidic IL under ultrasonic irradiation	2–6	64–97

Qian et al. (2016) reported an efficient and facile process for the preparation of a series of β-amino carbonyl compounds via Mannich reactions catalysed by caprolactam-based Brønsted acidic ILs under ultrasonic irradiation. [Capl][BF_4] was the most effective catalyst in the Mannich reaction, and good yields were obtained within 2–6 h (Table 4.4). The activity and stability of the catalyst were maintained even after using it for five times. The use of ultrasound can effectively shorten the reaction time and enhance the yields at ambient conditions.

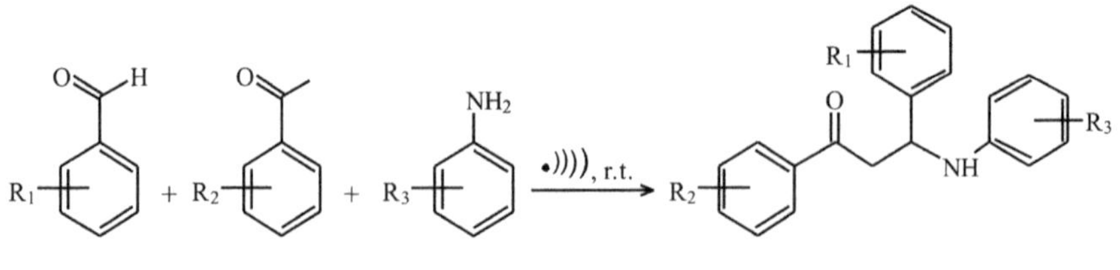

A novel series of 5-(4-(benzyloxy)-substituted phenyl)-3-((phenyl amino) methyl)-1,3,4-oxadiazole-2(3H)-thione Mannich bases were synthesized in good yields from the key compound 5-(4-(benzyloxy)phenyl)-1,3,4-oxadiazole-2(3H)-thione by aminomethylation with paraformaldehyde and substituted amines using molecular sieves and sonication (Nimbalkar et al., 2016).

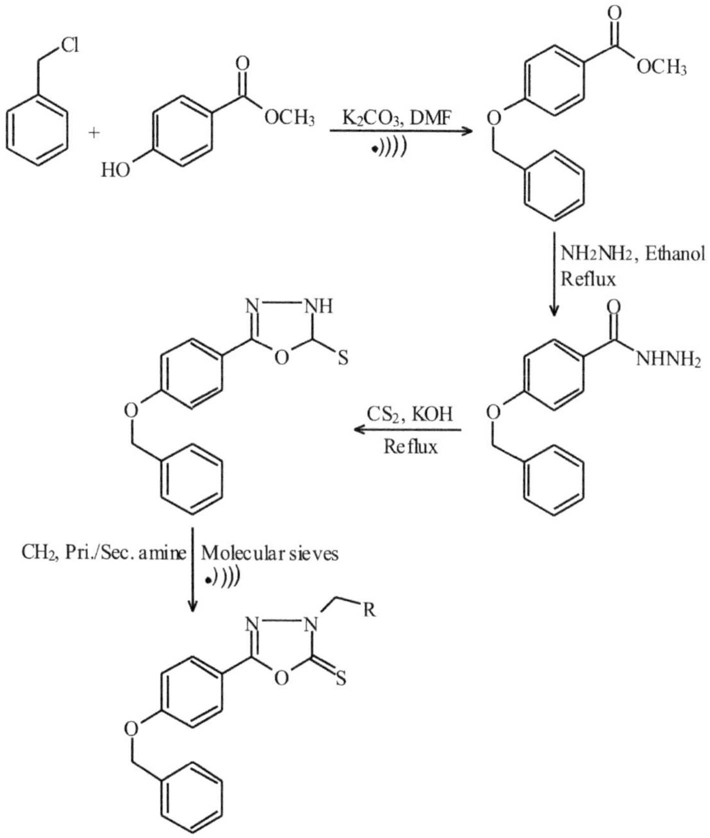

4.2.15 MICHAEL ADDITION REACTION

Michael reaction is the nucleophilic addition of a carbanion or another nucleophile to α,β-unsaturated carbonyl compounds. This is one of the most useful methods for the mild formation of C–C bonds (Little et al., 1995; Mather et al., 2006).

The use of ultrasound has found to be very effective for the Michael addition of indole to α,β-unsaturated carbonyl ketones. Ceric ammonium nitrate (CAN) is used as a catalyst in this reaction. The corresponding adducts were obtained in excellent yields with selectivity of substitution on the indole ring only at the 3-position without any *N*-alkylation products (Ji and Wang, 2003).

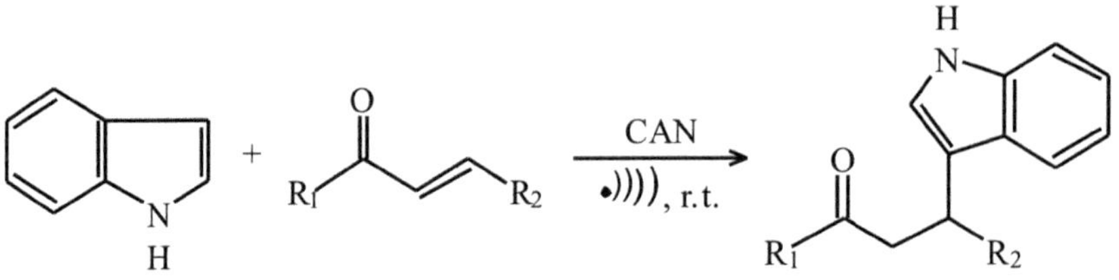

where CAN = Ceric ammonium nitrate.

Ultrasound-assisted hetero-Michael reaction has been studied by Arcadi et al. (2009). They described an alternative protocol for the Michael addition of thiols to 4-hydroxy-2-alkynoates. The reaction proceeds at room temperature in water under ultrasound irradiation. A sequential conjugate addition/lactonization reaction leads to important 4-amino-furan-2-one derivatives, when amines were used instead of thiols. An efficient protocol for the highly regioselective Michael addition of thiols to 4-hydroxy-2-alkynoates in water under ultrasound irradiation has also been reported. These reactions proceeded without the addition of any acid or base. Significant rate acceleration was observed in water as compared to organic solvents. A sequential conjugate addition/lactonization reaction leads to important 4-amino-furan-2-one derivatives on using amines in place of thiols. Ultrasound irradiation has shown significant advantages compared to the traditional stirring.

Long-chain dicationic ammonium salts were successfully used as efficient PTCs in the Michael addition reaction of various active methylene compounds to 2-cyclohexenone without solvent under ultrasonic irradiation (Oge et al., 2012). Sodium hydroxide (0.05 mmol) and a PTC (0.05 mmol) were added to a mixture of 2-cyclohexenone (1 mmol) and Michael donor (2 mmol). This reaction mixture was sonicated at 25 °C.

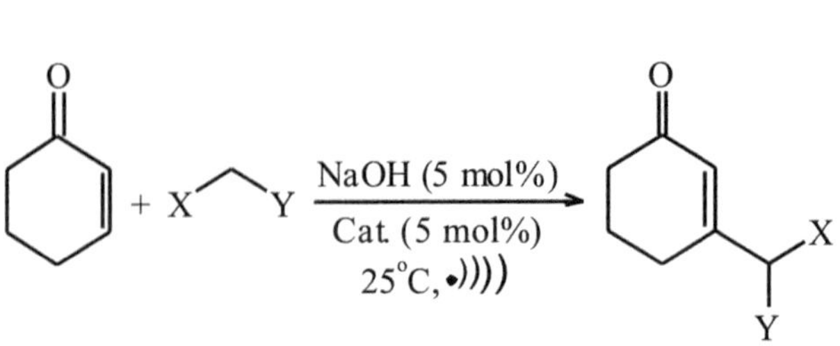

where X = Y = COOEt; X = Ph, Y = CN.

Imanzadeh et al. (2013) described an efficient, mild, nonexpensive, and eco-friendly protocol for the synthesis of *p*-toluenesulphonamide derivatives by aza-Michael addition reaction of *p*-toluenesulphonamide to fumaric esters using potassium carbonate under ultrasound irradiation. This method is simple, convenient, and products are also obtained in excellent yields.

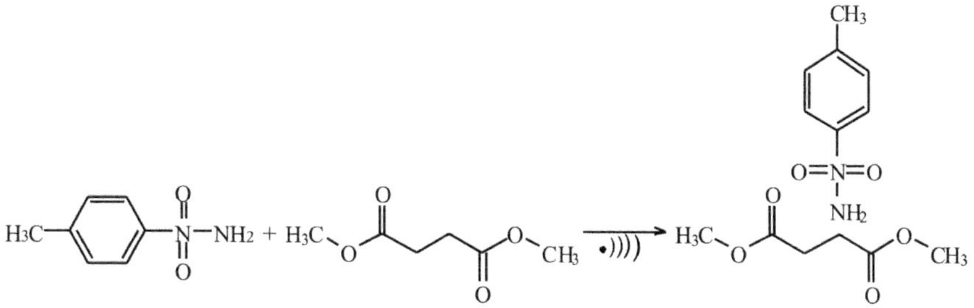

Imanzadeh and Gatar (2015) described solvent-free highly regioselective Michael addition of phenytoin to α,β-unsaturated esters in the presence of TBAB under ultrasound irradiation. α,β-unsaturated ester (2 mmol) was added to a well-ground mixture of DPH (5,5-diphenylhydantoin 2 mmol), (TBAB 2 mmol), and K_2CO_3 (2 mmol). Then these were mixed thoroughly with a glass rod, and this mixture was irradiated in the water bath of an ultrasonic cleaner at 25–70 °C.

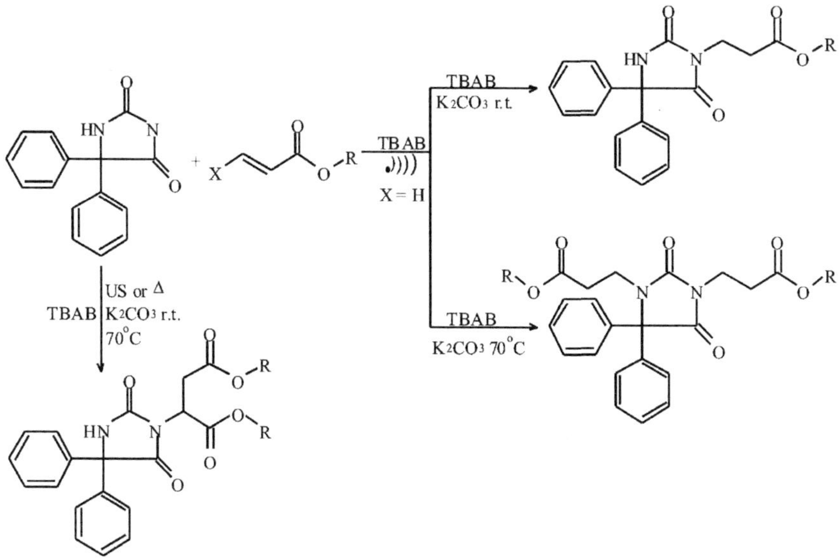

2-Pyrrolidinon-3-olates were synthesized via one-pot four-component reaction of 2-aminobenzothiazole, aromatic aldehydes, dimethyl acetylenedicarboxylate and morpholine/piperidine in the presence of Co_3O_4@ SiO_2 core–shell nanocomposite as a catalyst under ultrasound irradiation. Apart from other advantages this protocol provides advantage like reusability of the catalyst. The core-shell nanocomposite was also prepared using ultrasound irradiation (Ghasemzadeh and Abdollahi-Basir, 2016).

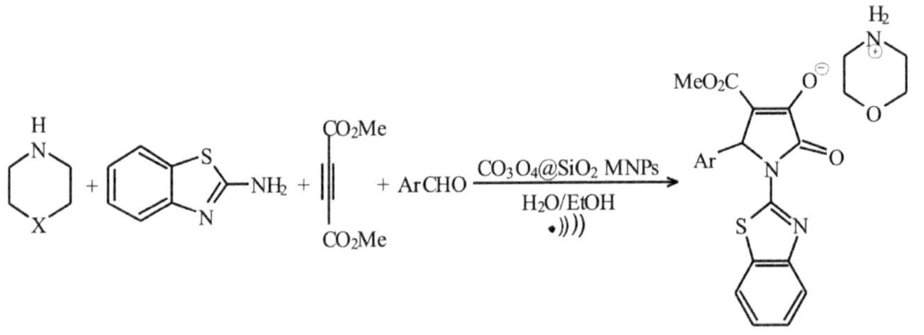

A simple, efficient, and green chemical ultrasound-assisted thio-Michael addition reaction between thiols and (*Z*)-azolactones was reported by Rocha et al. (2016), aiming to produce nonnatural amino acid derivatives by using chiral Zn[(L)-Pro]$_2$ as a heterogeneous catalyst. The product was obtained in good to excellent yields presenting high diastereoselectivity.

4.2.16 PINACOL COUPLING REACTION

A pinacol coupling reaction is an organic reaction in which a carbon–carbon covalent bond is formed between the carbonyl groups of an aldehyde or a ketone in the presence of an electron donor in a free radical process (Fittig, 1859).

The corresponding products of this reaction can be used as intermediates for the preparation of ketones and alkenes. Generally, the reaction is effected by treatment of carbonyl compounds with an appropriate metal reagent and/ or metal complex to give rise to the corresponding coupled product. The coupling products can have two newly formed stereocenters. As a consequence, efficient reaction conditions are required to control the stereochemistry of the 1,2-diols.

Anhydrous conditions and long reaction time are necessary to achieve high yield of the products. Apart from these, some of the reductants are expensive or toxic, the reduction conditions are critical, as well as an excess amount of metal is needed. The ultrasound irradiation was found effective in enhancing the reactivity of metal and is favorable for single-electron transfer reaction of the aldehydes resulting in diols.

The effect of ultrasound on pinacol coupling of benzaldehyde was studied by Meciarova et al. (2001) in aqueous medium mediated by magnesium, zinc, iron, nickel, and tin. It was observed that pinacol coupling of various aromatic carbonyl compounds yielded very good results in magnesium-mediated reaction. The use of ultrasonic irradiation reduces the reaction time and lower mole excess of the metal is required. Pinacol coupling proceeded well with zinc also, too, but no such reaction was observed with iron, however, some unexpected reaction (oxidation of benzaldehyde) was observed on using nickel and tin as metals. The conversion of starting material as well as the proportion of pinacol to corresponding benzyl alcohol was found to be strongly dependent on the steric environment surrounding the carbonyl group.

Zang et al. (2002) reported coupling reaction of aromatic aldehydes and ketones using Zn–$ZnCl_2$ in 50% aq. THF under ultrasound to afford pinacols in 5–77.5% yields.

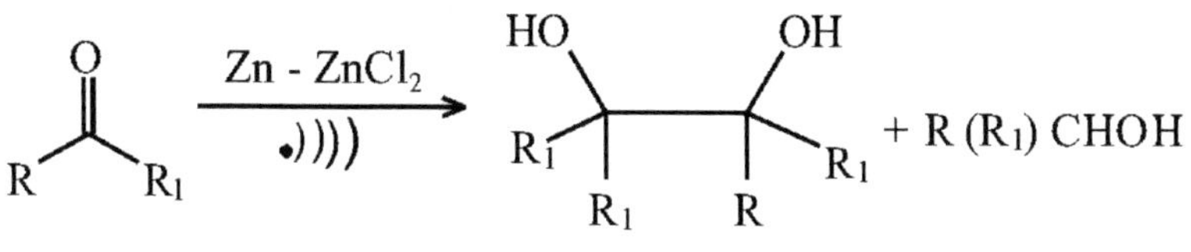

The Zn–$ZnCl_2$ reagent is not sensitive to oxygen. The coupling reaction of aromatic aldehydes afforded pinacols in 23.3–87.1% yields under ultrasound, by using $ZnCl_2$/Montmorillonite K10-Zn in 50% aqueous THF. But no pinacols were obtained from aromatic ketones such as acetophenone and 4-chloroacetophenone. No coupling reaction had occurred from benzaldehyde and 2, 4-dichlorobenzaldehyde without $ZnCl_2$ even for 8 h. The K10-$ZnCl_2$ could be recycled for three times without any significant loss in activity. When the catalyst is recycled first, second, and third time, and used again, the products were obtained in 87.1%, 86%, and 85% yield, respectively. No pinacol was obtained from benzaldehyde in the absence of ultrasound.

Zhang and Li (1999) reported pinacol coupling reaction of various aromatic aldehydes and ketones in aqueous NH_4Cl mediated by Mg. They could obtain pinacols in 40–90% yields. The reaction time was about 12–24 h and the molar ratio of ArCHO:Mg turning as high as 1:20. Toma and Mecarova (1999) employed Mg powder instead of Mg turning, which decreased the molar ratio of PhCHO:Mg to 1:10, shortened the reaction time to 10 min under ultrasound. They proved that reaction could be carried out in pure water (without the addition of ammonium chloride) in very good yields. Very good yields of pinacols are formed with zinc powder in aqueous NH_4Cl, but no reaction was observed with iron, nickel, and tin powders, even under sonication (Meciarová et al., 2002). Under ultrasound irradiation, complexes of $TiCl_4$–THF in CH_2Cl_2 could be reduced by Zn or Al to corresponding low-valent titanium complexes, which can reduce some aromatic aldehydes to corresponding pinacols in 33–98% yields within 4–25 min, but without ultrasound, the pinacols were obtained in lower yield. Pinacols were obtained in 57% yield using $TiCl_4$–THF–Zn under Ar stirring for 30 min (Li et al., 2000), and in 50% yield using $TiCl_4$–Et_2O–Al and stirring for 38 h (Li et al., 2003), but in the presence of $TiCl_4$–THF–Zn, and replacement of Ar by N_2, pinacols were obtained in 96% yield under ultrasonication for only 5 min. Under ultrasound, TiC14–THF–Al system provided pinacols in 90% yield for 20 min. Al was proved to be more diastereoselective (*dl/meso*) than Zn.

4.2.17 REFORMATSKY REACTION

Reformatsky reaction is a zinc-catalysed reaction between α-haloesters and aldehydes or ketones resulting in the formation of β-hydroxyesters. Disadvantages of this reaction include the use of zinc dust, which has low reactivity and also control of the resulting exothermic reaction. In that case, one has to activate the zinc dust to initiate the reaction. Yields of the Reformatsky reaction were found to be improved on using freshly prepared zinc powder, a heated column of zinc dust, copper–zinc couple, acid-washed zinc, and trimethylchlorosilane (Reformatsky, 1887).

The Reformatsky reaction is an organic reaction, which condenses aldehydes or ketones, with α-halo esters, using a metallic zinc to form β-hydroxy-esters (Reformatsky, 1887).

Han and Boudjouk (1982) found that there was a significant increase in the yield and rate of Reformatsky reaction under ultrasound. The best solvent was found to be dioxane. Neither ether nor benzene, the common Reformatsky solvents, produced good yields even after several hours of sonication.

Bang et al. (2002) reported Reformatsky reaction of indium, benzaldehyde with ethyl bromoacetate in various solvents. Among the several solvents tested, THF gave the best results in terms of conversion and reaction time. The reaction of benzaldehyde with ethyl bromoacetate resulted in ethyl 3-hydroxy-3-phenylpropanoate with 97 % yield in the presence of indium within 2 h; however, the yields were decreased in other solvents despite prolonged reaction time.

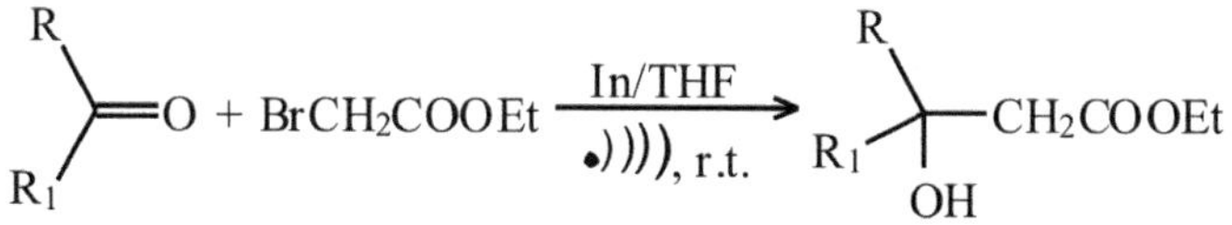

In the absence of ultrasonic waves, the reaction occurred much more slowly (17 h), and the yield of the β-hydroxyester was only 70 %. Ethyl iodoacetate gave similar results to that of ethyl bromoacetate. The reaction afforded the desired compounds in excellent yields (89–95 %) in case of aliphatic aldehydes, while aromatic aldehydes with the substituents, for example, methyl, chloro, bromo, monomethoxy, dimethoxy, nitro, or hydroxy showed little effects on the efficiency of the reactions. The desired products were also obtained in good to excellent yields in alkyl-substituted ethyl bromoacetate. Benzaldehyde reacted with ethyl 2-bromopropanoate and ethyl 2-bromo-2-methylpropanoate in the presence of indium to give β-hydroxyesters in 76 % and 82 % yields, respectively.

Reactions of sym-(keto)dibenzo-16-crown-5 with ethyl α-bromoesters, zinc dust, and iodine in dioxane and/or tetrahydrofuran under irradiation by HIU provided β-hydroxy lariat ether esters in 79–91 % yields (Ross and Bartsch, 2001). HIU-promoted Reformatsky reactions offer several advantages over conventional thermal reaction conditions.

Ross and Bartsch (2003) have reported the synthesis of β-hydroxyesters via the ultrasound-promoted Reformatsky reaction. They used nonactivated zinc dust and a catalytic amount of iodine. There are two methods for the improvement of the rate and yield of the

Reformatsky reactions. One is the use of low-intensity ultrasound (LIU) and another is using HIU.

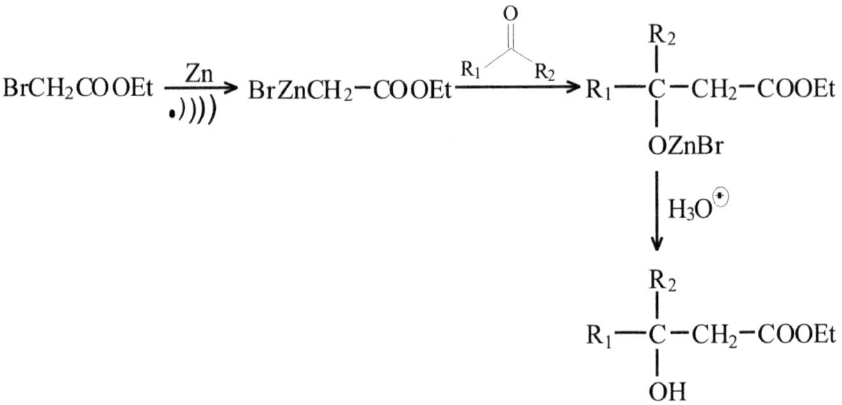

The use of a LIU using laboratory-cleaning bath can also improve the rates and yields of simple aldehydes and ketones with ethyl bromoacetate to a greater extent (Han and Boudjouk, 1982). However, in this case also, zinc dust has to be activated by washing with nitric acid.

Nathan and Richard (2003) reported Reformatsky reactions of a phenyl ketone, an α-bromoester, zinc dust, and a catalytic amount of iodine in dioxane under HIU irradiation from an ultrasonic probe. It gave high yields of β-hydroxyesters in shorter reaction times.

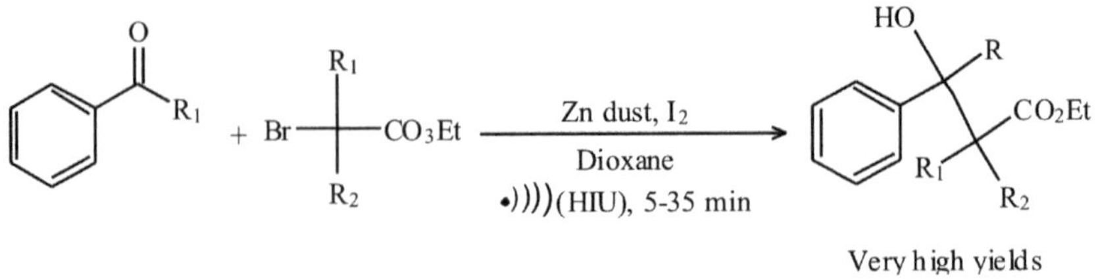

where R = Me, Et, i-Pr, t-Bu, Tr, Ph; R_1, R_2 = H, ME.

Reformatsky reactions of an imine, α-bromoester, zinc dust, and a catalytic amount of iodine in dioxane under HIU irradiation have been explored by Ross et al. (2004). They used direct immersion horn as the ultrasonic source.

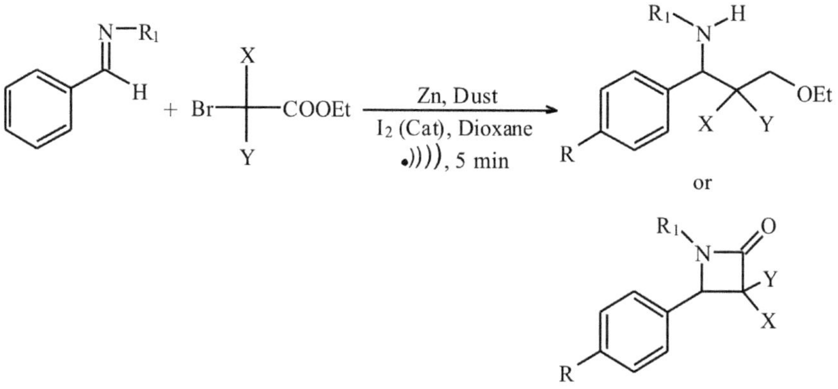

where R = OMe, Cl, CF_3, NMe_2; R_1 = Ph, p-$MeOC_6H_4$, p-ClC_6H_4.

A series of aldimines with a variety of sustituents were evaluated as potential electrophiles for this reaction with three α-bromoesters with differing steric demands. This method is quite successful for both enolizable and nonenolizable types of imines affording β-lactam in very good yields.

Raquel and Amalia (2012) reported the indium-mediated Reformatsky addition of bromoesters to glyoxylic oxime. Sonication of mixtures of oxime, indium powder, and ethyl 2-bromopropionate in tetrahydrofuran (THF) at room temperature for 6 h afforded 81% yield of products as a 3:2 mixture of syn/anti-isomers. Similarly, reaction of oxime and ethyl 2-bromophenylacetate provided the adduct in 88% yield, but as a 2:1 mixture of syn/anti-isomers. Interestingly, no reaction had occurred in the absence of sonication. Thus, ultrasonic irradiation was necessary to complete this reaction.

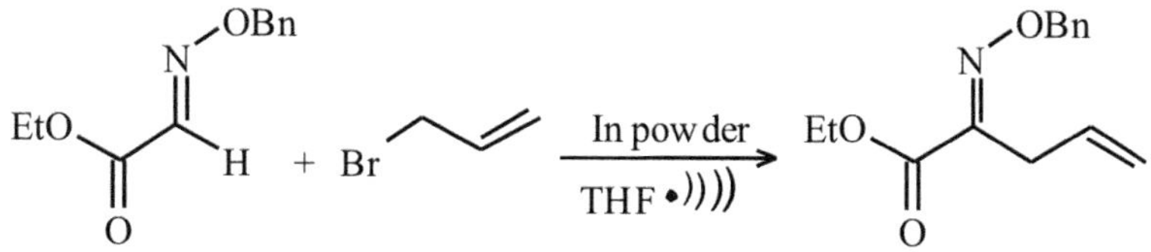

4.2.18 SONOGASHIRA COUPLING

Sonogashira reaction is a cross-coupling reaction used in organic synthesis to form carbon–carbon bonds. It employs a palladium catalyst to form a carbon–carbon bond between a terminal alkyne and an aryl or vinyl halide (Sonogashira, 2002). The Sonogashira cross-coupling reaction has been employed in a wide variety of areas, due to its usefulness in the formation of carbon–carbon bonds. The reaction can be carried out under mild conditions, such as room temperature in aqueous media and with a mild base. These conditions are favorable for extending the use of the Sonogashira cross-coupling reaction in the synthesis of complex molecules.

A copper-ligand free one-pot synthesis of benzo[b]furans via palladium acetate- catalysed tandem Sonogashira coupling and 5-endo-dig cyclization (inside the ring diagonal sp-hybridized carbon) was carried out under ultrasonic irradiation at ambient temperature (Palimkar et al., 2006). They applied the same method for synthesizing 2-substituted indoles via Sonogashira coupling and 5-endo-dig cyclization (Palimkar et al., 2008).

Li et al. (2009) synthesized 4,5-disubstituted-1,2,3-(NH)-triazoles in excellent yields by palladium- catalysed and ultrasonic-promoted Sonogashira coupling/1,3-dipolar cycloaddition of acid chlorides, terminal acetylenes, and sodium azide in one pot.

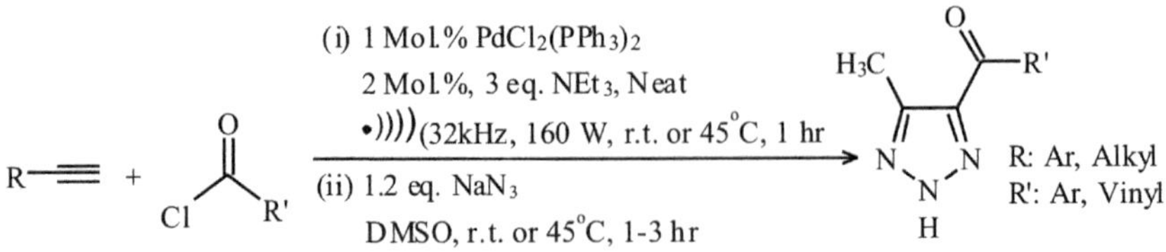

The ultrasound-mediated Pd nanoparticle- catalysed cross-coupling reaction was carried out between aryl iodides or bromides and terminal acetylenes in acetone as well as in IL [bbim]BF_4, under normal conditions of temperature and pressure (Gholap et al., 2005). It was interesting to observe that this reaction had worked well with a large number of substrates in the absence of any phosphine ligand or copper co-catalyst. However, the use of acetone as solvent furnished the desired coupled products in a shorter time. The yield was less in these conditions as compared to ILs, but it requires relatively more time to complete the reaction. It was observed that a catalyst can be recycled and reused with little loss of activity even after five runs. A control experiment was carried out under conventional conditions, and it was observed that no reaction took place even after several hours of stirring without ultrasound.

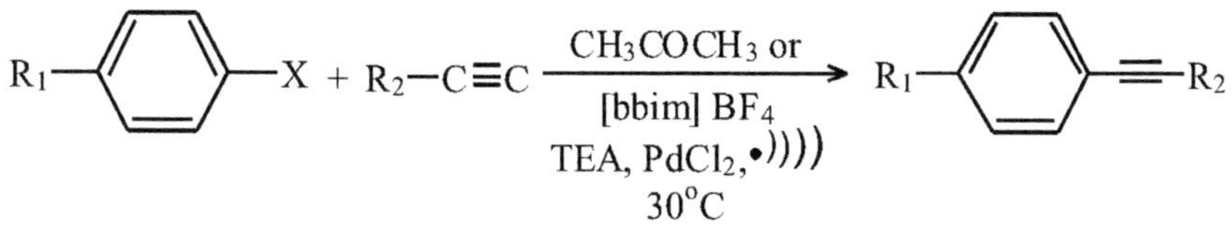

where R_1 = H, CH_3, NO_2, CHO; R_2 = Aryl, Cyclohexyl; X = I, Br.

This method was also carried out by Fu et al. (2008) for the Sonogashira coupling of ethynylferrocene with aryl iodides.

4.2.19 STRECKER SYNTHESIS

This is one of the important methods for the synthesis of amino acids via an aminonitrile intermediate. The Strecker synthesis is used for a series of chemical reactions for synthesizing an amino acid from an aldehyde and ketone (Strecker, 1850; 1854). The aldehyde is condensed with ammonium chloride in the presence of potassium cyanide to form an α-aminonitrile, which is subsequently hydrolyzed to give the desired amino acid.

Hanafusa et al. (1987) reported a direct reaction between an aldehyde, KCN, and NH_4Cl in acetonitrile in the presence of alumina under ultrasound exposure. Conventionally, this protocol normally leads to a mixture of products, but this can be made more specific under sonication.

The yield of the aminonitrile (the desired intermediate) is very poor on using benzaldehyde under stirring. The benzoin and hydroxynitrile were obtained as major products. They observed that alumina suspended in acetonitrile coupled with sonication improved the yield of target aminonitrile to about 90 %.

4.2.20 SUZUKI REACTION

Suzuki reaction is an organic coupling reaction catalysed by a palladium(0) complex, where the coupling partners are boronic acid and organohalide (Miyaura et al., 1979; Miyaura and Suzuki, 1979). Suzuki reaction was first investigated employing $Pd(OAc)_2$ as a catalyst in the absence of phosphine ligands. The ultrasound-assisted cross-coupling reaction between potassium trifluoroborate salts and aryl- or vinyltellurides for the preparation of

Z- or E-stilbenes has been reported by Cella and Stefani (2006). A series of alkenes with defined stereochemistry were obtained in good yields within 40 min under comparative conditions. It was observed that most of the starting material remained unchanged even after 24 h of magnetic stirring. However, the same reaction gives a yield of 63% in 18 h under reflux conditions. An important feature of this reaction is that it can even tolerate ester groups, which are sensitive to normal conditions.

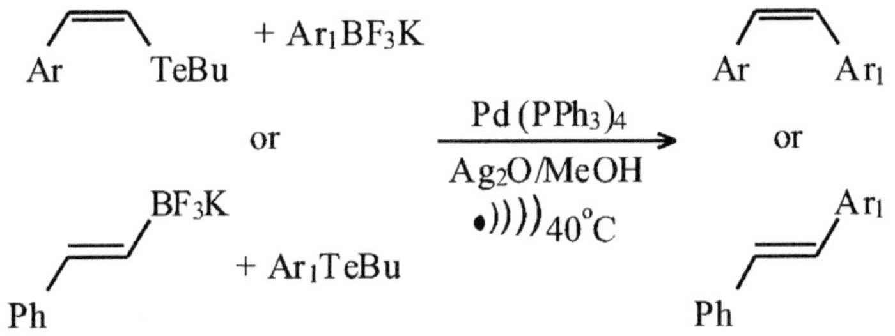

Guadagnin et al. (2008) have used this procedure with some modifications for the synthesis of 1,3-dienes, 1,3-enynes, and (Z)-(2-chlorovinyl)- alkenes.

Rajagopal et al. (2002) carried out Pd-catalysed Suzuki cross-coupling reactions of halobenzenes with phenylboronic acid in ILs under ultrasound irradiation. They used tetrafluoroborate bis-butylimidazol-2-ylidene-palladium(II) carbene complex instead of $Pd(OAc)_2$ and $[bbim]^+[BF_4]^-$/MeOH as a solvent system.

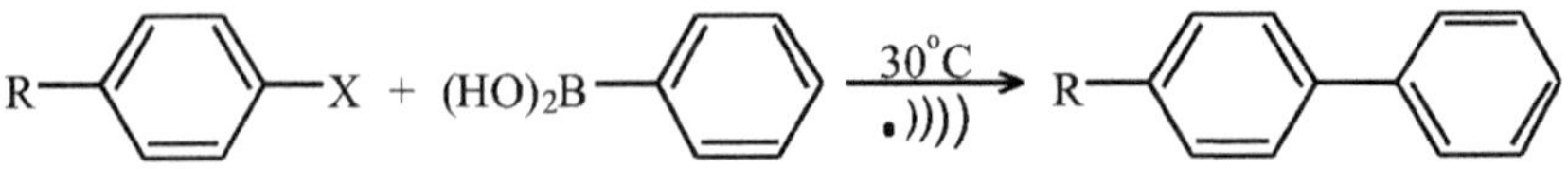

where X = I, Br, Cl; R = H, OCH_3, CH_3, Cl, NO_2.

Under these conditions, a series of biaryls were synthesized in good yields in shorter reaction times. This methodology is considered very important as aryl halides containing both the groups, electron-donating as well as electron-withdrawing groups, react readily. Even chlorobenzene undergoes the Suzuki reaction and afforded biaryls with 42–65% yields. The yield was moderate on using $Pd(OAc)_2$ as a catalyst. Chlorobenzenes are known to have poor reactivity under Suzuki conditions.

The modification allowed for recycling and using the palladium catalyst without any appreciable loss in its activity as a catalyst. A simple ultrasound-assisted procedure was reported by Martina et al. (2011) to synthesize solid-supported Cu(I) and Pd(II) catalysts. Cross-linked chitosan derivatives (CS-Cu, CS-Pd) were obtained by reacting hexamethylene isocyanate and chitosan in the presence of the corresponding metal salt. The in situ polymerization in water under sonochemical exposure was extremely fast and efficient. The Cu(I)-loaded polyurethane/urea-bridged chitosan was used for the azide/alkyne [3+2] cycloaddition, while those bearing Pd(II) complexes were used for Suzuki cross-couplings. In both these reactions, these novel catalysts can be reused several times with a little low activity.

Kulkarni et al. (2014) showed the use of an energy-efficient surface acoustic wave (SAW) device for driving closed-vessel SAW-assisted (CVSAW), ligand-free Suzuki couplings in aqueous media. The reactions were carried out on an mM scale with low- to ultra-low catalyst loadings, which were found to be driven by heating, resulting from the penetration of acoustic energy. The yields were uniformly high, and the reactions can be carried out without added ligand and in water.

The coupling reaction of aryl bromide and aryl boronic acid in water/DMF as solvent was studied using a palladium complex as a catalyst in the presence of ultrasound at room temperature by Said and Salem (2016). The effect on the reaction of a base and a solvent was also studied with and without ultrasound, and it was found that the rate of the reaction had increased.

4.2.21 ULLMANN REACTION

Ullmann reaction normally requires a 10-fold excess of copper and a longer reaction time of 48 h, but this amount can be reduced to only a 4-fold excess of copper and a reaction time of 10 h on the application of ultrasound (Lindley et al., 1986; 1987). The particle size of the copper was found to reduce from 87 to 25 μm. Only an increase in the surface area cannot explain the observed increase in reactivity and, therefore, it was suggested that the process of sonication also assists in the breakdown of intermediates and desorption of the products from the surface.

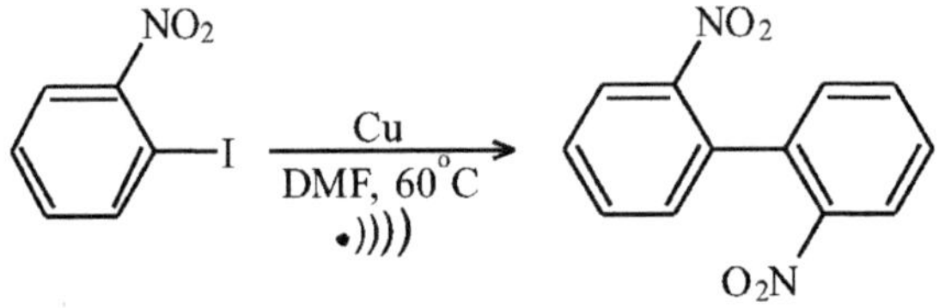

They revealed that the sonication is responsible for breaking down of intermediates or desorption of products from the surface as it also prevented the adsorption of copper on the walls of reaction vessels, which is quite a common problem encountered in conventional Ullmann reaction (Table 4.5). Such ionic reactions are accelerated by better mass transport and physical effects. They also called it false sonochemistry. It occurs, when the cavitation effects produce purely mechanical effects and this is most often in heterogeneous system

TABLE 4.5 Ullmann Reaction in the Presence of Ultrasound.

Product	Yield (%)
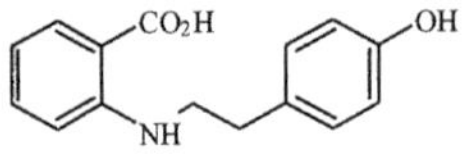	95
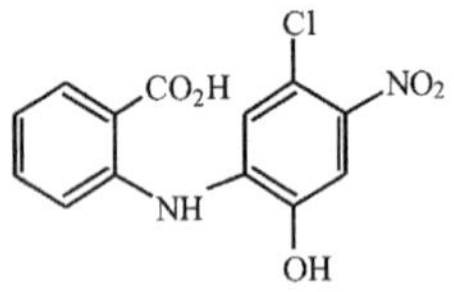	92
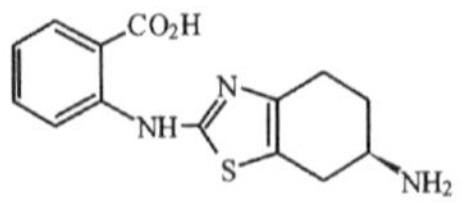	90
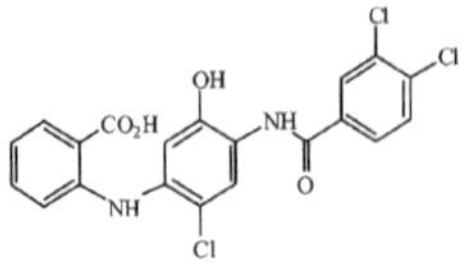	90
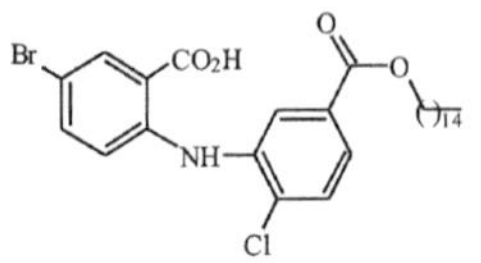	84

where rate enhancement derive from surface cleaning in heterogeneous catalysis, the enhancement of mass transfer and in liquid–liquid or solid–solid phase transfer reactions. On the other hand, true sonochemistry is the result of effects directly derived from the hotspots of cavitational collapse energy.

Smith and Dennis (1992) described the use of ultrasound in the reactions of phenols with bromoarenes in the presence of potassium carbonate and copper(I) iodide without an added solvent. It gave better yields of diaryl ethers (typically 70–90 %), at lower temperature (140 °C) than the traditional Ullmann reactions. The role of ultrasound here is primarily to break up the particles of the inorganic solids.

Docampo and Pellon (2003) studied Ullmann condensation between 2-chlorobenzoic acid and 2-aminopyridine derivatives using ultrasound. This reaction was conducted in the presence of anhydrous potassium carbonate and copper powder using DMF as a solvent. The ultrasound-assisted reaction required a shorter reaction time (20 min) with better yields as compared to conventional heating, that is, stirring for 6 h at reflux temperature.

Nelson and Horst (1991) reported that ultrasonic irradiation was found to promote the Ullmann coupling of picryl bromide at or below room temperature. In the presence of excess copper, a long-lived intermediate is formed that is quenched upon the work-up affording variable mixtures of trinitrobenzene and picric acid.

One-pot syntheses of 5*H*-[1,3]thiazolo[2,3-*b*] quinazolin-5-one, 12H-[1,3]benzothiazolo [2,3-*b*]quinazolin-12-one, and corresponding derivatives were developed using the copper-catalysed Ullmann condensation (Rolando et al., 2007). The use of ultrasonic irradiation enhanced yields and reduced the reaction time to minutes.

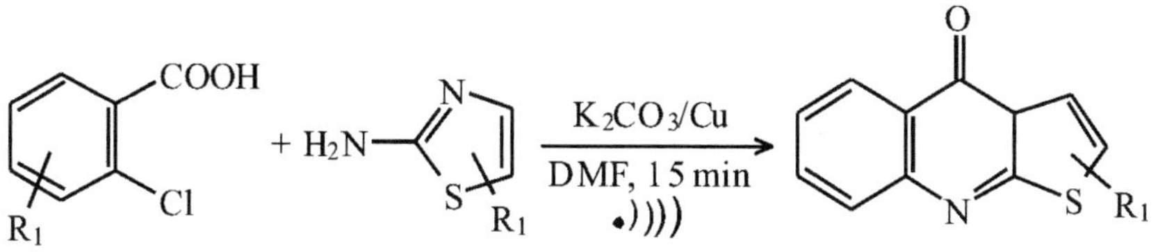

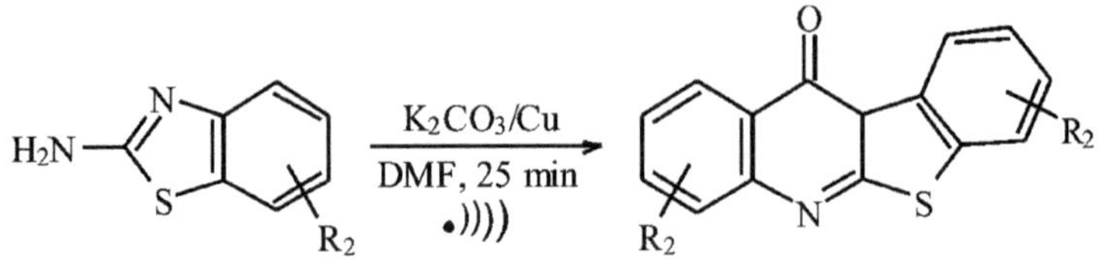

Michael et al. (2012) described the ultrasound-assisted Ullmann *N*-coupling of 2-halobenzoic acid with complex aryl- and alkylamines to give the corresponding products in good to excellent yields.

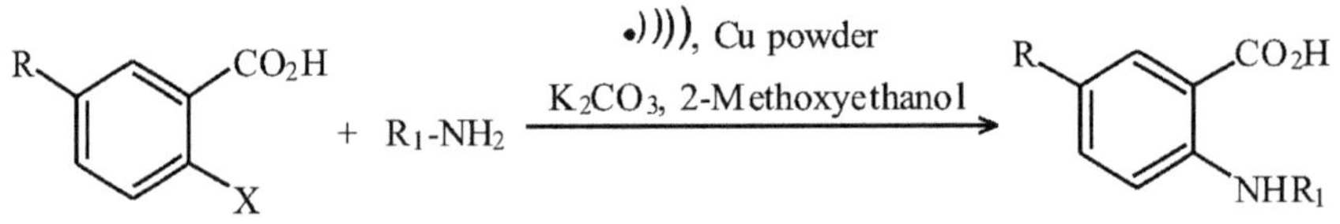

where R_1 = Aryl, Alkyl.

4.2.22 VILSMEIER–HAACK REACTION

The Vilsmeier–Haack reaction (also called the Vilsmeier reaction) is the reaction of a substituted amide with phosphorus oxychloride and an electron-rich arene to produce an aryl aldehyde or ketone. The reaction is named after Anton Vilsmeier and Albrecht Haack (Vilsmeier and Haack, 1927).

Ali et al. (2002) studied the Vilsmeier cyclization of acetanilides with Vilsmeier–Haack reagent under ultrasound irradiation. A solution of acetanilides was treated with Vilsmeier–Haack reagent (formed from $POCl_3$/DMF) in CH_2Cl_2 under ultrasound at 40 °C for 60 min to give corresponding 2-chloro-3-formylquinolines in good yields (65–85 %). On the other hand, the reaction under thermal condition resulted in the formation of the products with relatively lower yields (0–68 %). The desired formylated products were formed in 75–90 % yields under ultrasonic irradiation (Table 4.6). The acetophenone gave β-chlorocinnamaldehyde in 85 % yield, whereas 2-hydroxyacetophenone and 2,4-dihydroxyacetophenone provided corresponding 3-formylchromones in 82 % and 74 % yields, respectively.

TABLE 4.6 A Comparison Between Ultrasound and Thermal methods.

Substrate	Product	Yield (%)	
		US[a]	Thermal[b]
Acetanilide	2-Chloro-3-formyl quinoline	84	68
4-Methylacetanilide	6-Methyl-2-choro-3-formyl quinoline	85	50
4-Methoxyacetanilide	6-Methoxy-2-choro-3-formyl quinoline	82	58
3-Nitroacetanilide	7-Nitro-2-choro-3-formyl quinoline	65	0
4-Bromoacetanilide	6-Bromo-2-choro-3-formyl quinoline	68	35

[a]Ultrasound irradiation at 40 °C for 1 h.[b]Thermal reaction at 75 °C for 6 h.

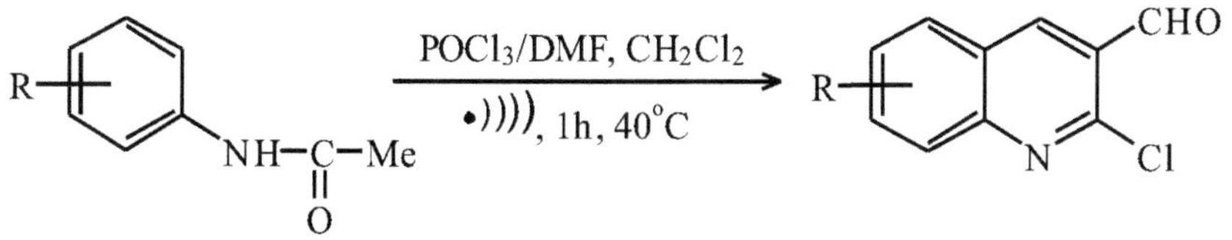

4.2.23 *WITTIG REACTION*

Wittig reaction or Wittig olefination is a chemical reaction of an aldehyde or ketone with a triphenyl phosphonium to give an alkene and triphenylphosphine oxide. This reaction was discovered by Georg Wittig (Wittig and Schollkopf, 1954). These reactions are most commonly used to couple aldehydes and ketones to singly substituted phosphine ylides with unstabilized ylides. This results in almost exclusively the Z-alkene product (Wittig and Haag, 1955).

Wittig reaction results in a novel type of annulation of an aromatic ring, when applied to *o*-quinones. Such reaction is called as Double-Wittig reaction. Yang et al. (1992) have reported that this process may be simplified by using ultrasound irradiation, which allowed the use of bases insensitive to moisture and air.

A comparison of solventless Wittig reactions to Wittig reactions under ultrasonication has been studied by Watanabe et al. (2005). Simple heating of a mixture of added components at 100 °C in an electric oven gives the corresponding olefins in good yields. Wittig reactions can be carried out in a biphasic medium under ultrasonication. In this case, the Wittig products can be isolated by simple evaporation of the organic phase without additional work-up.

Ramazani et al. (2015) reported ultrasound-promoted three-component reaction of *n*-isocyaniminotriphenyl-phosphorane, (*E*)-cinnamic acids, and biacetyl.

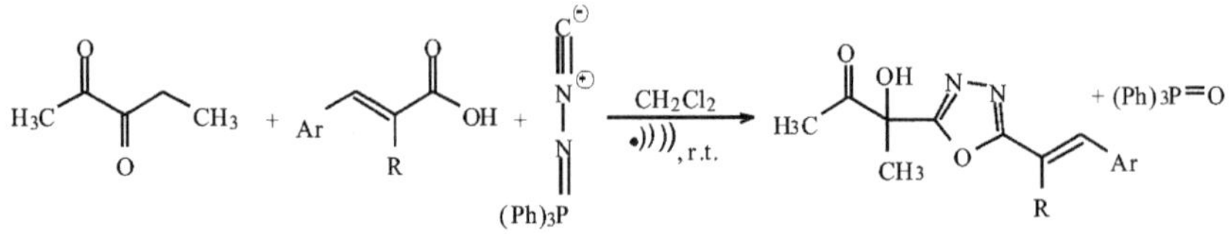

Fully substituted 1,3,4-oxadiazole derivatives were obtained in a one-pot three-component reaction of *n*-isocyaniminotriphenylphosphorane, biacetyl, and (*E*)-cinnamic acids in dichloromethane under ultrasound irradiation. This method produced the products in short reaction times (16–27 min) with excellent yields (91–97%).

4.3 PHASE-TRANSFER CATALYSIS

A PTC transfers an active species from one phase to the other, when two reactants are immiscible. Sometimes, a particular reaction is quite slow because of poor mass transfer. Thus, vigorous stirring is necessary for increasing the reaction rate. The presence of a PTC is also helpful in such reactions. Sonication either produces an extremely fine emulsion of immiscible liquids or assists in mass transfer and surface activation (in a solid/liquid system). These factors helped in enhancing PTC-catalysed heterogeneous reactions.

Regen and Singh (1982) cyclopropanated styrene in 96% yield in 1 h using a combination of both sonication and mechanical stirring. It was observed that the yield is reduced to 38% in 20 h with sonication alone because the power of the bath was not sufficient enough to disperse the solid reagent into the dense chloroform.

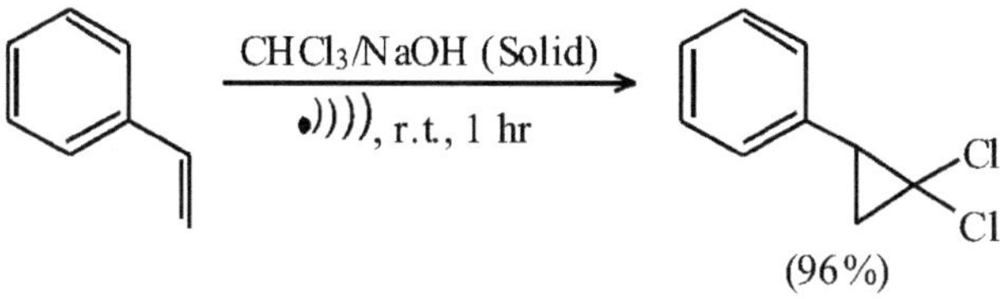

Davidson et al. (1987) observed only 19% yield of *N*-alkylated product of indole in 3 h using tert-butylammonium nitrate in toluene at 25 °C in the presence of solid KOH. This yield can be significantly improved by using ultrasound from 19% to 90% within 80 min.

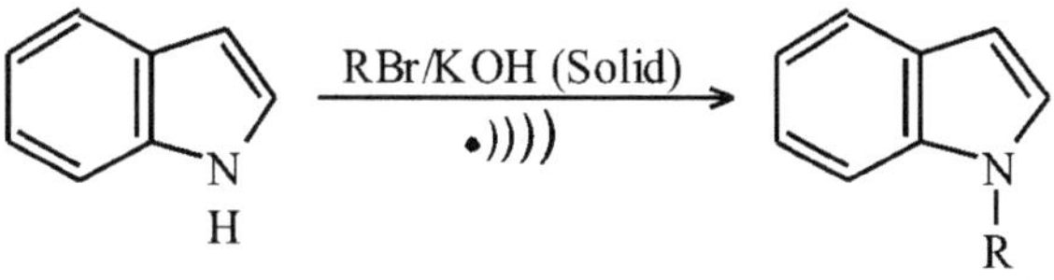

4.4 SUGARS

Deng et al. (2006) established that it is feasible to carry out a large number of conventional reactions in the synthesis and fictionalization of carbohydrates under ultrasound irradiation with better yields and reduced reaction times. It is very significant to note that the hindered disaccharide trehalose undergoes the reaction to afford the desired acetylated sugar in good yields. Tosylation and diol protection were also carried out successfully under sonochemical conditions.

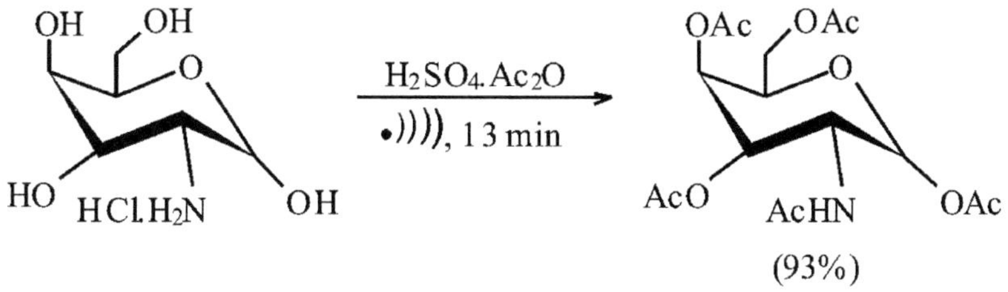

It was reported that the acyl group migrates within the carbohydrate scaffold in a few minutes (Deng and Chang, 2006). Otherwise, it takes hours or sometimes even days for the completion of this reaction under conventional conditions.

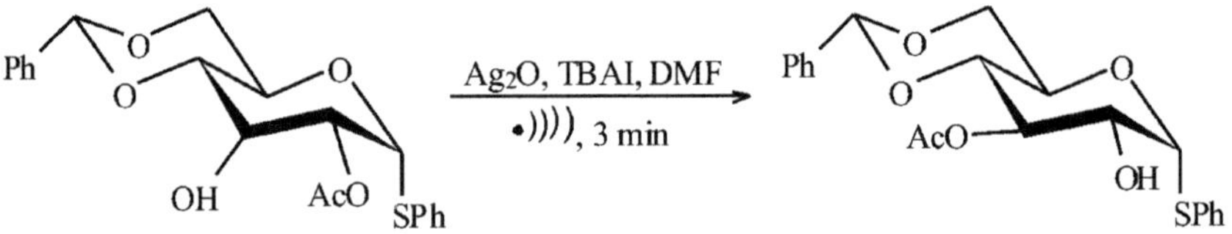

Tanifum and Chang (2008) have developed a new method for the synthesis of oligomannosides in good yields, in short reaction times and high stereoselectivity under ultrasound.

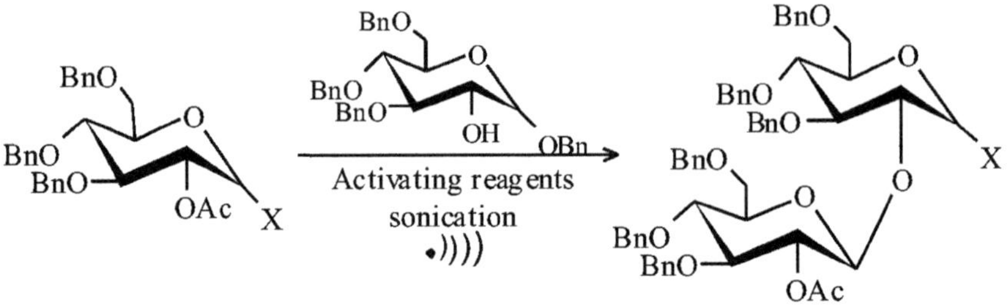

where X = SPh and OAc.

4.5 IONIC LIQUIDS

The increasing importance of sonochemistry as a green process and ILs as green solvents materials has resulted in their wide applications. The ILs find mild applications such as storage media for toxic gases, catalysts/solvents in organic syntheses, electrolytes, performance additives in pigments, and matrices because of their tailor-made nature. The existing conventional preparation of ILs requires excess solvents and hours or even days for completion in some cases. Microwaves were used for solving these problems. Although the reaction time was reduced from several hours to a few minutes and a large excess of alkyl halides and organic solvents was also avoided, the reaction is exothermic in nature and due to this, continuous microwave heating takes place leading to the formation of colored products. Ultrasound-assisted method was used to avoid this problem in dealing with these highly viscous materials, where efficient mixing is a challenge. Ultrasound-accelerated chemical reactions are well known and they proceed via the formation and adiabatic collapse of the transient cavitation bubbles. An efficient method for the preparation of ILs has been reported using ultrasound as the energy source by the simple exposure of neat reactants in a closed container to ultrasonic irradiation using a sonication bath. The current solvent-free approach requires shorter reaction time and lower reaction temperature in contrast to several hours needed under conventional heating conditions using an excess of reactants.

Namboodiri and Varma (2002) developed ultrasonic-assisted efficient method for the preparation of ILs using ultrasound as the energy source. In this process, they exposed neat reactants in a closed container using a sonication bath. One of the main advantages of this method is that the formation of the IL could be visibly monitored as the reaction contents turned from a clear solution to opaque and finally a clear viscous phase of solids.

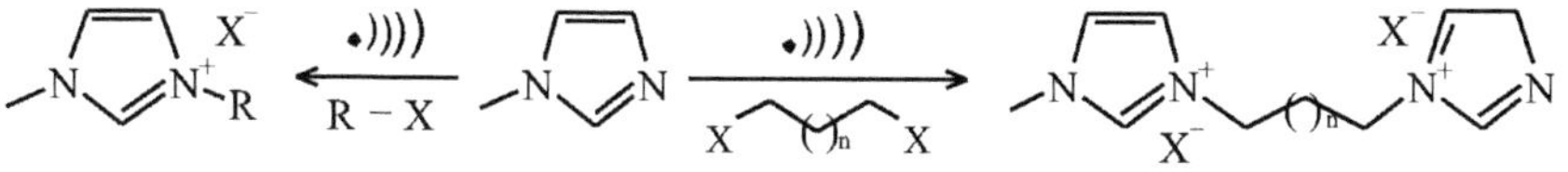

where $n=2$, 4, or 6; X = Cl, Br, I; R = nPr, n Bu, i-Bu, etc.

Imidazolium-based IL derivatives were synthesized using a facile and green ultrasound-assisted procedure (Ameta et al., 2015). A large number of ILs can be synthesized by a simple combination of different cations and anions. Ultrasound-assisted synthesis of ILs has an edge over classical conventional methods, that is, ecofriendly nature, simple product isolation, higher yield, shorter reaction times. The main advantages of this procedure as compared with conventional method were its milder conditions, shorter reaction time, higher yields and selectivity without the need for a transition metal or base catalyst. Introducing such an efficient synthetic protocol will reduce the cost of ILs, thus encouraging a wider use of these neoteric solvents.

Step I

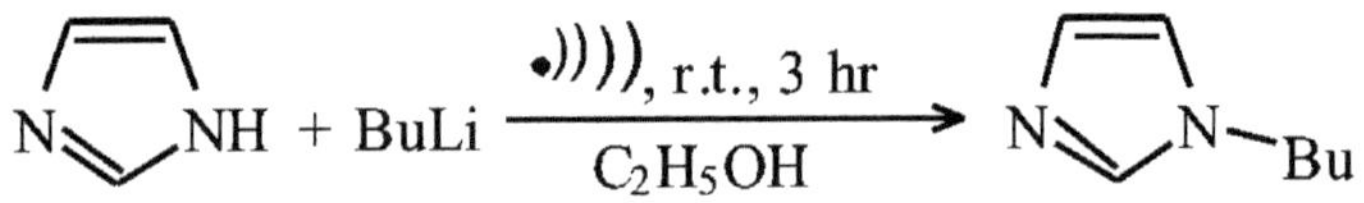

Step II

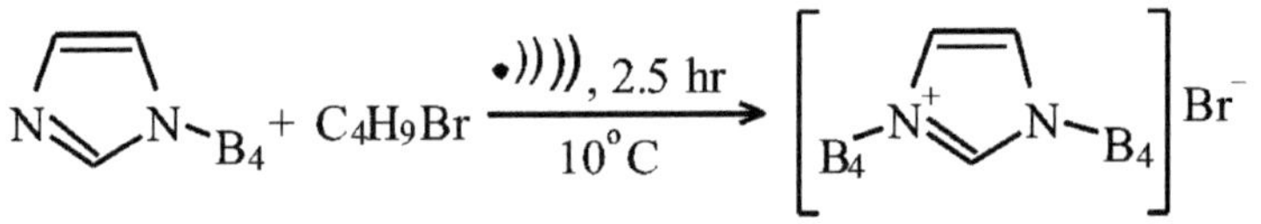

1,3-Dibutylimidazolian bromide

4.6 HETEROCYCLES

Some of the heterocycles are discussed in the following sections.

4.6.1 IMIDAZOLE DERIVATIVES

2-Arylbenzimidazole was synthesized by the reaction of substituted aldehydes with *o*-phenylenediamine catalysed by tetrabutylammonium fluoride (TBAF) in water under ultrasonic irradiation at room temperature (Joshi et al., 2010). It is a mild and chemoselective method for preparing such heterocycles. The use of quaternary ammonium fluoride salts is economical and these are relatively nontoxic reagents. Water is used as it is commonly accepted to be a green solvent in organic synthesis.

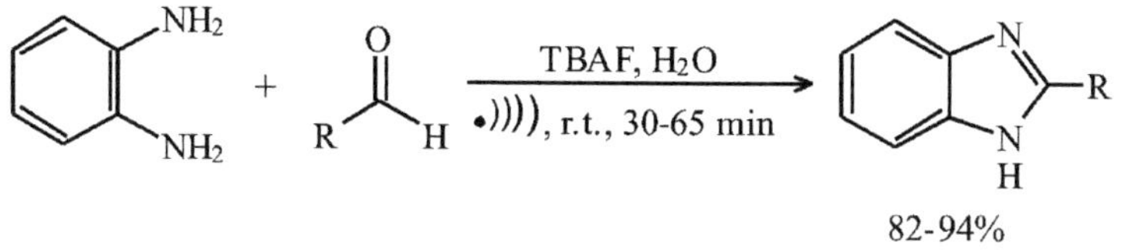

Arani and Safari (2011) reported a rapid and efficient synthesis of 5,5-diphenylhydantoin and 5,5-diphenyl-2-thiohydantoin derivatives with high yield (96–98 %) under mild conditions. These reactions were performed in DMSO/H_2O with ultrasonic irradiation and catalysed by KOH.

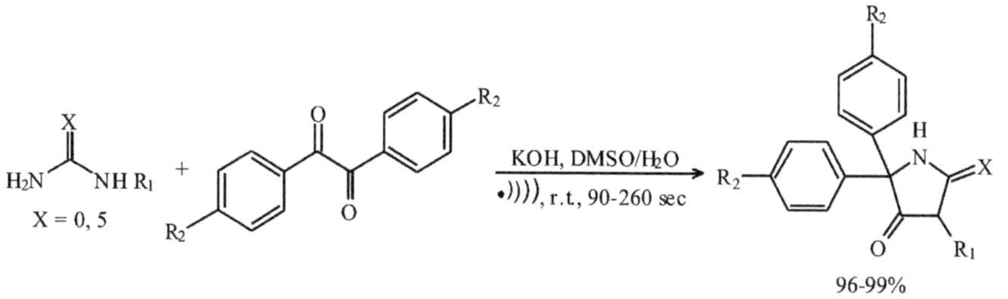

4.6.2 PYRAZOLE DERIVATIVES

A comparative study was made by Pathak et al. (2009) between four activating methods to obtain *N*-acetyl-pyrazolines. These include reflux, solvent-free conditions, microwave irradiation, and ultrasonic irradiation. Out of these four methods, microwave irradiation method was found to be the most effective, followed by ultrasound-assisted method. The reaction of 1,4-pentadien-3-ones with hydrazine and acetic acid in ethanol was completed within 10–25 min and afforded the products in good yields (79–95 %).

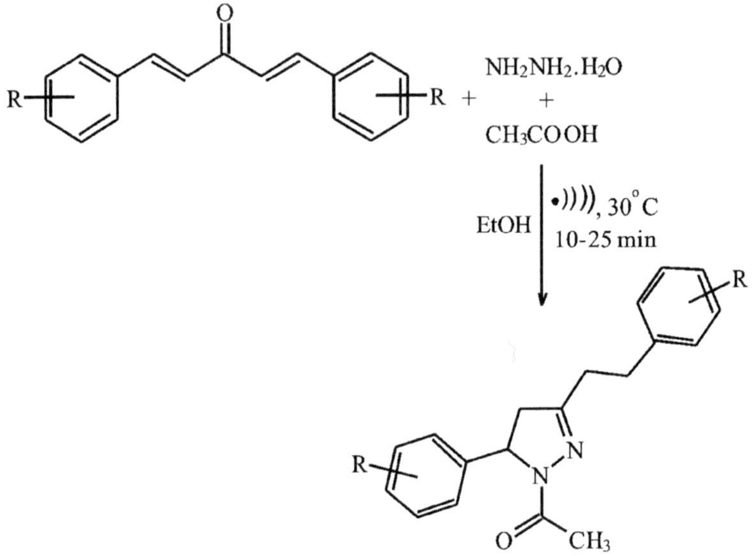

(76-91%)

An efficient and highly regioselective synthesis of ethyl 1-(2,4-dichlorophenyl)-1*H* pyrazole-3-carboxylates was reported under ultrasound irradiation (Machado et al., 2011). This reaction is completed in 10–12 min whereas it requires 2.5–3 h under conventional thermal conditions. An ultrasound-assisted and bismuth nitrate- catalysed eco-friendly route has been designed to synthesize a series of novel *N*-substituted pyrrole derivatives. (Bandyopadhyay et al., 2012) This is a simple method to prepare a series of *N*-substituted pyrroles with less nucleophilic polyaromatic amines.

4.6.3 ISOXAZOLE DERIVATIVES

Ultrasound-promoted synthesis of substituted pyrazoles and isoxazoles containing sulphone moiety has been investigated via the one-pot reaction of the carbanions of 1-aryl-2-(phenylsulphonyl)-ethanone with several different hydrazonyl halides in ethanol (Saleh and El-Rahman, 2009). These reactions were enhanced by ultrasonic irradiation, and the products were isolated in high yields (90–97%).

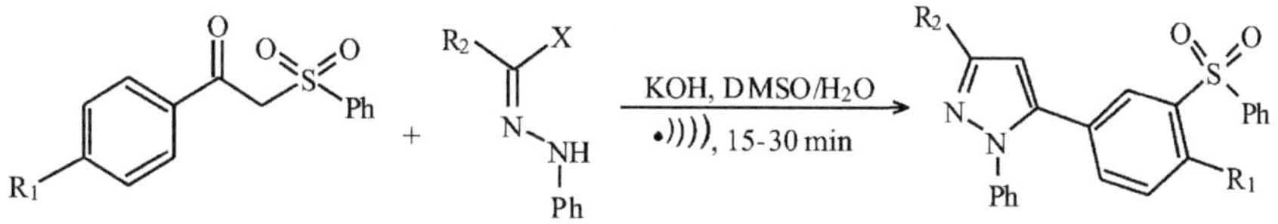

A benign and improved synthesis of 5-(2-chloroquinolin-3-yl)-3-phenyl-4,5-dihydroisoxazoline using acetic acid aqueous solution under ultrasound irradiation has been carried out by Tiwari et al. (2011). It was observed that sonochemical method gave better yields (87–90%) of the products in shorter times of 90–120 min than the corresponding thermal reactions (72–78%) and that too in 6–7 h.

4.6.4 OXAZOLE DERIVATIVES

Mirkhani et al. (2009) developed an improved sulphonated carbon-based solid acid as a novel, efficient, and reusable catalyst for chemoselective synthesis of 2-oxazolines and bis-oxazoline. It was found that reactions performed with a combination of the new catalyst and sonication was more efficient than those without ultrasonic irradiation.

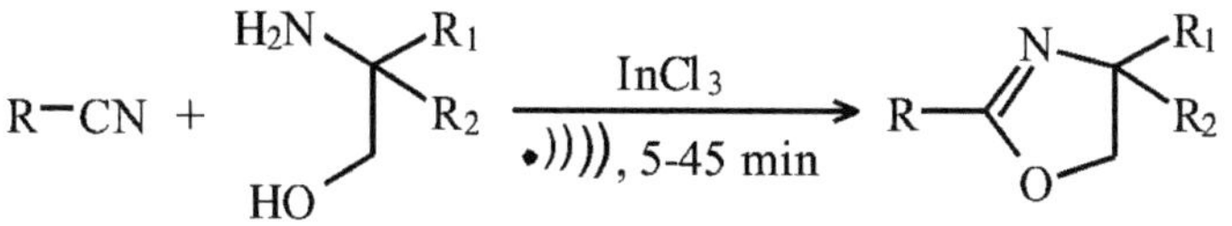

4.6.5 THIAZOLE DERIVATIVES

Spiro[indole-thiazolidinones] were prepared sonochemically via a three-component reaction in aqueous medium in the presence of cetyltrimethylammonium bromide as a PTC (Dandia et al., 2011). This reaction was completed in 40–50 min under ultrasound with the desired product in good to excellent yields (80–98%).

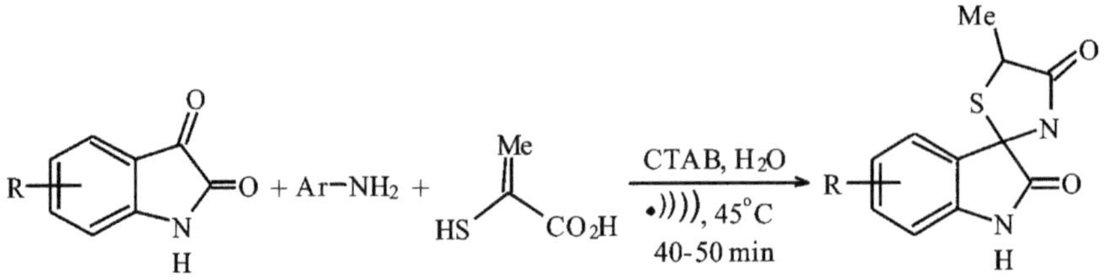

Ultrasound-promoted greener synthesis of 2-(3,5-diaryl-4,5-dihydro-1*H*-pyrazol-1-yl)-4-phenylthiazoles was reported by Venzke et al. (2011). This synthesis was completed in only 15 min in ethanol at room temperature and afforded the pure products in 47–93 % yields by simple filtration of the reaction mixture.

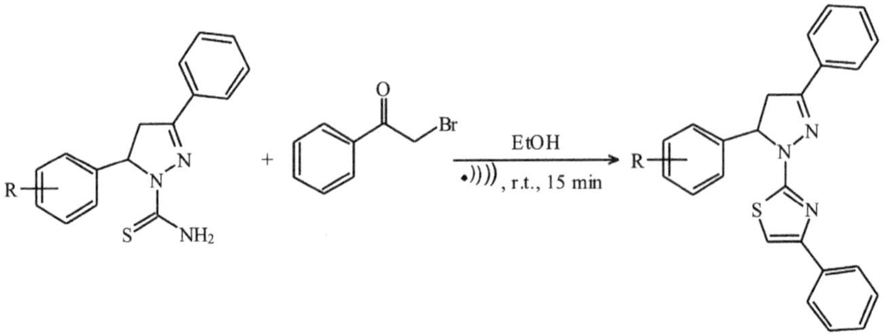

Thiazolidinones from piperonilamine were also obtained in good yields under sonication for 5 min. The corresponding conventional thermal reactions without ultrasound also furnished similar yields, but it required much longer times of 16 h (Neuenfeldt et al., 2011).

4.6.6 SELENAZOLE DERIVATIVES

An ultrasound-mediated preparation of 1,3-selenazoles was reported by Lalithamb et al. (2010). They reported an efficient preparation of desired products by the treatment of bromomethyl ketones with selenourea in acetone under ultrasonic irradiation for 5–10 min.

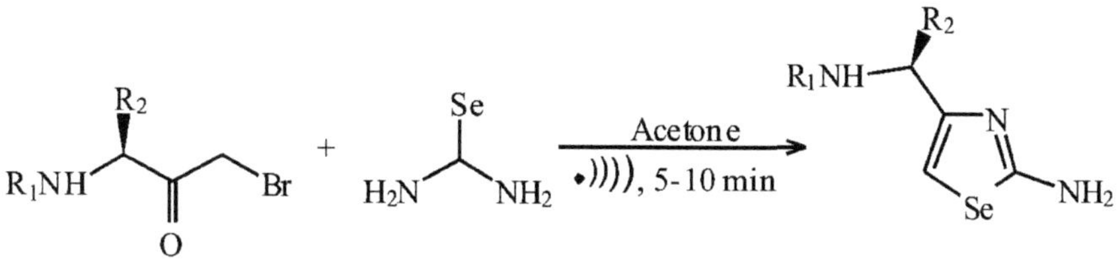

4.6.7 TRIAZOLE DERIVATIVES

Cravotto et al. (2010) used ultrasound with/without microwave for producing 1,2,3-triazole derivatives via copper- catalysed azide-alkyne cycloaddition reaction. It was reported that the best results were obtained, when azides and terminal alkynes were sonicated in the presence of Cu turnings in dioxane/H_2O at 70 °C or DMF at 100 °C. Substitution of water by DMF was required to prevent the formation of copper complexes so that the purification of the products is made simpler.

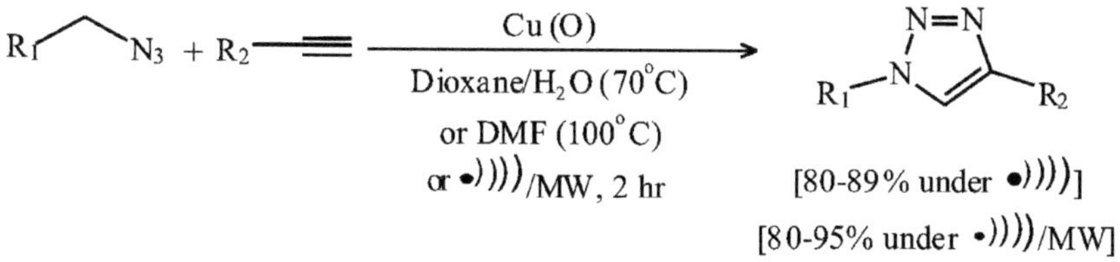

4.6.8 OXADIAZOLE DERIVATIVES

A comparison between conventional heating with microwave and ultrasound as ecofriendly energy sources for the synthesis of 2-substituted 1,2,3-triazoles and 4-amino-5-cyanopyrazole derivatives has been made by Al-Zaydi (2009). The products were obtained with excellent yields in this green chemical approach.

Ultrasound-promoted synthesis of 3-trichloromethyl-5-alkyl(aryl)-1,2,4-oxadiazoles was carried out by Bretanha et al. (2011), with excellent yields (84–98 %) and shorter reaction times of 15 min as compared to silent conditions.

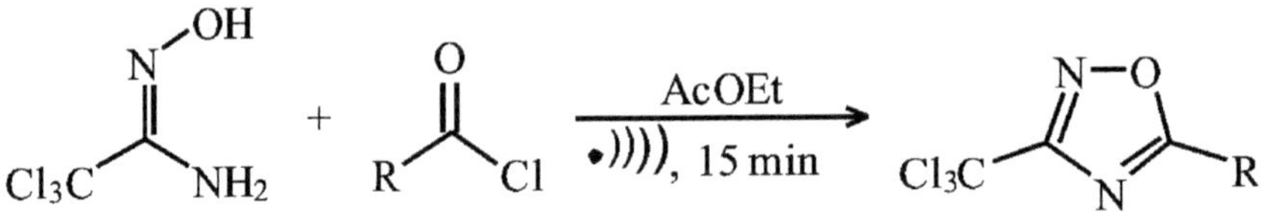

4.6.9 THIADIAZOLE DERIVATIVES

Some new 1,3,4-thiadiazole and bi(1,3,4-thiadiazole) derivatives incorporating pyrazolone moiety were synthesized by reacting 1-methyl-5-oxo-3-phenyl-2-pyrazolin-4-thiocarboxanilide with a series of hydrazonyl halides or *N,N'*-diphenyl-oxalodihydrazonoyl dichloride in the presence of triethylamine (TEA) (El-Rahman et al., 2009).

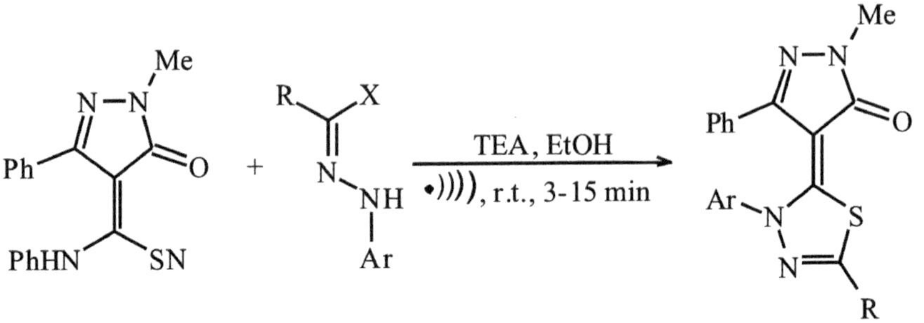

The reactions were carried out under both conventional and ultrasonic irradiation conditions. Improvement in rates and yields was observed, when reactions were carried out under sonication compared to classical condition.

4.6.10 TETRAZOLE DERIVATIVES

The reaction of a series of aromatic nitriles with sodium azide was catalysed by montmorillonite K-10 or kaolin clays in water or DMF as solvent, under ultrasound (Chermahini et al., 2010). The use of ultrasonic irradiation reduced the reaction times and increased the activity of a catalyst.

The amount of nitrile to sodium azide mole ratio, amount of catalyst, reaction time, and solvent type were also optimized. The versatility of this method was checked by using various nitriles, which showed reasonable yields of tetrazoles formation compared to conventional heating. It was found that using nitriles with electron-withdrawing groups result in both higher yields and lower reaction times. The catalysts could also be reused several times without significant loss of their catalytic activity. This methodology finds widespread use in organic synthesis for the preparation of tetrazoles. The advantages of this catalytic system are as follows: mild reaction condition, high product yields, easy preparation of the catalysts, nontoxicity of the catalysts, and simple and clean work-up of the desired products.

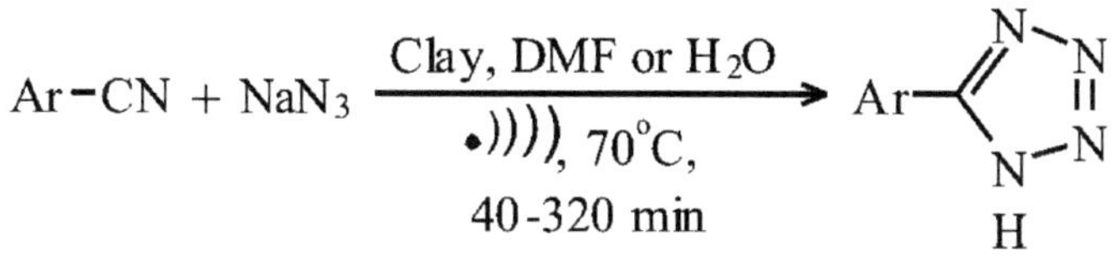

4.7 SONOCHEMICAL SWITCHING

Sonochemical switching is a unique feature of sonochemistry, which differentiates this technique from other methods enhancing chemical reactions. The extreme conditions within the bubble may lead to totally new reaction pathways, that is, via radicals generated in the vapor phase that would only have a transient existence in the bulk liquid. Such a switching has been observed in case of Kornblum–Russell reaction, where sonication favors a single-electron transfer pathway.

Dickens and Luche (1991) reported that 4-nitrobenzyl bromide reacts with 2-lithio-2-nitro-propane via a polar mechanism to give 4-nitrobenzaldehyde as a final product. When this reaction was carried out under sonication, it leads to the formation of a dinitro compound depending on the irradiation conditions and intensity of ultrasound.

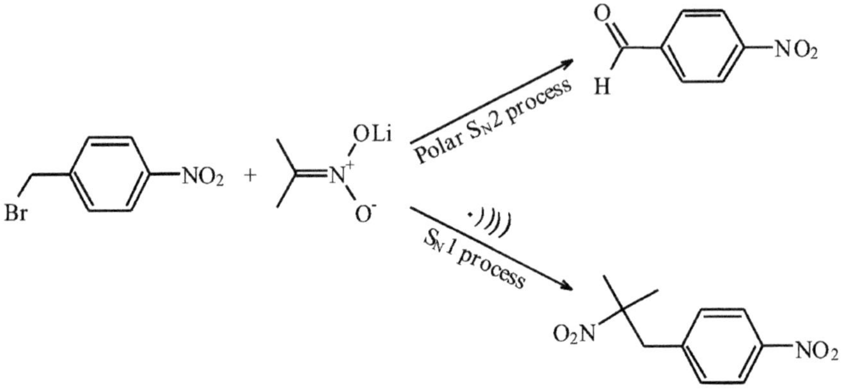

Ando and Kimura (1991) demonstrated a reaction between benzyl bromide and alumina-supported potassium cyanide in toluene. Benzyl cyanide was produced by nucleophilic displacement of the bromine by supported cyanide, but substituted toluene was obtained without sonication.

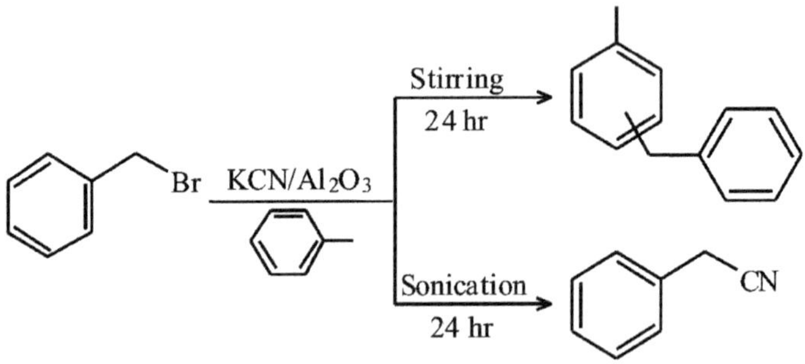

This reaction afforded diphenylmethane products via a Friedel–Crafts reaction between the bromo compound and the solvent, catalysed by Lewis acid sites on the surface of the solid phase reagent under stirring alone. On the contrary, sonication of the same reaction produced only the substitution product, that is, benzyl cyanide. It may be due to cavitations, which produce a structural change to the catalytic sites of the solid support, may be through masking them via cavitationally induced cyanide absorption. When nitric acid reacts with alcohol under ultrasound, the product obtained is acid. This reaction proceeds through radical cation. On the other hand, the product is nitrate ester under silent conditions.

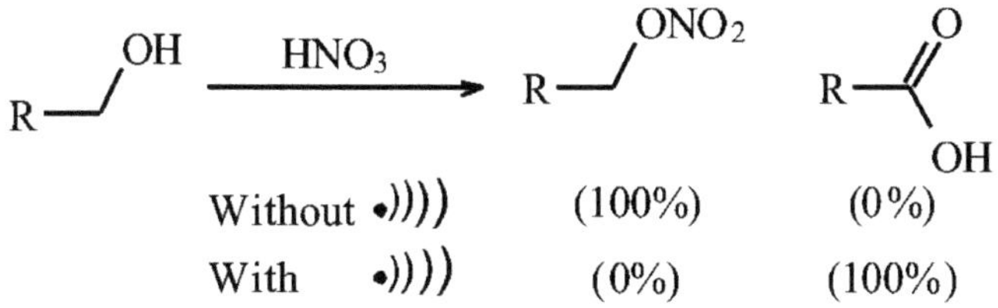

The reaction of *o*-allyl benzamides with Li/THF results in the formation of indanone in two steps (Einhom et al., 1988). The first step is accelerated by ultrasound which involves the ketyl radical anion. This anion cyclizes to 2-methylindanone and liberates an amide ion, which deprotonates the allyl moiety. Due to sonication, the ketyl is generated much more rapidly so that only the cyclization to 2-methylindanone occurs and other side reactions are omitted.

4.8 MISCELLANEOUS

It is necessary to develop some alternative methods for desulphurization of fossil fuel-derived oil, due to the requirement of stringent rules for ultra-low sulphur content of diesel fuels. Dibenzothiophene can be quantitatively oxidized by using appropriate oxidants and catalysts with the assistance of ultrasound irradiation within a few minutes. The removal efficiency of sulphur-bearing compounds in diesel fuels can reach or exceed 99% in a short contact time at ambient conditions (temperature and pressure) by the use of catalytic oxidation and ultrasound followed by solvent extraction. This approach is known as ultrasound-assisted oxidative desulphurization and it can be the basis for obtaining ultra-low sulphur-containing diesel oil (Mei et al., 2003).

Another application of ultrasound is allylation of carbonyl groups. Homoallylic alcohols can be obtained by sonicating a mixture of allylic halides, zinc, and aldehydes or ketones in aqueous THF (Petrier and Luche, 1985; Einhorn and Luche, 1987). However, the yields are not always satisfactory, but the replacement of the THF-water solvent by THF-aqueous NH_4Cl greatly improves this method. In this condition, stirring at room temperature is sufficient to provide activation.

One more application of ultrasound is alkylation. Bogolubsky et al. (2008) studied structurally and functionally diverse *N*-carbamoylamino acids, which were obtained through the alkylation of monosubstituted parabanic acids followed by hydrolysis of the intermediate products in very good yields and excellent purity.

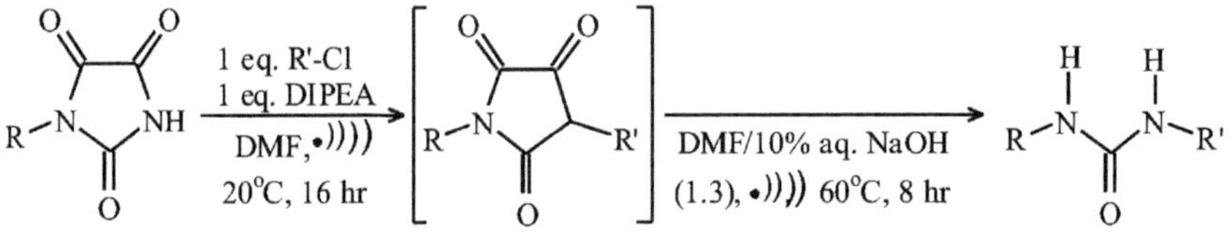

where R= Alkyl, Bn, Ph; R′= Benzyl, CHR″CO_2H.

Ultrasound has also been used for oxidation and reduction reactions. The use of ultrasonic irradiation accelerated the oxidation reaction of aromatic and heteroaromatic compounds. Ruthenium-catalysed oxidation of aromatic and heteroaromatic compounds has been reported. It was found that ultrasonic irradiation in a biphasic system consisting of substrate, CH_2Cl_2, H_2O, CH_3CN, $NaIO_4$ and catalytic amounts of $RuCl_3 \cdot nH_2O$, accelerates the oxidation reaction to afford the desired products in good yields (Tabatabaein et al., 2007).

Various mono-, di- and α,β-unsaturated cyanides were reduced with a Cu–Al alloy in NaOD–D_2O and THF to the corresponding deuterated aliphatic amines, such as nonylamines, putrescine, and 1,6-hexanediamine, with high deuterium content in the presence of ultrasound (Tsuzuki et al., 1996).

Salvador et al. (1993) studied the zinc-mediated reduction of α,β-unsaturated ketones in acetic acid. They reported that this reaction proceeds very efficiently under sonochemical conditions. Different α-enone systems gave two kinds of products: olefins and allylic alcohols. The steroids with the common 4-en-3-one system are efficiently reduced to D3-steroids whereas 16-ene-20-one gave the corresponding allylic alcohol systems under similar conditions.

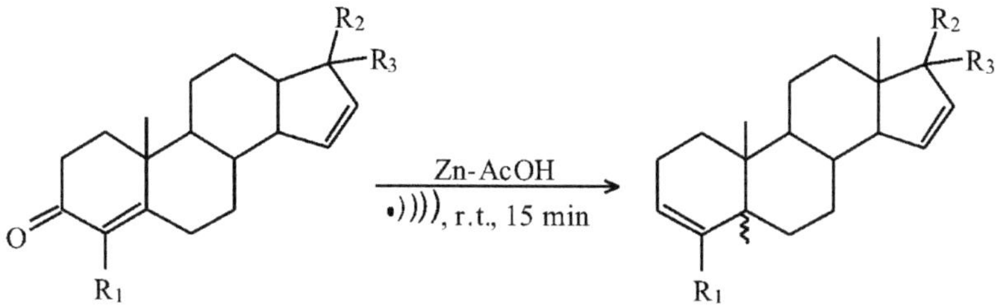

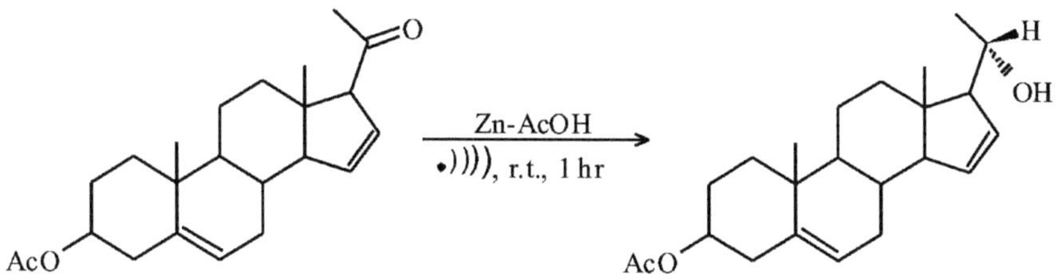

A convenient one-pot cyclocondensation method for benzothiazole and benzimidazole syntheses was reported by Yuan and Guo (2011). Some benzothiazoles and benzimidazoles were prepared from aromatic *o*-diamines or *o*-aminothiophenol and aldehydes using chlorotrimethylsilane in DMF as promoter and water scavenger under ultrasonic irradiation in good yields (84–97 %).

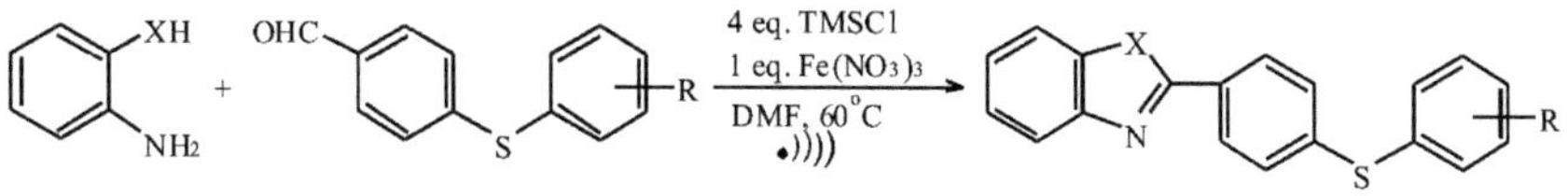

where X = S, NH.

Xiong et al. (2013) reported a simple and an efficient method for the selective cleavage of terminal acetonides. When terminal acetonides in the presence of wide range of functional group are treated with silica-supported boron trifluoride as a catalyst under ultrasound irradiation, it furnished the corresponding diols in reasonable good yields (82–95 %). It was interesting to note that acid labile *p*-methoxybenzyl group (protecting group) remained as such under these conditions of deprotection.

Sonochemistry is gaining significant importance in organic synthesis based on laboratory results and the availability of scale-up systems. Ultrasound is an efficient and practically benign means of activation in synthetic chemistry and it has been employed for decades with varied success. This high-energy input not only enhances mechanical effects in heterogeneous processes but also induces new reactivity thereby leading to the formation of unexpected chemical species. It has made sonochemistry a unique technique supported by the remarkable phenomenon of cavitation, which is currently the subject of intense research and it has already yielded some thought-provoking results. Although applications for sonochemistry can be found in many areas, sonochemical processes are most widely developed for heterogeneous reactions. It is a multidisciplinary field where chemists, physicists, chemical engineers, and mathematicians must cooperate to develop a better understanding of the processes taking place within the collapsing bubbles to develop some newer dimensions in its applications.

The challenges of scale-up still represent a current concern of nonconventional technologies. A series of commercially available ultrasonic reactors can be readily adapted for scaling up a particular process, even operating below the ultrasonic threshold.

Because organics are invariably linked to biomaterials and pharmaceuticals, sonochemical reactions conducted on larger scales in these areas will have a wide and long-term impact without any debate. Besides improving numerous organic reactions, sonication can also initiate crystal nucleation, switching a chemical reaction, etc. Both sonosynthesis and sonocrystallization are likely to be very useful in drug discovery. Chemists, in general, but organic synthetic chemists, in particular, must explore their researches further using this harmless and greener sound energy. The usefulness of sonochemistry has now no limitations and it will continue to expand its arena encompassing, Green chemistry, photochemistry, electrochemistry, bio-imaging, biosensors, industrial chemistry, biotechnology, etc.

REFERENCES

Abimannan, P.; Rajendran, V. *Iran. J. Chem. Eng.* **2016,** *13*, 54–62.

Adams, R.; Marvel, C. S. *Org. Synth.* **1941,** *1*, 94–95.

Ali, M. M.; Sana Tasneem, S.; Rajanna, K. C.; Saiprakash, P. K. *Synth. Commun.* **2002,** *32*(9), 1351–1356.

Al-Zaydi, K. M. *Ultrason. Sonochem.* **2009,** *16*, 805–809.

Ameta, G.; Pathak, A. K.; Ameta, C.; Ameta, R.; Punjabi, P. B. *J. Mol. Liq.* **2015,** *211*, 934–937.

An, G.; Ji, X.; Han, J.; Pan, Y. *Synth. Commun.* **2011,** *41*, 1464–1471.

Ando, T.; Kimura, T. *Ultrasonic Organic Synthesis Involving Nonmetal Solids, Advances in Sonochemistry;* Mason, T. J., Ed.; JAI Press: London, 1991; p 211.
Arani, N. M.; Safari, V. J. *Ultrason. Sonochem.* **2011,** *18*, 640–643.
Arcadi, A.; Alfonsi, M.; Marinelli, F. *Tetrahedron Lett.* **2009,** *50*, 2060–2064.
Babukantharaju, V. V. S.; Tantry, S. J. *Int. J. Pept. Res. Ther.* **2005,** *11*(2), 131–137.
Bandyopadhyay, D.; Mukherjee, S.; Granados, J. C.; Short J. D.; Banik, B. K. *Eur. J. Med. Chem.* **2012,** *50*, 209–215.
Bang, K.; Lee, K.; Park, Y. K.; Lee, P. H. *Bull. Korean Chem. Soc.* **2002,** *23*, 1272–1276.
Batovska, D.; Parushev, S.; Slavova, A.; Bankova, V.; Tsvetkova, I.; Ninova, M.; Najdenski, H. *Eur. J. Med. Chem.* **2007,** *42*, 87–92.
Baylis, A. B.; Hillman, M. E. D. German Patent 2,155,113, 1972.
Bhavani, S.; Ashfaq, M. A.; Rambabu, D.; Rao, M. V. B.; Pal, M. *Arabian J. Chem.* **2016**. DOI: 10.1016/j.arabjc.2016.02.002.
Biginelli, P. *Gazz. Chim. Ital.* **1893,** *23*, 360–413.
Blaise, E. E. *C. R. Chim.* **1901,** *132*, 478–480.
Bogolubsky, A. V.; Ryabukhin, S. V.; Pakhomov, G. G.; Ostapchuk, E. N.; Shivanyuk, A. N.; Tolmachev, A. A. *Synlett* **2008,** 2279–2282.
Bretanha, L. C.; Teixeira, V. E.; Ritter, M.; Siqueira, G. M.; Cunico, W.; Pereira, C. M. P.; Freitag, R. A. *Ultrason. Sonochem.* **2011,** *18*, 704–707.
Calvino, V.; Picallo, M.; Lopez-Peinado, A. J.; Martin-Aranda, R. M.; Duran-Valle, C. J. *J. Appl. Surf. Sci.* **2005,** *252*(17), 6071–6074.
Cannizzaro, S. *Liebigs Ann. Chem. Pharm.* **1853,** *88*, 129–130.
Carey, F. A.; Sundberg, R. J. *Advanced Organic Chemistry Part B: Reactions and Synthesis*, 3rd ed.; Plenum: New York, 1993; p 55.
Cason, J.; Rinehart, K. L.; Thorston, S. D. *J. Org. Chem.* **1953,** *18*, 1594.
Caulier, T. P.; Reisse, J. *J. Org. Chem.* **1996,** *61*, 2547–2548.
Cella, R.; Stefani, H. A. *Tetrahedron* **2006,** *62*, 5656–5662.
Chen, L.-H.; Chung, T.-W.; Narhe, B. D.; Sun, C.-M. *ACS Comb. Sci.* **2016,** *18*, 162–169.
Chermahini, A. N.; Teimouri, A.; Momenbeik, F.; Zarei, A.; Dalirnasab, Z.; Ghaedi, A.; Roosta, M. *J. Heterocycl. Chem.* **2010,** *47*, 913–922.
Coelho, F.; Almeida, W. P.; Veronese, D.; Mateus, C. R.; Silva Lopes, E. C.; Rossi, R. C.; Silveira, G. P.C.; Pavam, C. H. *Tetrahedron* **2002,** *58*(37), 7437–7447.
Cravotto, G.; Demetri, A.; Nano, G. M.; Palmisano, G.; Penoni, A.; Tagliapietra S. *Eur. J. Org. Chem.* **2003,** *22*, 4438–4444.
Cravotto, G.; Fokin, V. V.; Garella, D.; Binello, A.; Boffa, L.; Barge, A. *J. Comb. Chem.* **2010,** *12*, 13–15.
Curtius, T. *Ber. Dtsch. Chem. Ges. Berlin* **1890**, *23*, 3023–3033.
Curtius, T. *J. Prakt. Chem.* **1894,** *50*, 275–294.
Dandia, A.; Singh, R.; Bhaskaran, S.; Samant, S. D. *Green Chem.* **2011,** *13*, 1852–1859.
Davidson, R. S.; Safdar, A.; Spencer, J. D; Robinson B. *Ultrasonics* **1987,** *25*, 35–39.
De-la-Torre, P.; Osorio, E.; Alzate-Morales, J. H.; Caballero, J.; Trilleras, J.; Astudillo-Saavedra, L.; Brito, I.; Cárdenas, A.; Quiroga, J.; Gutiérrez, M. *Ultrason. Sonochem.* **2014,** *21*(5), 1666–1674.
Deng, S.; Chang, C. W. T. *Synlett* **2006,** 756.
Deng, S.; Gangadharmath, U.; Chang, C. W. T. *J. Org. Chem.* **2006,** *71*, 5179–5185.

Deshmukh, R. R.; Rajagopal, R.; Srinivasan, K. V. *Chem. Commun.* **2001,** 1544–1545.

Dhawanpalli, R. S.; Kale, H. B.; Khan, P. A. *J. Innovations Pharm. Biol. Sci.* **2015,** *2*(2), 209–217.

Dickens, M. J.; Luche, J.-L. *Tetrahedron Lett.* **1991,** *32,* 4709–4712.

Diels, O.; Alder, K. *Justus Liebigs Ann. Chem.* **1928,** *460,* 98–122.

Diels, O.; Alder, K. *Ber. Dtsch. Chem. Ges. Berlin* **1929,** *62,* 554–562.

Docampo, L. M.; Pellon, R. F. *Synth. Commun.* **2003,** *33,* 1777–1782.

Drahl, C. *Chem. Eng. News* **2010,** *88*(22), 31–33.

Einhorn, C.; Luche, J. L. *J. Organomettalic Chem.* **1987,** *322,* 177–183.

Einhorn, J.; Einhorn, C.; Luche, J.-L. *Tetrahedron Lett.* **1988,** *29,* 2183–2184.

El-Rahman, N. M. A.; Saleh, T. S.; Mady, M. F. *Ultrason. Sonochem.* **2009,** *16,* 70–74.

Entezari, M. H.; Shameli, A. A. *Ultrason. Sonochem.* **2000,** *7,* 169–172.

Estager, J.; Leveque, J. M.; Turgis, R.; Draye, M. *Tetrahedron Lett.* **2007,** *48*(5), 755–759.

Fernando, C.; Wanda, A. P.; Demetrius, V.; Cristiano, M. R.; Elizandra, L. S. C.; Rodrigo, R. C.; Gabriel, S. C. P.; Cesar, P. H. *Tetrahedron* **2002,** *58,* 7437–7447.

Fittig, R. *Justus Liebigs Ann. Chem.* **1859,** *110,* 23–45.

Fu, N.; Zhang, Y.; Yang, D.; Chen, B.; Wu, X. *Catal. Commun.* **2008,** *9,* 976–979.

Fuentes, A.; Marinas, J. M.; Sinisterra, J. V. *Tetrahedron Lett.* **1987,** *28,* 4541–4544.

Ghadami, M.; Jafari, A. A. *Environ. Chem. Lett.* **2015,** *13*(2), 191–196.

Ghasemzadeh, M. A.; Abdollahi-Basir, M. H. *Green Chem. Lett. Rev.* **2016,** *9*(3), 156–165.

Gholap, A. R.; Venkatesan, K.; Pasricha, R.; Daniel, T.; Lahoti, R. J.; Srinivasan, K. V. *J. Org. Chem.* **2005,** *70,* 4869–4872.

Ghotbinejad, M.; Khosropour, A. R.; Mohammadpoor-Baltork, I.; Moghadam, M.; Tangestaninejad, S.; Mirkhani, V. *RSC Adv.* **2014,** *4,* 8590–8596.

Girase, P. S.; Khairnar, B. J.; Nagarale, D. V.; Chaudhari, B. R. *Eur. J. Pharm. Med. Res.* **2016,** *3*(3), 333–336.

Guadagnin, R. C.; Suganuma, C. A.; Singh, F. V.; Vieira, A. S., Cella, R.; Stefani, H. A. *Tetrahedron Lett.* **2008,** *49,* 4713–4716.

Hagu, H.; Salmar, S.; Tuulmets, A. *Ultrason. Sonochem.* **2007,** *14*(4), 445–449.

Han, B. H.; Boudjouk, P. *J. Org. Chem.* **1982,** *47,* 5030–5032.

Hanafusa, T.; Ichihara, J.; Ashida, T. *Chem. Lett.* **1987,** *16,* 687–690.

Hathaway, B. A. *J. Chem. Educ.* **1987,** *64,* 367–368.

Heck, R. F.; Nolley, J. P. *J. Org. Chem.* **1972,** *37*(14), 2320–2322.

Hemantha, H. P.; Chennakrishnareddy, G.; Vishwanatha, T. M.; Sureshbabu, V. V. *Synlett* **2009,** 407–410.

Imanzadeh, G.; Rezaee-Gatar, S. *Arkivoc* **2015,** *5,* 121–133.

Imanzadeh, G.; Kazemi, F.; Zamanloo, M.; Mansoori, Y. *Ultrason. Sonochem.* **2013,** *20*(2), 722–728.

Iwamato, K.; Hamay, M.; Hashimoto, N.; Kimura, H.; Suzuki, Y.; Sato, M. *Tetrahedron Lett.* **2006,** *47,* 7175–7177.

Javed, T.; Mason, T. J.; Phull, S. S.; Baker, N. R.; Robertson, A. *Ultrason. Sonochem.* **1995,** *2,* S3–4.

Ji, J. S.; Wang, Y. S. *Synlett* **2003,** 2074.

Joshi, R. S.; Mandhane, P. G.; Dabhade, S. K.; Gill, C. H. *J. Chin. Chem. Soc.* **2010,** *57,* 1227–1231.

Kaiser, C.; Weinstock, J. *Org. Synth.* **1988,** *6,* 910–912.

Kakaei, S.; Kalal, H. S.; Hoveidi, H. *J. Sci., Islamic Repub. Iran* **2015,** *26*(2), 117–123.
Kappe, C. O. *Tetrahedron* **1993,** *49*(32), 6937–6963.
Keipour, H.; Hosseini, A.; Khalilzadeh, M. A.; Ollevier, T. *Lett. Org. Chem.* **2015,** *12*, 645–650.
Khaligh, N. G.; Mihankhah, T. *Chin. J. Catal.* **2013,** *34*, 2167–2173.
Knoevenagel, E. *Ber. Dtsch. Chem. Ges.* **1898,** *31*(3), 2596–2619.
Kulkarni, K.; Friend, J.; Yeo, L.; Perlmutter, P. *Ultrason. Sonochem.* **2014,** *21*, 1305–1309.
Lalithamba, H. S.; Narendra, N.; Naik, S. A.; Sureshbabu, V. V. *Arkivoc* **2010,** *ix*, 77–90.
Lee, A. S.-Y.; Cheng, R.-Y. *Tetrahedron Lett.* **1997,** *38*, 443–446.
Li, J. T.; Chen, G. F.; Wang, J. X.; Li, T. S. *Synth. Commun.* **1999a,** *29*, 965–971.
Li, J. T.; Li, T. S.; Li, L. J.; Cheng, X. *Ultrason. Sonochem.* **1999b,** *6*, 199–201.
Li, T. Y.; Cui, W.; Liu, J. G.; Zhao, J. Z.; Wang, Z. M. *Chem. Commun.* **2000,** 139–140.
Li, J. T.; Zhang, H. J.; Meng, L. H.; Li, L. J.; Yin, Y. H.; Li, T. S. *Ultrason. Sonochem.* **2001,** *8*, 93–95.
Li, J. T.; Yang, W. Z.; Wang, S. X.; Li, S. H.; Li, T. S. *Ultrason. Sonochem.* **2002,** *9*, 237–239.
Li, J. T.; Yang, J. H.; Li, T. S. *Youji Huaxue (Chin. J. Org. Chem.)* **2003,** *23*(12), 1428.
Li, J. T.; Wang, D.; Zhang, Y.; Li, J.; Chen, B. *Org. Lett.* **2009,** *11*, 3024–3027.
Li, J. T.; Sun, S. F.; Sun, M. X. *Ultrason. Sonochem.* **2011,** *18*(1), 42–44.
Lindley, J.; Lorimer, P. J.; Mason, J. T. *Ultrasonics* **1986,** *24*, 292–293.
Lindley, J.; Mason, J. T.; Lorimer, P. J. *Ultrasonics* **1987,** *25*, 45–48.
Little, R. D.; Masjedizadeh, M. R.; Wallquist, O.; McLoughlin, J. I. *Org. React.* **1995,** *47*, 315–552.
Liu, C.-J.; Wang, J.-D. *Molecules* **2010,** *15*(4), 2087–2095.
Lossen, W. *Justus Liebigs Ann. Chem.* **1872,** *161*(2–3), 347–362.
Lossen, W. *Justus Liebigs Ann. Chem.* **1875,** *175*(3), 313–325.
Machado, P.; Lima, G. R.; Rotta, M.; Bonacorso, H. G.; Zanatta, N.; Martins, M. A. P. *Ultrason. Sonochem.* **2011,** *18*, 293–299.
Mahata, P. K.; Barun, O.; Ila, H.; Junjappa, H. *Synlett* **2002,** 1345–1347.
Mamaghani, M.; Dastmard, S. *Ultrason. Sonochem.* **2009,** *16*, 445–447.
Mandhane, P. G.; Joshi, R. S.; Nagargoje D. R.; Gill, C. H. *Tetrahedron Lett.* **2010,** *51*(23), 3138–3140.
Mannich, C.; Krosche, W. *Arch. Pharm.* **1912,** *250*, 647–667.
Martina, K.; Leonhardt, S. E. S.; Ondruschka, B.; Curini, M.; Inello, A.; Cravotto G. *J. Mol. Catal. A: Chem.* **2011,** *334*, 60–64.
Martín-Aranda, R. M.; Ortega-Cantero, E.; Cervantes, M. L. R.; Bañares-Muñoz, M. A. *J. Chem. Technol. Biotechnol.* **1997,** *80*(2), 234–238.
Mather, B.; Viswanathan, K.; Miller, K.; Long, T. *Prog. Polym. Sci.* **2006,** *31*(5), 487–531.
McNulty, J.; Steere, J. A.; Wolf, S. *Tetrahedron Lett.* **1998,** *39*, 8013–8016.
Mecarova, M.; Toma, S. *Green Chem.* **1999,** *1*, 257–260.
Meciarová, M.; Toma, S.; Babiak, P. *Chem. Pap.* **2001,** *55*, 302–307.
Meciarova, M.; Polácková, V.; Toma, S. *Chem. Pap.* **2002,** *56*, 208–213.
Mei, H.; Mei, B. W.; Yen, T. F. *Fuel* **2003,** *82*(4), 405–414.
Michael, R. R.; Krishnaraj, A.; Muthupandian, A. K.; Sambandam A. *Org. Prep. Proced. Int.* **2012,** *44*, 271–280.
Mirkhani, V.; Moghadam, M.; Tangestaninejad, S.; Mohammadpoor-Baltork, I.; Mahdavi, M. *Monatsh. Chem.* **2009,** *140*, 1489–1494.

Miyaura, N.; Suzuki, A. *Chem. Commun.* **1979,** *19*, 866–867.

Miyaura, N.; Yamada, K.; Suzuki, A. *Tetrahedron Lett.* **1979,** *20*(36), 3437–3440.

Mizoroki, T.; Mori, K.; Ozaki, A. *Bull. Chem. Soc. Jpn.* **1971,** *44*(2), 581–584.

Nakano, T.; Irifune, S.; Umano, S.; Inada, A.; Ishii, Y.; Ogawa, M. *J. Org. Chem.* **1987,** *52*, 2239–2244.

Namboodiri, V. V.; Varma, R. S. *Org. Lett.* **2002,** *4*, 3161–3163.

Nathan, A. R.; Richard, A. B. *J. Org. Chem.* **2003,** *68*(2), 360–366.

Nelson, K; Horst, G. A. *Synth. Commun.* **1991,** *21*, 293–305.

Neuenfeldt, P. D.; Duval, A. R.; Drawanz, B. B.; Rosales, P. F.; Gomes, C. R. B.; Pereira, C. M. P.; Cunico, W. *Ultrason. Sonochem.* **2011,** *18*, 65–67.

Nimbalkar, U. D.; Tupe, S. G.; Vazquez, J. A. S.; Khan, F. A. K.; Sangshetti, J. N.; Nikalije, P. G. *Molecules* **2016,** *21*, 484–497.

Oge, A.; Mavis, M. E.; Yolacan, C.; Aydogan, F. *Turk. J. Chem.* **2012,** *36*(11), 137–146.

Ozturkcan, S. A.; Turhan, K.; Turgut, Z. *Chem. Pap.* **2012,** *66*, 61–66.

Palimkar, S. S.; Kumar, H. P.; Lahoti, J. R.; Srinivasan, V. K. *Tetrahedron* **2006,** *62*, 5109–5115.

Palimkar, S. S.; More, S. V.; Srinivasan, V. K. *Ultrason. Sonochem.* **2008,** *15*, 853–862.

Pathak, V. N.; Joshi, R.; Sharma, J.; Gupta, N.; Rao, V. M. *Phosphorus, Sulfur, and Silicon* **2009,** *184*, 1854–1865.

Pelit, E.; Turgut, Z. *Ultrason. Sonochem.* **2014,** *21*(4), 1600–1607.

Pétrier, C.; Luche, J. L. *J. Org. Chem.* **1985,** *50*, 910–912.

Pinto, D. C. G. A.; Silva, A. M. S.; Levai, A.; Cavaleiro, J. A. S.; Patonay, T.; Elguero, J. *Eur. J. Org. Chem.* **2000,** *14*, 2593–2599.

Polackova, V.; Tomova, V.; Elecko, P.; Toma, S. *Ultrason. Sonochem.* **1996,** *3*, 15–17.

Popp, F. D. *J. Org. Chem.* **1960,** *25*, 646–647.

Pourabdi, L.; Osati, F.; Mojtahedi, M. M.; Abaee, S. M. J. *Sulfur Chem.* **2016,** *37*, 1–9.

Qian, H.; Wang, K.; Juejie, Z. *Lett. Org. Chem.* **2016,** *13*(2), 143–147.

Rajagopal, R.; Jarikote, V. D.; Srinivasan, K. V. *Chem. Commun.* **2002,** 616–617.

Ramazani, A.; Rouhani, M.; Nasrabadi, F. Z.; Gouranlou, F. *Phosphorus, Sulfur, and Silicon Relat. Elem.* **2015,** *190*(1), 20–28.

Ramesh, G.; Sunil, J.; Pramod, K. *Org. Chem.: Ind. J. Res.* **2016,** *12*(3), 1–8.

Rao, H. S. P.; Rafi, S.; Padmavathy, K. *Tetrahedron* **2008,** *64*, 8037–8043.

Raquel, G. S.; Amalia, M. E. *Ultrason. Sonochem.* **2012,** *19*, 916–920.

Reeves, R. L. Condensations Leading to Double Bonds. In *Chemistry of Carbonyl Group;* Patai S., Ed.; Wiley: New York, 1966; pp. 580–593.

Reformatsky, S. *Ber. Dtsch. Chem. Ges.* **1887,** *20*(1), 1210–1211.

Regen, S. L.; Singh, A. *J. Org. Chem.* **1982,** *47*, 1587–1588.

Rocha, M. P. D.; Oliveira, A. R.; Albuquerque, T. B.; da Silva, C. D. G.; Katla, R.; Domingues, N. L. C. *RSC Adv.* **2016,** *6*, 4979–4982.

Rodriguez, J. M.; Dolors, P. M. *Tetrahedron Lett.* **2011,** *52*(21), 2629–2632.

Rolando, F. P.; Maite, L. D.; Mirta L. F. *Synth. Commun.* **2007,** *37*(11), 1853–1864.

Roos, G. H. P.; Rampersadh, P. *Synth. Commun.* **1993,** *23*, 1261–1266.

Ross, N. A.; Bartsch, R. A. *J. Heterocycl. Chem.* **2001,** *38*(6), 1255–1258.

Ross, N. A.; Bartsch, R. A. *J. Org. Chem.* **2003,** *68*, 360–366.

Ross, N. A.; MacGregor, R. R.; Bartsch, R. A. *Tetrahedron* **2004,** *60*, 2035–2041.

Saadatjoo, N.; Golshekan, M.; Shariati, S.; Azizi, P.; Nemati, F. *Arabian J. Chem.* **2012**. DOI: org/10.1016/j.arabjc.2012.11.018.

Safari, J.; Zarnegar, Z.; Ahmadi, M.; Seyyedi, S. *J. Saudi Chem. Soc.* **2015,** *19*, 628-633.

Said, K.; Salem, R. *Adv. Chem. Eng. Sci.* **2016,** *6*, 111–123.

Saïd, K.; Majed, K.; Younes, M.; Ridha, B. S. *Mediterr. J. Chem.* **2011,** *1*(1), 13–18.

Saleh, T. S.; El-Rahman, T. S. *Ultrason. Sonochem.* **2009,** *16*, 237–242.

Salvador, R. A. J.; Sa e Melo, L. M.; Neves, S. A.; Campos, A. S. *Tetrahedron Lett.* **1993,** *34*, 357–361.

Shekouhy, M.; Hasaninejad, A. *Ultrason. Sonochem.* **2012,** *19*, 307-313.

Singhal, S.; Joseph, J. K.; Jain, S. L.; Sain, B. *Green Chem. Lett. Rev.* **2010,** *3*(1), 23–26.

Smith, K.; Dennis, J. *J. Chem. Soc. Perkin Trans.* **1992,** *1*, 407–408.

Soengas, R. G.; Silva, A. M. S. *Synlett* **2012,** *23*, 873–876.

Sonogashira, K. *J. Organomet. Chem.* **2002,** *653*, 46–49.

Strecker, A. *Ann. Chem. Pharm.* **1850,** *75*(1), 27–45.

Strecker, A. *Ann. Chem. Pharm.* **1854,** *91*(3), 349–351.

Suresh; Sandhu, J. S. *Org. Med. Chem. Lett.* **2013,** *3*, 2–6.

Tabatabaeian, K.; Mamaghani, M.; Mahmoodi, O. N.; Khorshidi, A. *Catal. Commun.* **2007,** *9*, 416–420.

Tanifum, C. T.; Chang, C. W. T. *J. Org. Chem.* **2009,** *74*, 634–644.

Tiwari, V.; Parvez, A.; Meshram, J. *Ultrason. Sonochem.* **2011,** *18*, 911–916.

Tsuzuki, H.; Harada, T.; Mukumoto, M.; Mataka, S.; Tsukinoki, T.; Kakinami, T.; Nagano, Y.; Tashiro, M. *J. Labelled Compd. Radiopharm.* **1996,** *38*, 385–393.

Vasantha, B.; Hemantha, H. P.; Sureshbabu, V. V. *Synthesis* **2010,** 17, 2990–2996.

Venzke, D.; Flores, A. F. C.; Quina, F. H.; Pizzuti, L.; Pereira, C. M. P. *Ultrason. Sonochem.* **2011,** *18*, 370–374.

Villacampa, M.; Maria Perez, J.; Avendana C.; Menendzz, J. C. *Tetrahedron* **1994,** *50*(33), 10047–10054.

Vilsmeier, A.; Haack, A. *Chem. Ber.* **1927,** *60*, 119–122.

Wang, F.-D.; Yue, J.-M. *Eur. J. Org. Chem.* **2005,** *12*, 2575–2579.

Watanabe, M.; Morais, G. R.; Mataka, S.; Ideta, K.; Thiemann, T. *J. Chem. Sci.* **2005,** *60*(8), 909–915.

Wittig, G.; Haag, W. *Chem. Ber.* **1955,** *88*(11), 1654–1666.

Wittig.,G.; Schollkopf, U. *Chem. Ber.* **1954,** *87*(9), 1318–1330.

Wohler, F.; Liebig, J. *J. Ann. Pharm.* **1832,** *3*, 249–282.

Xiong, J.; Yan, S.; Ding, N.; Zhang, W.; Li, Y. *J. Carbohydr. Chem.* **2013,** *32*, 184–192.

Yang, C.; Yang, D. T. C.; Harvey, R. G. *Synlett* **1992,** 799–780.

Yuan, Y.-Q.; Guo, S.-R. *Synth. Commun.* **2011,** *41*, 2169–2177.

Zang, H. J.; Li, J. T.; Ning, N.; Wei, N.; Li, T. S. *Ind. J. Chem.* **2002,** *41B*, 1078–1080.

Zhang, W. C.; Li, C. J. *J. Org. Chem.* **1999,** *64*, 3230–3236.

Zhang, Z.; Zha, Z.; Gan, C.; Pan, C.; Zhou, Y.; Wang, Z.; Zhou, M.-M. *J. Org. Chem.* **2006,** *71*, 4339–4342.

Zhang, D.; Fu, H.; Shi, L.; Pan, C.; Li, Q.; Chu, Y.; Yu, W. *Inorg. Chem.* **2007,** *46*, 2446–2451.

Zhang, G.; Huang, Z.; Zou, J. *Chin. J. Chem.* **2009,** *27*, 1967–1974.

CHAPTER 5

INORGANIC, COORDINATION AND ORGANOMETALLIC COMPOUNDS

KIRAN MEGHWAL[1], SHARONI GUPTA[2], CHETNA GOMBER[3]

[1]*Department of Chemistry, M. L. Sukhadia University, Udaipur, India*
E-mail: meghwal.kiran1506@gmail.com

[2]*Department of Chemistry, M. L. Sukhadia University, Udaipur, India*
E-mail: sharoni290490@gmail.com

[3]*Department of Chemistry, JIET Group of Institutions, Jodhpur, India*
E-mail: chirag11gomber@gmail.com

CONTENTS

5.1 INTRODUCTION

One of the potential and widely studied areas in the field of sonochemistry is the synthesis of inorganic compounds. The physical and chemical effects generated after collapse of the bubbles are responsible for the formation of these materials and in some of the cases these may also result in the generation of some unusual materials with novel properties. Normally, preparation of inorganic materials is carried out in liquid phase using 20–40 kHz ultrasound irradiation. Here, either a horn- or bath-type transducer is used with/without the presence of Ar or some other inert gases. Smaller-sized particles could be formed under ultrasonic conditions as compared to traditional chemical processes. This process assists in the generation of inorganic materials without using any toxic precursors and common additives as required in case of conventional methods. Thus, sonochemical process could be considered as an eco-friendly or green chemical-processing technique.

It is an established fact that the processing techniques and the experimental parameters are deciding factors for the size and morphology of the materials generated. Many conventional methods are known including co-precipitation, hydrothermal, sol-gel, combustion, microemulsion, microwave assisted, photochemical, mechanochemical, solid state, electrochemical, gas reaction, pyrolysis and so on. Most of these methods require high temperature, air, or use temperature-sensitive organometallic precursors/environmental pollutants. Sonochemical- or ultrasound-assisted synthesis seems to be an alternative and effective mean as compared to these techniques. The use of ultrasound promotes a reaction in a liquid medium due to the dynamics of bubbles that involves its formation, growth, and implosive collapse. Such an implosive collapse generates drastic intensity for a short time that is sufficient to induce either chemical or physical transformations. Sonochemical processes are generally environmentally benign in nature and applicable to synthesis of inorganic compounds, which are selective in achieving amorphous/crystalline products with uniform size distribution of particles.

The reactivity of a number of inorganic solids is affected on exposure to ultrasonic irradiation. A large increase in the rates of intercalation of a wide range of compounds into various layered inorganic solids has been observed by Suslick and Grinstaff (1990). It has been revealed that ultrasound has multiple effects on the morphology and surface characteristics of inorganic solids. It creates substantial surface damage resulting into increased surface area and causes increased particle aggregation (Green and Thompson, 1987).

Roger et al. (1989) investigated ultrasound-assisted alkyl-metal synthesis and found that sonochemically activated potassium metal is a powerful and stoichiometric-reducing reagent. The corresponding metallates $[M(C_5Me_5)(CO)_n]^- K^+$ quickly form the dimers $M(C_5Me_5)(CO)_n]_2$ (M=Fe, Ru, n=2; M = Mo, n=3). They carried out reduction on a small scale (1–20 mmol) with a 100- or 600-W acoustic power generator and a titanium horn directly immersed in the solution. These reduction reactions were completed within 15–20 min. In situ alkylation of these metallates was done with one equivalent of an electrophile RX (where R = CH_3, CH_3CH_2, CH_3OCH_2, CH_3OCH_2), which provided a facile synthesis of metal alkyl $M(C_5Me_5)(CO)_nR$.

Nagarjun and Tomar (2009) applied ultrasound for the synthesis of ternary oxide $AgMO_2$ (M = Fe, Ga). They observed that crystalline α-$AgFeO_2$ can be obtained from the alkaline solutions of silver and iron hydroxides by sonication for 40 min. When β-$NaFeO_2$ was sonicated with aqueous silver nitrate solution for 60 min, β-$AgFeO_2$ (orthorhombic) was ion-exchanged product. Similarly, sonication of β-$NaGaO_2$ with aqueous silver nitrate solution for 60 min resulted in the formation of olive green-coloured α-$AgGaO_2$. It was confirmed that the ion exchange was complete through sonication.

A rapid ultrasound-assisted extraction procedure for the determination of total mercury, inorganic mercury (IM), and methyl mercury (MM) was reported by Krishna et al. (2005) in various environmental matrices such as animal tissues, samples of plant origin, and coal-fly ash. The mercury contents were estimated by cold vapor atomic absorption spectrometry. $SnCl_2$ was used as reducing agent in estimation of IM whereas total mercury was determined after the oxidation of MM through UV irradiation. Various operational parameters such as composition of extractant (HNO_3 and thiourea), time, and amplitude of sonication were found to be different for different matrices. Quantitative recovery of total mercury was achieved using mixtures of nitric acid and urea. These were optimized at 5% HNO_3 and 0.02% thiourea, 10% HNO_3 and 0.02% thiourea, 20% HNO_3 and 0.2% thiourea for fish tissues, plant matrices, and coal-fly ash samples, respectively. This ultrasound-assisted extraction procedure reduces the time required for sample treatment significantly for the extraction of mercury species. It was observed that extracted mercury species were quite stable even after 24 h of sonication. This method can be successfully applied for the determination of Hg in samples of lichens, mosses, coal-fly ash, and coal samples.

Olivine-structured $LiFePO_4$/C composite powder was synthesized by Jugovic et al. (2008) via ultrasound-assisted method. Lithium iron phosphate is of utmost importance as it is used as storage cathode for the next generation of rechargeable lithium batteries. Polyvinyl alcohol (PVA) solution was used as the source of in situ-formed carbon. Electrochemical characteristic of this composite was also evaluated by using galvanostatic charge/discharge tests. Carbon film with a thickness of several nanometers can be observed on the thin particle edges as well as small carbon agglomerates (typical size <6 nm) at certain spots on the substrate surface.

5.2 OXIDES

Antonelli and Ying (1995) synthesized hexagonally packed mesoporous TiO_2 via a modified sol-gel synthesis. Mesoporous TiO_2 has been widely known because of its high surface area and large uniform pores, as these properties have vital importance in catalysis and solar cell applications. A variety of mesoporous TiO_2 have been synthesized using surfactants as templates through conventional methods (Kluson et al., 2000), but conventional synthesis of mesoporous TiO_2 takes longer times, sometimes even several days for completion. Wang et al. (2000) reported a sonochemical synthesis of mesoporous TiO_2 with wormhole-like framework structures. Long-chain amine and titanium isopropoxide (TIP) were used as the structure-directing agent and precursor. This has shortened the synthesis time to a considerable extent. Mesoporous anatase with a broad pore size distribution was obtained on injecting TIP directly into deionized water in a cell under continuous sonication without using any surfactants (Huang et al., 2000).

Yu et al. (2002) also synthesized three-dimensional and thermally stable mesoporous TiO_2 without using any surfactants via treatment with high-intensity ultrasound irradiation for a short period of time. These monodispersed TiO_2 sol particles were generated initially by ultrasound-assisted hydrolysis of acetic acid-modified TIP. Then, the mesoporous spherical or globular particles were produced by controlled condensation and agglomeration of these sol nanoparticles under high-intensity ultrasound irradiation. It was indicated that these mesoporous TiO_2 particles retain mesoporosity with a narrow pore size distribution and high surface area up to 673 K. It was found thermally stable, may be due to its thick inorganic walls, consisting of TiO_2 nanoparticles. This synthetic method is environmentally benign. The photocatalytic activity of mesoporous TiO_2 for the oxidation of acetone in air was also observed. Negligible activity of as-prepared mesoporous TiO_2 was attributed to its amorphous structure;

however, calcined mesoporous TiO_2 shows better activity than commercial photocatalyst P25.

Nano-sized noble metal particles show high activity because of their unique optical, magnetic, and electric properties. These particles have high catalytic activities for the degradation of some contaminants because of their very large surface-to-volume ratios. Zhang et al. (2010) developed a simple, facile, and rapid sonochemical method to synthesize Fe_3O_4/Ag composites. These composites can be recycled six times by magnetic separation without any major loss of activity. The catalytic activity of this as-prepared composite was also investigated for rhodamine B degradation.

Fe_3O_4 nanoparticles were also prepared by Li and Gaillard (2009) using the conventional chemical co-precipitation of Fe(II) and Fe(III) chlorides (Fe(II)/Fe(III) ratio=0.5) with 1.5 M NaOH. The black precipitate was magnetically separated and washed. These Fe_3O_4 microspheres were modified with (3-aminopropyl)triethyoxysilane (APTES) and sonicated for 30 min. Fe_3O_4/APTES composite (0.05 g) microspheres were dispersed in 100 mL of 2.9 mM $Ag(NH_3)_2^+$ solution. After the adsorption of $Ag(NH_3)_2^+$, the reaction mixture was irradiated with a high-intensity ultrasonic probe (20 kHz) in ambient air for 1 h. The resulting dark material was separated by a magnet. It was indicated that Fe_3O_4 and Ag were present mainly as separated phases in Fe_3O_4/Ag composites. The catalytic activity of as-prepared composite catalysts was found to decrease during the degradation reaction in the recyclable experiments. This may be due to the fact that a small portion of crystal is lost during the process of catalytic reaction. It was found that only 0.033 μg mL^{-1} Ag was in the processed water indicating that Ag was stable enough to remain attached to the Fe_3O_4/APTES. Therefore, it was concluded that this catalyst is stable.

ZnO has been considered as the reference standard among various semiconducting metallic oxides used for photocatalysis, due to its high photocatalytic activity, excellent chemical and mechanical stability, lack of toxicity, and low cost (Miquel et al., 2010). The photocatalytic activity of ZnO can be improved by changing its size and morphology (Bokhimi et al., 2008), doping (Kang et al., 1996), use of composite materials (Omidi et al., 2013), and so forth. ZnO doped with transition metals such as Co, Ni, Cu, Mn, and Fe has been used to improve photocatalytic activity. Doping of the ZnO matrix is possible by using various methods (Suwanboon et al., 2014). Cd-doped ZnO was synthesized with different amounts of Cd dopant using sonochemical method. The photocatalytic activity of as-prepared ZnO was evaluated by the degradation of methylene blue and methyl orange in solution as model systems. It was interesting to note that all the as-synthesized Cd-doped ZnO samples were more active photocatalytically than pure ZnO for degrading these dyes.

Both 0.005 mol $Zn(NO_3)_2{\cdot}6H_2O$ and 0–3% mol $Cd(NO_3)_2{\cdot}4H_2O$ were dissolved separately in 100 mL deionized water with vigorous stirring until complete dissolution. Then it is followed by drop-wise addition of NaOH solution (3 M) until pH becomes 10. The solutions were then mixed and the mixtures were sonicated for 5 h at 120/480 W, 35 kHz. The precipitates obtained were isolated by filtration, and washed with water and ethanol several times. All samples were dried and collected. X-ray powder diffraction (XRD) patterns of Cd-doped ZnO were found very similar to those of pure ZnO, indicating that Cd doping did not change the crystalline structure of ZnO. A slight increase in the lattice constant may due to the stress developed on the substitution of Zn atoms (0.074 nm) by Cd atoms (0.092 nm) (Lupan et al., 2011).

Titanium dioxide is a commonly used photocatalyst because of its unique properties of high optical activity, low cost, nontoxicity, and chemical stability and, therefore, semiconducting titanium dioxide is widely used in the purification of air and water. Composites of TiO_2 and SiO_2 were used as they have high photocatalytic and thermal photocatalysts (Mohseni et al., 2013). As titanium is nontoxic, it is combined with materials such as ZnO and Ag, and used to eliminate harmful bacteria (Lo and Huang, 2010). The absorption of visible light is very important for photocatalysts. So these days, a combination of two photocatalytic nanocomposite components is used to overcome this problem (Chong et al., 2010). The nontoxic titanium produced nano composites as triplet photocatalysts, $TiO_2/SiO_2/Ag$ by sonochemical method, and these composites were used for removing paint and controlling water pollution.

Copper oxide nanoparticles were synthesized by Reza et al. (2010) via a sonochemical process using copper nitrate and sodium hydroxide in the presence of PVA as a starting precursor. The product was calcined at various temperatures ranging between 400 and 700 °C. High-purity CuO nanoparticles were obtained by this process. It was also reported that crystallization and particle size of CuO were strongly dependent on the reaction time and calcination temperature.

Thermogravimetric (TG) and differential-thermal analyses of copper oxide nanoparticles (sonicated for 30 min) were performed. There are two weight losses in the temperature ranges of 180–250 °C and 500–700 °C. The first weight loss is originated from evaporation of PVA and deionized water in mixed solution. The second weight loss is due to the oxidation of metal copper in air resulting to the crystallization of CuO. XRD data indicated that good powder of CuO particles can be obtained by sonochemical process only. The time of sonication and calcination temperature were the major factors that decide the formation and crystallization of CuO nanoparticles (Wongpisutpaisana et al., 2011).

Liu et al. (2009) successfully synthesized porous Cu_2O microcubes via in situ sonochemical route at an ultrasound power of 90% for 3 h in the presence of polyvinylpyrrolidone. They used $CuCl_2 \cdot 2H_2O$, $C_6H_{12}O_6 \cdot H_2O$, trisodium citrate, and NaOH as precursor materials. They showed that the as-prepared porous Cu_2O microcubes possessed good electrochemical and photocatalytic properties. It can degrade organic dye pyronin B and restrain the propagation of *Escherichia coli*. It was indicated that the as-obtained Cu_2O crystals have good crystallinity and large size. The degradation effect was much better as porous Cu_2O microcubes have a bigger surface area in comparison with smooth Cu_2O microcubes. It was observed that the solution was opaque due to the rapid propagation of *E. coli* in the absence of porous Cu_2O microcubes but the cultivating liquid containing porous Cu_2O microcubes was still transparent, which indicates that the propagation of *E. coli* has been restrained. The concentration of *E. coli* gradually reduced with the increase in the amount of porous Cu_2O mirocubes from 0 to 200 mgl^{-1}.

A single-step sonochemical synthesis of hybrid vanadium oxide/polyaniline nanowires has been carried out by Malta et al. (2008) starting from crystalline V_2O_5 and aniline in aqueous medium. This method of synthesis provides high-power ultrasounds on heterogeneous solid–aqueous phases, ultimately leading to nanowires with the dimension 5–10 mm length. The oxidative polymerization within the inorganic matrix goes along with morphological changes and monomer intercalation. The electronic conductivity of these hybrid nanowires reaches 0.8 $S \cdot cm^{-1}$ at room temperature.

Zhang et al. (2007) synthesized polycrystalline CeO_2 nanorods (5–10 nm diameter and 50–150 nm length) via ultrasonication using polyethylene glycol (PEG) as a structure-directing agent at room temperature. The effect of various operation reaction parameters was investigated such as content and molecular weight of PEG, KOH concentration, pH, and the time of sonication. They also observed that the content and molecular

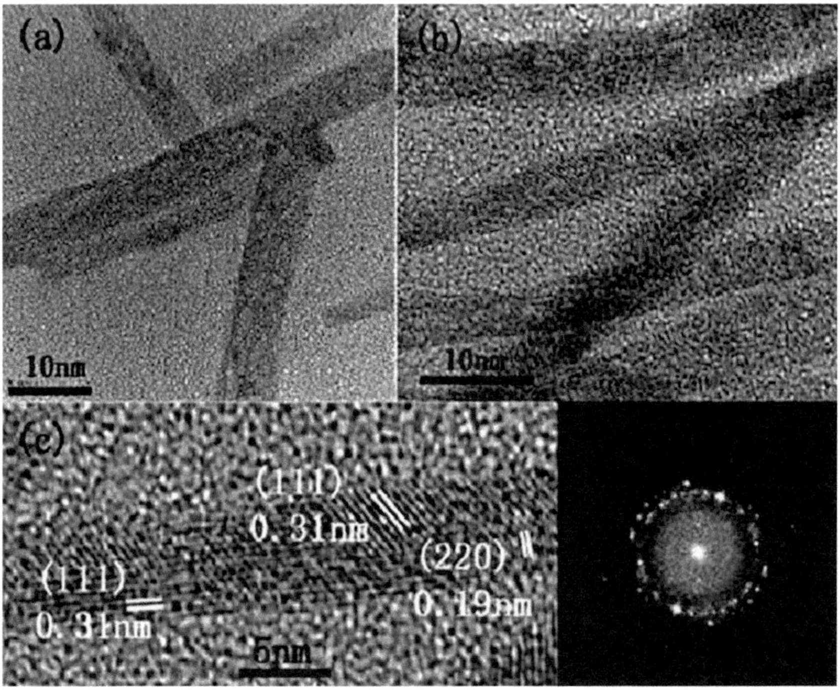

FIGURE 5.1 TEM images of CeO_2 nanorods. (a and b) HRTEM image of the single nanorod (c)

Source: Zhang et al., 2007, with permission.

weight of PEG as well as sonication time played a crucial role in the formation of one-dimensional CeO_2 nanorods. High-resolution transmission electron microscopy (HRTEM) revealed that the CeO_2 nanorods were composed of many tiny interconnected grains at various orientations (Fig. 5.1).

Activated barium hydroxide is used as a catalyst as it is nontoxic, easy to handle, low cost, and has a high catalytic activity (Pinto et al., 2000). Activated barium hydroxide has been used as a catalyst for the synthesis of 1,5-diarylpenta-2,4-dien-1-ones under ultrasound exposure (Xin et al., 2009). This methodology offers several advantages over other conventional methods with simple procedure such as excellent yields, shorter reaction times, and milder conditions. 1,5-Diarylpenta-2,4-dien-1-ones are important intermediates and raw materials widely used as precursors to different drugs, nonlinear optical materials, and biological activities.

Conventional synthesis of 1,5-diarylpenta-2,4-dien-1-ones takes a longer time (20 h) via Claisen–Schmidt condensation and poor product yields (55–81%) (Batovska et al., 2007).

Patil and Gogate (2016) synthesized vanadium pentoxide supported on alumina as a heterogeneous catalyst. They used two different approaches:

1. Conventional synthesis which is based on ozonation
2. Combination of ultrasound with ozonation

In both these cases, commercially available vanadium sulfate hydrate was used as a precursor. It was observed that nanoparticles of synthesized catalyst with ultrasound resulted in reduced particle size and higher stability. As-obtained catalyst was used in the reaction for synthesis of *n*-hexyl acetate.

$LiCoO_2$ compound exhibits two different phases. These are:

- Low-temperature phase (LT-$LiCoO_2$ and pseudo-spinel structure)
- High-temperature phase (HT-$LiCoO_2$ and a-$NaFeO_2$-type structure)

Nanosized HT-$LiCoO_2$ powders were synthesized sonochemically in an aqueous solution of lithium hydroxide containing cobalt hydroxide at ~80 °C without high-temperature treatment (Kim and Kim, 2008). The effects of various parameters such as LiOH concentration, oxidation conditions, time of ultrasound irradiation, and temperature on the formation of the nano-sized HT-$LiCoO_2$ phase were observed. The HT-$LiCoO_2$ aggregates have a mean particle diameter of ~20 nm.

A facile and surfactant-free method was reported by Xie et al. (2012) for synthesizing double-pyramid-like ZnO architectures via an ultrasound-assisted route. On ultrasound irradiation, uniform octahedral ZnC_2O_4 precursors were first formed, which were then transformed to macro-/mesoporous ZnO with the original shape after calcination at 350 °C. As-obtained hierarchical ZnO double-pyramids were having ZnO nanoparticles (~35 nm size and pores of 20–100 nm). It was revealed that the presence of intrinsic defects in the porous ZnO accounts for the narrowed band gap (3.16 eV). This ZnO has large specific surface area and therefore, exhibits extraordinary photocatalytic activity to degrade organic pollutants.

Nano-sized $CoAl_2O_4$ pigments are important as a colouring agent in glaze and bulk tile compositions. These were successfully synthesized by avoiding mechanical stirring via hydrothermal process with ultrasonic irradiation (Kim et al., 2012). A difference was also observed by Kim et al. in physicochemical and optical properties of the $CoAl_2O_4$ pigments prepared by an ultrasonic-assisted-hydrothermal method. It was reported that ultrasonic-assisted $CoAl_2O_4$ pigments presented a narrow particle size distribution with vivid blue colour, and better thermal stability so as to allow their use for ceramic inks processed at high temperature. Nanosized powders with better physicochemical and optical properties were achieved by the application of ultrasonic irradiation during the hydrothermal process.

Yahyavi et al. (2015) synthesized Ni–Co/Al_2O_3–MgO nanocatalysts by taking different ratios of Al/Mg ratios via sonochemical method. These catalysts were used in CH_4/CO_2 reforming. Ni–Co/Al_2O_3–MgO nanocatalyst with Al_2O_3/MgO = 1.5 was also prepared via impregnation method to know the effect of ultrasonic irradiation. It was observed that the sonochemical method gives better dispersion, smaller particle size, higher specific area, uniform morphology and higher catalytic efficiency in comparison with the sample obtained by impregnation. Among all the prepared samples via sonochemistry with different ratios, the Ni–Co/Al_2O_3–MgO nanocatalyst with Al_2O_3/MgO = 5 showed good catalytic performance.

Manganese dioxide was obtained via an ultrasound–microwave-intensified precipitation reaction with the soft template of triblock polymer P123 (Wang et al., 2014). By combining these methods, microwave heating and ultrasonic dispersion, MnO_2 particles were obtained with loose sphere-network structure with tiny pores of 4–5 nm diameter inside the spheres along architectural mesopores and macropores among the secondary particles. This loose network became more solid with bigger particles and less surface area on replacing microwave heating or ultrasonic dispersing with conventional hot-plate heating or magnetic stirring. It was reported that MnO_2 prepared with intensified precipitation had the highest discharge specific capacitance of ~214 Fg^{-1} and it showed almost no capacitance decline even after 1000 cycles but some acceptable decline after 5000 cycles. The MnO_2 sample had a spherical morphology and the diameter was ~200 nm. These spheres had a loose network structure and were composed of nanorods (~5 nm in diameter and 100 nm length) (Fig. 5.2).

X-ray diffraction patterns of sample indicated that the sample of MnO_2 was in a partially crystalline state and it is in α-MnO_2 form; however, this sample changed to Mn_2O_3 with high crystallinity on calcining at 520 °C.

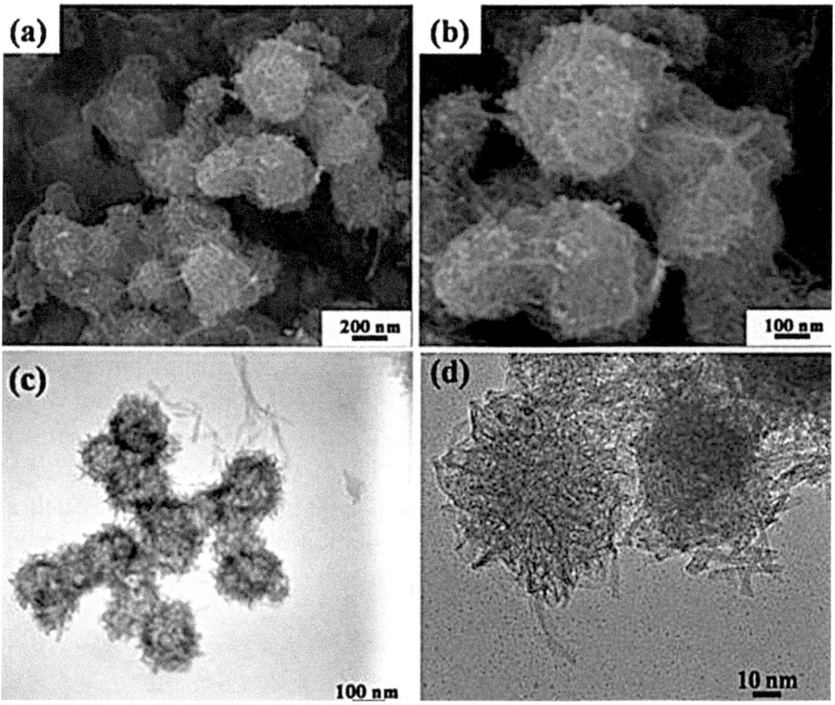

FIGURE 5.2 SEM (a, b) and TEM images of MnO_2 sample (c, d)
Source: Wang et al., 2014, with permission.

Ultrafine powders of chromium oxide (Cr_2O_3) and manganese oxide (Mn_2O_3) have been prepared at room temperature by the sonochemical reduction of an aqueous solution containing ammonium dichromate [$(NH_4)_2Cr_2O_7$] and potassium permanganate ($KMnO_4$), respectively (Dhas et al., 1997). The yield of the sonochemical reduction has been enhanced by raising the reaction temperature or by using an aqueous solution of ethanol (0.1 M). As-obtained powders are nano-sized (50–200 nm), and the surface area varies from 35 to 48 $m^2 \cdot g^{-1}$. Amorphous Mn_2O_3 and Cr_2O_3 powders were converted into crystalline form by heating them at 600 and 900 K, for 4 h, respectively.

5.3 SULPHIDES

Dhas and Suslick (2005) first reported the sonochemical preparation of bulk MoS_2 using $Mo(CO)_6$ and elemental sulfur in 1,2,3,5-tetramethylbenzene.

Later, mixed Co–Mo and Ni–Mo sulfides supported on Al_2O_3 were also prepared using a similar approach (Dhas et al., 2001). Catalytic activity of these materials was found to be three to five times higher than that of the conventional catalysts. Thus, there is a good potential in sonochemical method for improving the performance of hydrodesulfurization catalysts.

Uzcanga et al. (2005) developed a sonochemical reaction between $(NH_4)_6Mo_7O_{24}$ and CH_3COSH in aqueous solution that leads to the formation of hollow MoS_2 microspheres. It was observed that the walls of these hollow spheres have short- and low-stacked MoS_2 fringes. These fringes provide this material a very high intrinsic catalytic activity in the thiophene hydrodesulfurization reaction. The sonochemically prepared MoS_2 consists of open spherical aggregates of 0.2–2 μm in diameter with walls composed of smaller size particles (fig. 5.3a), while it was clearly shown by transmission electron microscopy (TEM) (fig. 5.3b) that the large aggregates as well as surrounding smaller particles are hollow. Such a morphology is like the shape of the cavitation bubbles in the sonicated solution. It seems to be first report of ultrasound-assisted formation of inorganic hollow-based spheres.

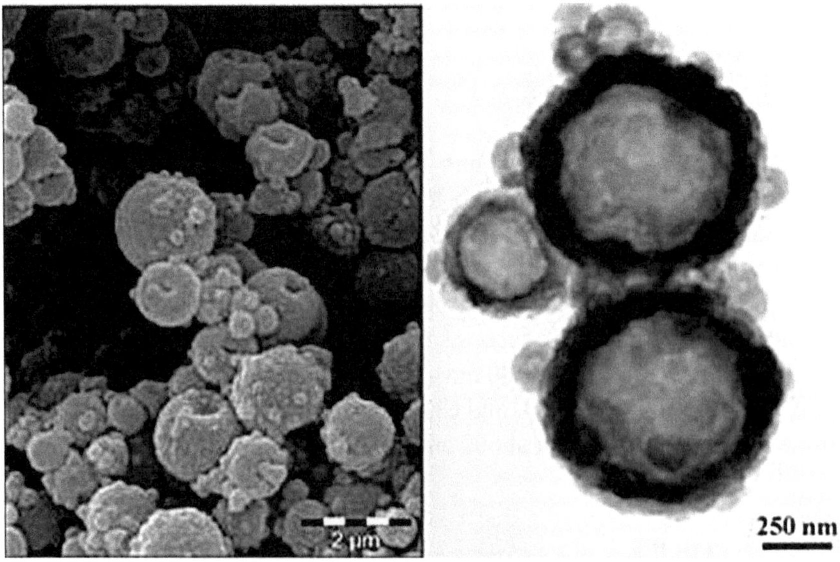

FIGURE 5.3 Sonochemically prepared MoS_2 microspheres: Scanning electron micrograph (a) and Low-magnification transmission electron micrograph (b)
Source: Uzcanga et al., 2005, with permission.

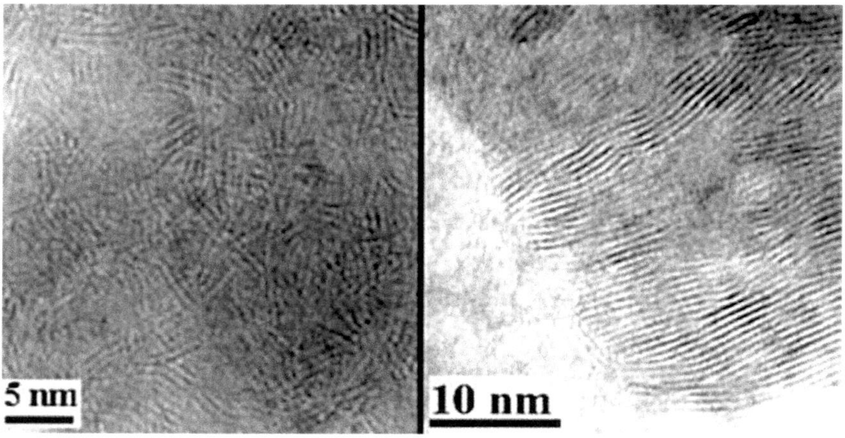

FIGURE 5.4 High-resolution electron microscopy images of MoS_2 (Sonochemically prepared (left) and MoS_2 ex-ATM (right).
Source: Weber et al., 1995, with permission.

It was found that sonochemical preparation of MoS_2 in aqueous solution affects not only the shape of agglomerates but also the properties of MoS_2 crystallites. The XRD pattern of this MoS_2 indicates that there is lower stacking of the slabs than the sample obtained by thermal decomposition. This fact is further supported by high-resolution electron micrographs (fig. 5.4). It was also noticed that the slabs are much shorter in the sonochemically prepared sample and less ordered as compared to ex-ATM sample. In spite of this difference in sonochemically and thermally decomposed ATM, molybdenum is present in fully sulfidized form, with Mo $3d_{3/2\text{-}5/2}$ doublet with the respective binding energies 229.2 and 232.6 eV corresponding to the Mo(IV) species in the MoS_2 sulfide (Weber et al., 1995).

Wang et al. (2006) synthesized PbS hollow nanospheres of diameter 80–250 nm using surfactant-assisted sonochemical route. The structural characterization of PbS nanoparticles indicates that these particles have diameters of about 12 nm and the shells are hollow spheres. The formation of the hollow nanostructure was explained by a vesicle-template mechanism, where sonication and surfactant play important roles. Uniform silica layers were successfully coated onto the hollow spheres via a modified Stober method to enhance their performance for some promising applications. The morphology and microstructure of PbS hollow nanospheres are

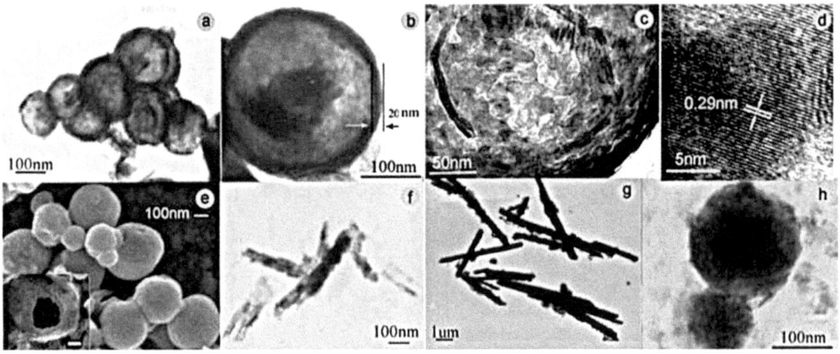

FIGURE 5.5 TEM, HRTEM and SEM images of the PbS hollow spheres (a–e)
Source: Wang et al., 2006, with permission.

clearly demonstrated by TEM and scanning electron microscopy (SEM) images (Fig. 5.5), where the strong contrast between the dark edge and the relatively bright center is due to their hollow nature. The diameter of the hollow sphere is in the range of 80–200 nm and the wall thickness was around 20 nm (approximately) from the enlarged TEM image of a single hollow sphere. Spherical nanostructures with a smooth exterior are clear from the SEM image. The diameter of the sphere is about 80–250 nm, which was comparable to the TEM results. SEM indicates that the spheres are not very compact and some of them may be destroyed by intensive post sonication. PbS nanocrystals with diameters of about 12 nm were observed, indicating that the walls of the PbS nanospheres are really constructed by small particles.

The synthesis of single-crystalline Sb_2S_3 nanorods by the ionic liquid-assisted sonochemical method (ILASM) was carried out by Salinas-Estevan et al. (2010) (Fig. 5.6). Antimony(III) chloride, thioacetamide, and absolute ethanol were used as starting materials and 1-butyl-3-methylimidazolium tetrafluoroborate ([BMIM][BF_4]) was used as ionic liquid.

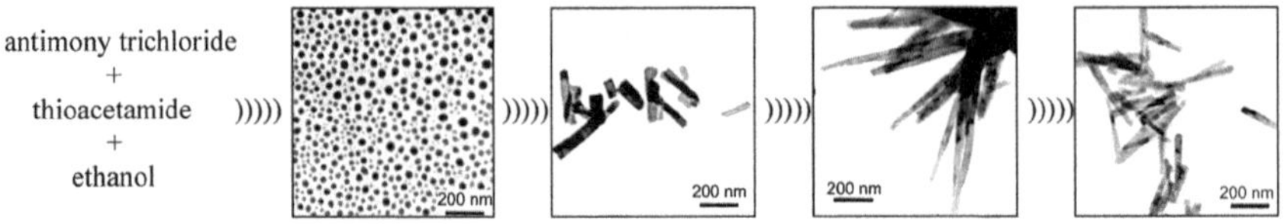

FIGURE 5.6 Synthesis of Sb_2S_3 nanorods
Source: Wang et al., 2003, with permission).

It was reported that the ionic liquid played an important role in the final morphology of Sb_2S_3. Single-crystalline Sb_2S_3 nanorods were prepared in the presence of [BMIM][BF_4], whereas round Sb_2S_3 structures were formed in the absence of [BMIM][BF_4], under otherwise identical conditions.

$CuInS_2$ (CIS) nanoparticles were synthesized successfully via a new copper precursor [bis(acetylacetonato)copper(II)], [$Cu(acac)_2$], at room temperature by ultrasonic method (Amiri et al., 2014). The effect of sulfur source, solvent, and reaction time was investigated on product morphology and particle size. $CuInS_2$ nanoparticles were prepared and coated on FTO. Later, the coated FTO was sintered so that a compact and dense $CuInS_2$ film was produced and photovoltaic characteristics were measured.

Wang et al. (2002) prepared bismuth sulfide nanorods by sonochemical method from an aqueous solution of bismuth nitrate and sodium thiosulfate in the presence of complexing agents. Bismuth sulfide nanorods with different diameters and lengths could be obtained by using different complexing agents, including ethylenediaminetetraacetic acid, triethanolamine and sodium tartrate. Bi_2S_3 nanorods have also been successfully prepared by using thioacetamide as the sulphur source. When 20% *N,N*-dimethylformamide was used as the solvent, higher yield was observed and smaller sizes of Bi_2S_3 nanorods were obtained. The TEM image revealed that the product consists of needle-shaped nanorods. The diameter of the nanorods ranges from 20 to 30 nm, and the length is about 200–250 nm. This rod-type morphology of the product was possibly due to the inherent chain-type structure of bismuth sulfide.

It is known that Bi_2S_3 crystallizes with a lamellar structure with linked Bi_2S_3 units forming infinite bands, which in turn are connected via weaker Van der Waals interactions. It seems that the formation of Bi_2S_3 may have been originated from the preferential directional growth of Bi_2S_3 crystallites.

The sonication of $InCl_3$ with thioacetamide in an aqueous solution has led to different products, depending on the sonication temperature (Avivi et al., 2001). On one hand, sonication at 0 °C yielded In_2O_3 as the major product and In_2S_3 as a minor component. On the other hand, when the reaction was carried out at room temperature, nanocrystalline In_2S_3 was obtained as the sole product. The products of sonication at 0 °C were obtained in the amorphous form.

Amorphous WS_2 has been prepared by ultrasound irradiation of $W(CO)_6$ solution in diphenylmethane (DPhM) in the presence of a slight excess of sulfur at 90 °C under argon (Nikitenko et al., 2002). On heating, the amorphous powder at 800 °C under argon yields WS_2 nanorods and their packings. The average size of WS_2 nanorods was found to be 3–10 nm and 1–5 mm thickness and length, respectively.

Yin et al. (2003) obtained ZnS/PS microspheres through sonochemical treatment containing zinc acetate and thioacetamide in an ethanol solution. It was observed that the composite particles were optically hollow due to the large refractive index contrast between the core and shell materials. Hollow ZnS spheres were formed by annealing the core-shell spheres in a muffle oven under the stream of 3:1 nitrogen and hydrogen gas at 600 °C. PS microspheres obtained by emulsion polymerization were smooth, and monodisperse with diameter was about 265 nm. TEM image showed that ZnS shell appears as a dark ring around the PS core. It was found that the particles slightly agglomerated.

However, it was found that the coating on the PS microspheres was smooth and uniform after the microspheres were treated in the ethanol solution of zinc acetate and thioacetamide by sonochemical treatment. The thickness of ZnS shells was about 67.5 nm. When the PS microspheres were treated in ethanol solution for 3 h, smooth and integral coating on PS beads was obtained. If the treatment time was carried out <3 h, only part of PS microspheres was covered with ZnS and nonintegral coating was formed. They indicated that ZnS nanoparticles were formed first and then adsorbed onto the surface of PS microspheres. The diameter of hollow ZnS spheres was about 220 nm. In order to protect the ZnS from oxidation, the composite was calcined in a muffle oven under the stream of 3:1 nitrogen gas and hydrogen gas.

Slurry of amorphous silica microspheres, zinc acetate, and thioacetamide in an aqueous medium for 3 h under ambient air and ultrasonic irradiation yields zinc sulphide coated on silica (Dhas et al., 1999). XRD of the initial zinc sulfide–silica (ZSS) powder yields diffraction peaks corresponding to the ZnS phase. TEM image of ZSS showed that the porous ZnS nanoparticles (diameter 1–5 nm) coated the silica (SiO_2) surface as thin layers or nanoclusters, depending on the reactant concentration. Some structural changes occurred in the siloxane network and surface silanol groups of SiO_2 on ultrasonic deposition of ZnS as evident from IR. The optical absorption of porous ZnS shows a broad band at around 610 nm, which was attributed to the unusual surface-state transition.

The absorption energy of the surface state transition was lower than the band gap of the ZnS particles and probably, it stems from the dangling surface bonds or defects. On the other hand, ZSS does not show the surface-state transition of ZnS, may be due to the strong surface interaction with SiO_2. It was proposed that the coating process takes place via ultrasonic-cavitation-induced initial grafting of zinc acetate onto the silica surface, followed by the displacement of acetate ion by in situ-generated S^{2-} species.

Regular stibnite (Sb_2S_3) nanorods with diameters of 20–40 nm and lengths of 220–350 nm have been successfully synthesized by Wang et al. (2003), using a sonochemical method under ambient air from an ethanolic solution containing antimony trichloride and thioacetamide. It was revealed that the Sb_2S_3 nanorods crystallize in an orthorhombic structure and predominantly increase in size along the (001) crystalline plane. High-intensity ultrasound irradiation plays an important role in the formation of these Sb_2S_3 nanorods. The sonochemical formation of stibnite nanorods can be divided into four steps. These are:

- Ultrasound-induced decomposition of the precursor, which leads to the formation of amorphous Sb_2S_3 nanospheres
- Ultrasound-induced crystallization of these amorphous nanospheres and generation of nanocrystalline irregular short rods
- A crystal growth process, giving rise to the formation of regular needle-shaped nanowhiskers
- Surface corrosion and fragmentation of the nanowhiskers by ultrasound irradiation, resulting in the formation of regular nanorods

The optical properties of the Sb_2S_3 amorphous nanospheres, irregular short nanorods, needle-shaped nanowhiskers, and regular nanorods were investigated by diffuse reflection spectroscopic measurements, and the band gaps were measured to be 2.45, 1.99, 1.85 and 1.94 eV, respectively.

5.4 COORDINATION COMPOUNDS

Reversible coordination networks were prepared by Paulusse et al. (2007) combining diphenylphosphinite telechelic polytetrahydrofuran with $[RhCl(COD)]_2$ or $[IrCl(COD)]_2$ in chloroform. Both these systems resulted in stable gel formation at concentrations above 50 and 30 $g{\cdot}L^{-1}$ for the rhodium(I) and iridium(I) networks, respectively. It was reported

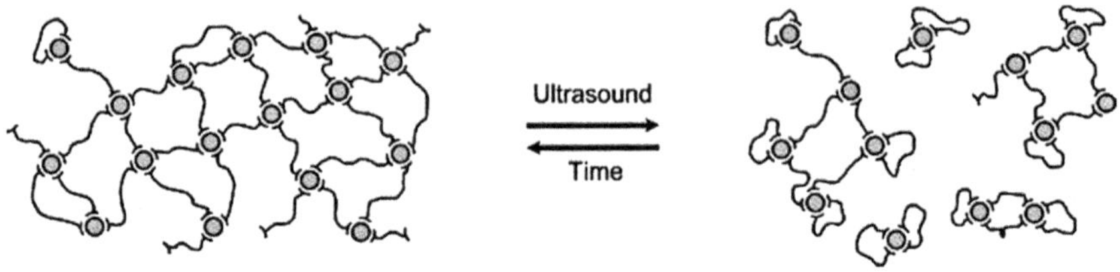

FIGURE 5.7 Effects of sonication on the coordination networks, resulting in loss of the gel fraction.
Source: Paulusse et al., 2007, with permission.

that ultrasonication of the rhodium(I) gel caused liquefaction after 3 min followed by regelation, just after 1 min of stopping sonication. On the contrary, iridium(I) gel was also liquefied after 3 min of sonication, but its regelation took about 1.5 h at room temperature and required more than 10 days at −20 °C (fig. 5.7).

It was proposed that sonication of the gels resulted in ligand exchange, which changes the network topology keeping the coordination chemistry intact. The fraction of metal centers in active cross-links decreases on sonication and, as a result, the gel fraction is reduced to zero. Such an effect offers opportunities to use ultrasound in the activation of dormant transition metal catalysts.

There is the transient formation of coordinatively unsaturated metal complexes during reversibly breaking coordinative bonds in the rhodium(I) and iridium(I) coordination networks. These species are known to be excellent catalysts in a number of reactions and therefore, this coordination networks will prove to be promising systems to probe the effect of mechanical forces on catalytic activity and selectivity.

Co-B catalysts show good practice for hydrogen evolution via hydrolysis of alkaline sodium borohydride solutions and, therefore, Cogkuner et al. (2014) investigated this and proposed a sonochemical approach for synthesis of Co-B catalysts and hydrolysis of alkaline $NaBH_4$ solutions. Application of ultrasound on synthesizing process improved both the properties intrinsic and extrinsic properties of Co-B catalysts such as surface area, particle size, pore volume, spectral, crystal, and pore diameter. It was observed that Co-B catalysts prepared by this sonochemical approach have smaller particle size, higher surface area, and higher pore volume as compared to the Co-B catalysts prepared by routine co-precipitation synthesis. It was also indicated that the advantages of alkaline $NaBH_4$

solution sonohydrolysis provide superficial effects on hydrogen evolution, that is, maximum hydrogen generation rate with minimizing activation energy (Fuentes et al., 1987).

Liu et al. (2011) employed ultrasonic irradiation to modify activated carbon (AC) by sodium hypochlorite, and tested its ability to remove cobalt(II) from aqueous solutions. Modified AC show enhanced adsorption capacity for Co(II) and it increases with the increasing NaOCl impregnating concentration. A combination of adsorption and chemical precipitation was considered responsible for Co(II) removal. Liu et al. (2011) reported that enhanced Co(II) removal by the ultrasonic-assisted NaOCl modified carbons was due to improved cation exchange capacity and higher content of adjacent aliphatic functional groups providing more complexation sites for cobalt.

Dharmarathna et al. (2012) developed a rapid and direct sonochemical method to synthesize cryptomelane-type manganese octahedral molecular sieve (OMS-2) materials. Such materials have very high surface area of 288 ± 1 m^2g^{-1} and small particle sizes in the range of 1–7 nm under nonthermal conditions. Reaction time was reduced by 50% using a cosolvent system such as water/acetone apart from achieving higher surface area. The following factors were identified as most important in the formation of the pure cryptomelane phase: (i) reaction time, (ii) temperature, and (iii) acetone concentration. They reported that OMS-2 materials synthesized using sonochemical methods were having greater amounts of defects and, as a result, show excellent catalytic performances for oxidation of benzyl alcohol than to OMS-2 synthesized using reflux methods and commercial MnO_2.

Li et al. (2013) investigated the adsorption of chromium(VI) and uranium(VI) nitrate on Ca–Al hydrotalcite (Ca–Al LDHs) compounds. These materials were synthesized in a modified co-precipitation method under ultrasonic irradiation. Different factors affecting synthesis of compounds were also investigated such as temperature, reaction time, calcination and aging conditions. It was reported that Ca–Al LDHs has the highest capacity absorption for both Cr(VI) and U(VI), 104.82 ± 0.02 and 54.79 ± 0.02 mg g^{-1} for rate of adsorption up to 98.78 ± 0.02% and 90.28 ± 0.02%, respectively. It suggests that Ca–Al LDHs are efficient in removing these U(VI) ions; however, they show fast adsorption and slow release as compared to Cr(VI).

The sonochemical synthesis of Mn–ferrite nanoparticles was reported by Goswami et al. (2013) using acetates as precursors. Mn–ferrite oxide

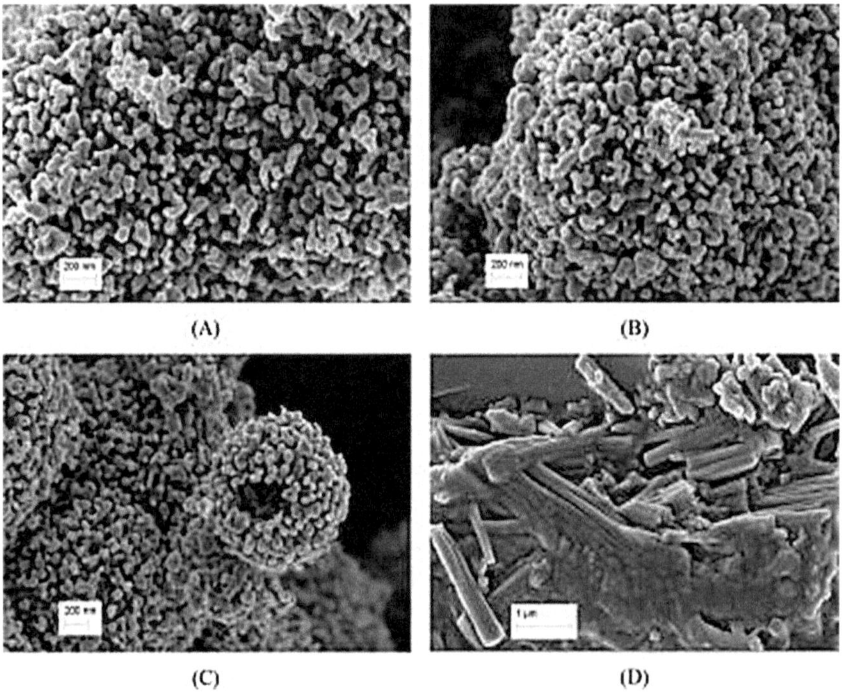

FIGURE 5.8 Field-emission scanning electron microscopy (FE-SEM) images of Mn–ferrite formed with external calcination.
Source: Goswami et al., 2013, with permission.

precursors were formed by sonication followed by their external calcination. It was observed that pH does not play any dominant role in this synthesis. Collisions between metal oxide particles induced by shock waves generated by transient cavitation were not sufficient to cross the activation energy barrier so as to form ferrite. Calcination temperature is an important parameter that influences the magnetic properties of ferrites. It was observed that calcination at 950 °C leads to the formation of rods with grain growth that introduces large shape anisotropy. Ferrite particles have a spherical shape, more or less, with a narrow size distribution as evident from FE-SEM images. The particle size was found to increase with calcination temperature (fig. 5.8). High surface activity at the nanometer size range may be due to some agglomeration.

The preparation of micro-hexagonal rods of a new 1D polymeric lead(II) complex containing the Pb_2-$(\mu$-$N_3)_2$ motif, $[Pb(dmp)(N_3)_2]_n$

(where dmp=2,9-dimethyl-1,10-phenanthroline) by the sonochemical method has been reported by Hanifehpour et al. (2012). A single-crystalline material was obtained by applying a heat gradient applied to a solution of the reagents. The single-crystal XRD analyses show that the coordination number for the Pb(II) ions is six, PbN_6, with stereochemically active electron lone pairs and that the coordination sphere being hemidirected. It also show that the chains of $[Pb(dmp)(N_3)_2]_n$ interact with each other through the π–π stacking interactions creating a 3D framework.

Mizukoshi et al. (1999) prepared platinum nanoparticles in an aqueous system using high-intensity ultrasound (200 kHz, 6 $W\cdot cm^{-2}$). The particles formed in the presence of a surfactant (sodium dodecyl sulfate, SDS) were stable, homogeneously spherical, and relatively monodispersed with an average 2.6 nm diameter. The rate of formation of the platinum nanoparticles was 26.7 $\mu M\ min^{-1}$ in the Pt(II)–SDS system. Reducing species generated near and/or in the hot bubbles, which were sonochemically induced in the media, would react with the $[PtCl_4]^{2-}$ complexes to form the platinum nanoparticles.

Three kinds of the reducing species were proposed to be formed in the sonicated system and these are:

- Radicals formed from the thermal decomposition of SDS at the interfacial region between the cavitation bubbles and bulk solution
- Radicals formed via reactions of the hydroxyl radicals or hydrogen atoms with SDS
- Hydrogen atoms

During the reduction of the Pt(II) ion, the effectiveness of while first these processes are in the order: Second is more effective and then third is only slightly effective, whereas in the cases of gold and palladium nanoparticles, the hydrogen atoms were the main reductive species.

A nano-sized mixed ligand Cd(II) coordination polymer, $\{[Cd(bpa)(4,4'\text{-bipy})_2(H_2O)_2](ClO_4)_2\}_n$ (where bpa = trans-1,2-bis(4-pyridyl)ethane and 4,4′-bipy = 4,4′-bipyridine, has been synthesized by a sonochemical method (Ranjbar and Morsali, 2011). This grows in one dimension by two different bridging ligands, 4,4′-bipy and bpa. CdO nanoparticles were obtained by direct calcination at 500 °C and decomposition in oleic acid at 200 °C of this nano-sized coordination polymer.

Moosavi et al. (2012) investigated the growth of silver bromide nanoparticles on polyester fiber by sequential dipping steps in alternating bath of potassium bromide and silver nitrate under ultrasound irradiation. They studied the effects of ultrasound irradiation, concentration, and sequential dipping steps in growth of the AgBr nanoparticles. Particle sizes and morphology of nanoparticle were found to depend on power of ultrasound irradiation, sequential dipping steps, and concentration. Moosavi et al. (2012) reported a decrease in the particles size in these systems, on increasing the sonication power. An increase in sequential dipping steps and concentration also led to an increase in particle size.

A new methodology for the synthesis of pillared clays with the Al_{13}–Fe and Al_{13}–Fe–Ce polymers in solid state was adapted by using microwaves and ultrasound in the aging of the olygomeric solution, in the precipitation of the corresponding sulfates, and in the intercalation of the polymer with the powdered clay (Olaya et al., 2009). This methodology reduces significantly the water consumption in the intercalation process, and the time for the synthesis of both the polymer in the solid state, and the pillared clay. A higher percentage of Fe and Ce, associated with Al_{13}–Fe and Al_{13}–Fe–Ce sulphates, was dissociated during the exchange process with nitrate. A monoclinical structure of Al_{13} and Al_{13}–Fe sulfates was observed, whereas the Al_{13}–Fe–Ce sulfates present two crystalline phases, monoclinical and cubical. It was shown that the pillared clays, synthesized with this methodology, have more intense and remarkably homogenous pillaring signals, and better textural properties also than the clays pillared by other conventional method of synthesis. The catalytic activity was assessed for the phenol oxidation in dilute aqueous medium at 25 °C and at atmospheric pressure. As-synthesized solids showed similar catalytic properties than those prepared by the conventional method.

Arenas et al. (2013) obtained hybrid nanocomposites based on anatase titania:polypyrrole (TiO_2:PPy) from a simple, one-step, ultrasonic (UT)-assisted synthesis. The properties of these crystalline nanocomposites were compared with those of others fabricated using cold-assisted synthesis without any UT assistance, which required a hydrothermal treatment to yield crystalline anatase titania in the nanocomposite at low temperature (130 °C) and that too in a short time (3 h). The SEM results demonstrated that the UT-assisted synthesis is an easy method to obtain anatase TiO_2:PPy nanocomposites with controlled morphology using low energy. As absorption in the visible region was observed (λ_{max} = 670 nm), which indicates that these nanocomposites are promising candidates for harvesting energy in solar cells. Devices based on these nanocomposites were also evaluated for their electrical properties, where an increase in the photocurrent was observed with the nanocomposites obtained a from the UT-assisted synthesis.

A new nanostructured Bi(III) supramolecular compound, {[Bi_2(4,4′-Hbipy)$_{1.678}$(4,4′-H bipy)$_{0.322}$(μ-I)$_2$$I_{5.678}$] (4,4′-bipy)}, (where 4,4′-bipy = 4,4′-bipyridine}, was synthesized by a sonochemical method (Soltanzadeh & Morsali, 2010). Its nanostructure and crystal structure were investigated using SEM, powder XRD, IR spectroscopy, and elemental analysis. The thermal stability of bulk compound and nano-sized particles was studied by TG and differential thermal analyses. Bi_2O_3 and BiI_3 nanostructures were obtained by calcinations of these nanostructures at 400 °C under air and nitrogen atmospheres, respectively.

Fard-Jahromi and Morsali (2010) synthesized nanostructures of two new Pb(II) two-dimensional coordination polymers, [Pb(μ-4-pyc)(μ-NCS)(μ-H_2O)]$_n$ and [Pb(μ-4-pyc)(μ-N_3)(μ-H_2O)]$_n$, where (4-Hpyc = 4-pyridinecarboxylic acid), via the sonochemical method. These compounds were structurally characterized and consisted of two-dimensional polymeric units. $Pb_2(SO_4)O$ and PbO nanoparticles were obtained by calcination of the nanostructures at 600 °C. This method of preparation of coordination polymers may have some advantages such as it takes place in shorter reaction times, produces better yields, and also coordination polymers of nanosize. Reduction of the particle size of the coordination polymers to a few dozen nanometers results in lower thermal stability as compared to the single crystalline samples.

A new nano-sized lead(II) coordination polymer of maleic acid (H_2Mal), [Pb(μ_7-Mal)]$_n$ has been synthesized by Aboutorabi and Morsali (2010) via sonochemical method. The compound was structurally characterized by single-crystal XRD. Thermal stability of nano and bulk samples of compound was compared with each other. Pure-phase micro-sized lead (II) oxide was produced after the calcination of nano-sized compound at 600 °C. The morphology of compound is also quite interesting. It is a nano flower-like structure formed by nano-plates with sizes of about 40–70 nm. XRD shows that the complex in the solid state is a 2D polymeric network.

Okitsu et al. (2005) investigated the rate of sonochemical reduction of Au(III) to produce Au nanoparticles in aqueous solutions containing 1-propanol. This reaction was found to be strongly dependent on the ultrasound frequency. The size and distribution of the Au nanoparticles can also be correlated with the rate of Au(III) reduction, which in turn is influenced by the applied frequency. They suggested that the rate of Au(III) reduction as well as the size distribution of Au particles was governed by the chemical effects of cavitation and therefore, the rate is not significantly affected by the physical effects accompanying ultrasound-induced cavitation.

Straw-like nanostructure of a new mixed-ligand Zn(II) two-dimensional coordination polymer, $\{[Zn(\mu\text{-}4,4'\text{-bipy})(\mu\text{-}3\text{-bpdb})(H_2O)_2](ClO_4)_2\cdot 4,4'\text{-bipy}\cdot 3\text{-bpdb}\cdot H_2O\}_n$, (where 4,4′-bipy = 4,4′ bipyridine and 3-bpdb = 1,4-bis(3-pyridyl)-2,3-diaza-1,3-butadiene), was synthesized by a sonochemical method (Khanpour et al., 2010). This compound consisted of two-dimensional polymeric units. ZnO nanoparticles were obtained by calcination of this compound at 500 °C under atmospheric air.

5.5 ORGANOMETALLIC COMPOUNDS

The field of organometallic sonochemistry was pioneered by Renaud (1950), who reported the preparation of organolithium, organomercury, and organomagnesium compounds in undried solvents in simple cleaning bath. Based on the formation or reaction step, organometallic sonochemistry is divided into two categories: homogeneous and heterogeneous processes. Homogeneous process requires a high-intensity ultrasound to alter chemical reactions in a single liquid, whereas heterogeneous process uses ultrasound to alter chemical reactions in two-phase system, where the mass transfer takes place from the bulk solution to metal surface.

5.5.1 HOMOGENEOUS SONOCHEMISTRY

5.5.1.1 CLUSTRIFICATION

The clustrification of $Fe(CO)_5$ to $Fe_3(CO)_{12}$ in alkane solvents during sonolysis was observed, together with the formation of finely divided iron

(Suslick et al., 1981; 1983). The proposed mechanism during sonolysis of $Fe(CO)_5$ has been proposed as:

$$Fe(CO)_5 \xrightarrow{\bullet))))} Fe(CO)_{5-n} + n\ CO \quad (n = 1\text{-}5)$$

During cavitation, the initial multiple dissociative loss of CO from $Fe(CO)_3$ takes place.

$$Fe(CO)_3 + Fe(CO)_5 \longrightarrow Fe_2(CO)_8$$

Formation of $Fe(CO)_8$ was confirmed by ligand-trapping studies.

$$2Fe(CO)_4 \longrightarrow Fe_2(CO)_8$$

Dimerization of $Fe(CO)_4$ takes place in the localized cavitation site.

The reaction of putative $Fe_2(CO)_8$ with $Fe(CO)_5$ proceeds through initial dissociation in analogy to the matrix isolation following first-order kinetics.

$$Fe_2(CO)_8 + Fe(CO)_5 \longrightarrow Fe_3(CO)_{12} + CO$$

Sonolysis of $Fe_2(CO)_9$ yields $Fe(CO)_5$ and finely divided iron; on the contrary, $Fe_2(CO)_9$ is not generated during the sonolysis of $Fe(CO)_5$ to $Fe_3(CO)_{12}$. These reactions are summarized in Table 5.1.

TABLE 5.1 Some Clustrifications.

Reactant	Product	References
$Fe(CO)_5$	$Fe_3(CO)_{12}$, Fe	Suslick et al. (1981)
$Fe_2(CO)_9$	Fe, $Fe(CO)_5$	Suslick et al. (1983)

5.5.1.2 *LIGAND SUBSTITUTION*

Sonication of $Fe(CO)_5$ in the presence of phosphines or phosphites yields:

$$Fe(CO)_{5-n}L_n \ (n = 1, 2, \text{and } 3).$$

$$Fe(CO)_5 \xrightarrow{\bullet))))} Fe(CO)_{5-n} + n\ CO \quad (n = 1\text{-}5)$$

$$Fe(CO)_4 + L \longrightarrow Fe(CO)_4L$$

$Fe(CO)_4L$ does not sonochemically convert to $Fe(CO)_3L_2$.

$$Fe(CO)_3 + L \longrightarrow Fe(CO)_3L$$

$$Fe(CO)_3 + CO \longrightarrow Fe(CO)_4$$

$$Fe(CO)_3L + L \longrightarrow Fe(CO)_3L_2$$

Sonochemical ligand substitution has been observed to occur with a variety of metal carbonyls. Ligand substitution (L)* reaction in homogeneous organometallic sonochemistry is represented in Table 5.2.

TABLE 5.2 Some of the Ligand Substitution Reactions.

Reactants	Products	References
$Fe(CO)_5+L$	$Fe(CO)_4L$, $Fe(CO)_3L_2$, $Fe(CO)_2L_3$	Suslick et al. (1981, 1983)
$FeCp(CO)_2I+L$	$FeCp(CO)(L)I$	Suslick and Wang
$Cr(CO)_6+L$	$Cr(CO)_5L$, $Cr(CO)_4L_2$, $Cr(CO)_3L_3$	Suslick et al. (1981, 1983)
$Mo(CO)_6+L$	$Mo(CO)_5L$, $Mo(CO)_4L_2$	Suslick et al. (1981, 1983)
$W(CO)_6+L$	$W(CO)_5L$, $W(CO)_4L_2$	Suslick et al. (1981, 1983)
$Mn_2(CO)_{10}+L$	$Mn_2(CO)_8L_2$	Suslick et al. (1981, 1983)

L: Phosphines and phosphites

π-Allyl lactone (tricarbonyl) iron complexes are useful intermediates to synthesize lactones and lactam (Geoffroy and Wrighton, 1979).

Similar reaction is observed thermally with $Fe(CO)_4$ (Tetrahydrofuran)-$Fe(CO)_5$ and $Fe_3(CO)_{12}$ do not undergo this type of reaction.

5.5.1.3 SECONDARY SONOCHEMICAL REACTIONS

Metal carbonyls in halocarbon solvents undergo halogenation. The rate of this reaction is solvent independent.

$$R_3CX \xrightarrow{\bullet))))} R_3C^{\bullet} + X^{\bullet}$$

$$2R_3C^{\bullet} \longrightarrow R_3CCR_3$$

$$2X^{\bullet} \longrightarrow X_2$$

$$M_2(CO)_{10} + 2X^{\bullet} \longrightarrow 2\,M(CO)_5X$$

$$M_2(CO)_{10} + X_2 \longrightarrow 2\,M(CO)_5X$$

Secondary sonochemical reaction of $Co_2(CO)_8$ in *n*-alkanes (C_5H_{12} through $C_{10}H_{22}$) yields $Co_2(CO)_6(H_2C_2)$ and $Co_2(CO)_{10}(H_2C_2)$ with a small amount of $Co_4(CO)_{12}$, as it is formed due to pyrolysis of $Co_2(CO)_8$. The presence of (H_2C_2) is confirmed by isotopic labeling.

5.5.1.4 SONOCATALYTIC REACTIONS

Isomerization of 1- pentene to *cis*- and *trans*-2-pentene can be catalysed by metal carbonyls upon sonication.

$$M(CO)_n \longrightarrow M(CO)_m + (n-m)CO$$

$$M(CO)_n + 1\text{-alkene} \longrightarrow M(CO)_x\ (1\text{-alkene}) + (m-x)CO$$

$$M(CO)_x(1\text{-alkene}) \longrightarrow M(CO)_x(H)\ (\pi\text{-allyl})$$

$$M(CO)_x(H)\ (\pi\text{-allyl}) \longrightarrow M(CO)_x(2\text{-alkene})$$

$$M(CO)_x(2\text{-alkene}) + 1\text{-alkene} \longrightarrow M(CO)_x(1\text{-alkene}) + 2\text{-alkene}$$

These reactions give the formation of a hydrido-π-allyl intermediate and alkene rearrangement through hydride migration, which forms more stable 2-alkene complex. Some of the sonocatalytic reactions are shown in Table 5.3.

TABLE 5.3 Some of the Sonocatalytic Reactions.

Reactant	Product	References
$Fe_x(CO)_y$ + 1-alkene	*cis*-, *trans*-2-alkene	Suslick et al. (1981, 1983)
$Ru_x(CO)_y$ + 1-alkene	*cis*-, *trans*-2-alkene	Suslick et al. (1983)
$Mo_x(CO)_y$ + 1-alkene	*cis*-, *trans*-2-alkene	Suslick et al. (1983)
$Co_x(CO)_y$ + 1-alkene	*cis*-, *trans*-2-alkene	Suslick et al. (1983)

5.5.2 HETEROGENEOUS SONOCHEMISTRY

Sonochemistry has been utilized to accelerate reaction rates involving metals through surface activation, which is accomplished in the following three different ways:

- Sonication during the reaction
- Pretreatment of the metal before using it in a conventional reaction
- Conversion of a metal to a different and more reactive form

5.5.2.1 MERCURY

Organomercury refers to the group of organometallic compounds containing mercury. The Hg–C bond exhibits stability towards air and moisture, but it is sensitive to light.

The use of ultrasound in liquid–liquid heterogeneous system was limited to mercury at initial level. Fry and Bujanauskas (1978) observed that the reduction of 2,4-dibromo-2-4-dimethylpentan-3-one by finely dispersed mercury in ketonic solvents provides a method to synthesize 4-isopropylidene-5,5-dimethyl-1,3-dioxolan ring system. Here, first oxyallyl cation is formed.

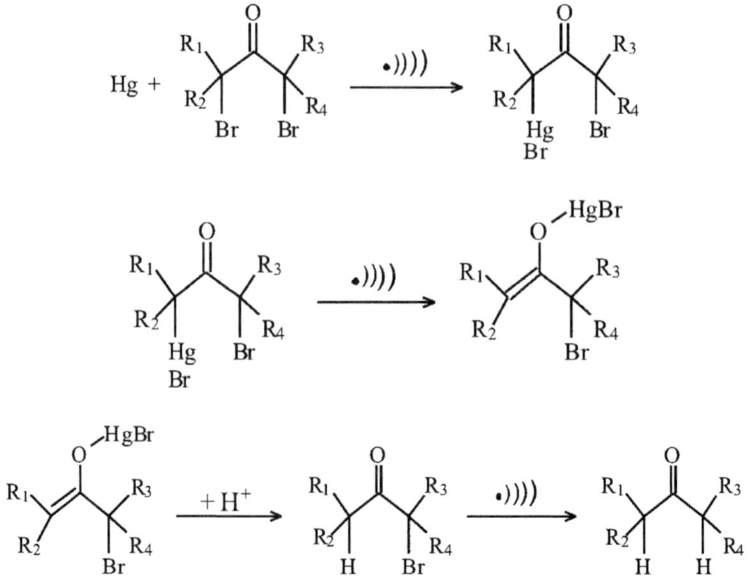

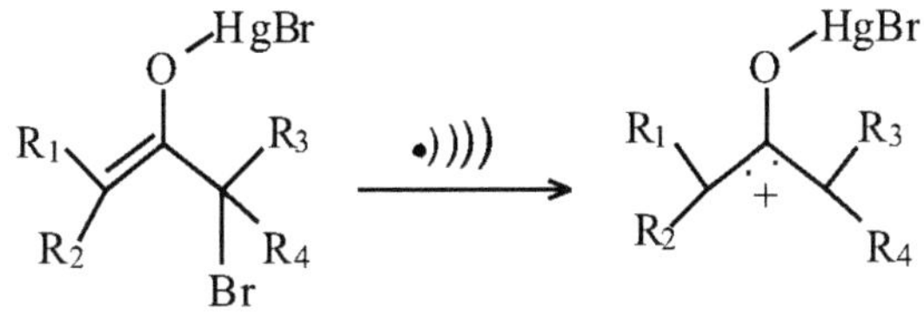

Oxyallyl cation

This oxyallyl cation reacts with some alcohol or carboxylic acid to yield the final product.

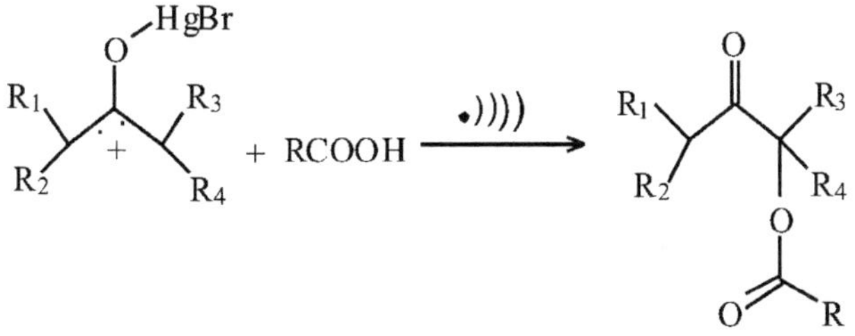

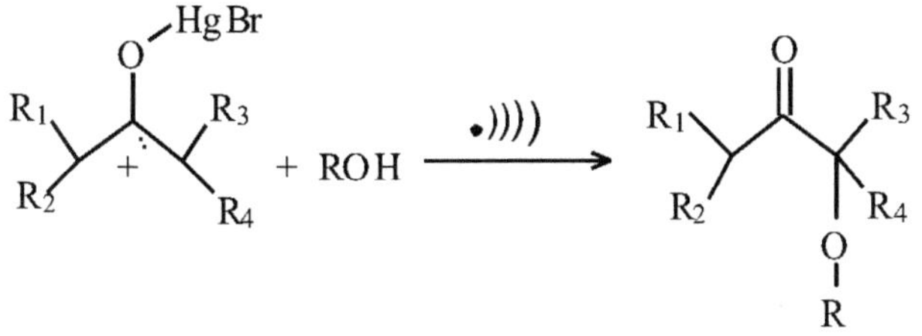

These reactions represent synthesis of 1,3-dioxolan, which involves nucleophilic attack on the mercurial oxyallyl cation. The use of ultrasound results in rate enhancement (Fry and Ginsburg, 1979) in this system. Electrochemical reduction by dispersed mercury is explained by the given reaction:

$$C_6H_5(H)(Br)CSC(Br)(H)C_6H_5 + Hg \xrightarrow{\bullet))))} \text{t- } C_6H_5(H)C{=}C(H)C_6H_5$$

Thiobenzaldehyde or 1,2-diphenylthiirane is not reported to be formed during the reaction.

Sonochemical reactions of nucleophiles with α,α′-dibromoketones and mercury have been reported by Fry et al. (Fry and Hen, 1979; Fry and Lefor, 1979; Fry et al., 1979).

Al-Juaid et al. (1996) studied the preparation and thermal stabilities of organomercury compounds containing the bulky ligands such as $(Me_3Si)_3C$ or $(PhMe_2Si)_3C$.

$$HgR'Br + MgRX \xrightarrow{\bullet))))} HgR'R \; [R^1 = C(SiMe_3)_3$$

where R = Me, ^{i}Pr, Bu, t-Bu or Ph.

$$HgR^2Cl + LiR \xrightarrow{\bullet))))} HgR^2R$$

where $R^2 = C(SiMe_2Ph)_3$; R = Me, Bu or Ph.

$$HgR^2Cl + Mg(CH_2Ph)Cl \xrightarrow{\bullet))))} HgR^2R$$

where $R^2 = C(SiMe_2Ph)_3$; $R = CH_2Ph$

It was observed that the replacement of one R group in HgR″Cl by some bulky R′ or R″ group leads to a significant increase in thermal stability. The compound HgR″Cl does not react further with LiR″ in tetrahydrofuran, but with LiR′, it gives HgR′R″.

Hashemi et al. (2012) synthesized a new nano-sized mercury(II) coordination polymers $[Hg(4\text{-bpdb})(SCN)_{2n}]$; (4-bpdb = 1,4-*bis*(4-pyridyl)-3,4-diaza-1,3-butadiene) by a sonochemical method. Such coordination polymers are suitable precursors for the preparation of nanoscale materials without the need conditions such as high temperature, longer time, and control of pressure.

$$Hg(NO_3)_2 + 2\text{-bpdh}[L]^* + Hg(SCN)_2 \xrightarrow{\bullet))))} Hg(4\text{-bpdb})(SCN)_2]_n$$

Ranjbar et al. (2013) carried out reaction between 2,9-dimethyl-1.10-phenanthroline and $Hg(CH_3COO)_2$ in the presence of halide ions (KX, where X = I, Br) under ultrasonic irradiations. This reaction leads to the formation of $HgI_2(C_{14}H_{12}N_2)$ and $HgBr_2(C_{14}H_{12}N_2)$, respectively.

The positive charge on Hg(II) is neutralized by the halide ions serving as ligands.

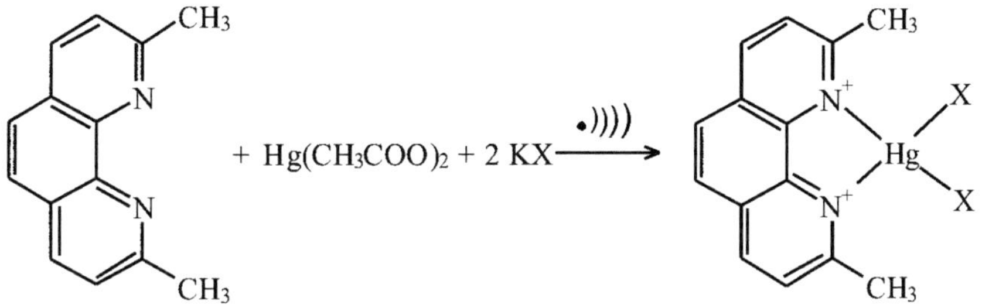

5.5.2.2 ZINC

Kitazume and Ishikawa (1981) synthesized triflurotrimethyl carboninols from trifluoromethylation of carbonyl compounds with trifluoromethyl zinc iodide under sonication using 35 W (32 kHz) for irradiation 30 min.

$$RR'C = O + Zn + CF_3I \xrightarrow{\bullet))))} RR'(F_3C)COZnI \xrightarrow{H^+} RR'C(OH)CF_3$$

TABLE 5.4 Organometallic Sonochemistry of Mercury.

Reactant	Product	References
$(R_2BrC)_2CO$+$(H_3C)_2CO$	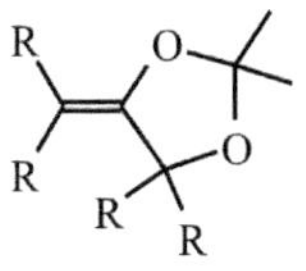	Fry et al. (1978)
$C_6H_5(H)(Br)CSC(Br)(H)C_6H_5$	t- $C_6H_5(H)C$=$C(H)C_6H_5$	Fry et al. (1981)
$(R_2BrC)_2CO$+ROH	$(HR_2C)CO[C(OR')R_2]$	Fry et al. (1981)
$Hg(NO_3)_2$+2-bpdh+$Hg(SCN)_2$	$Hg(4\text{-bpdb})(SCN)_2]_n$	Al-Juaid et al. (1996)
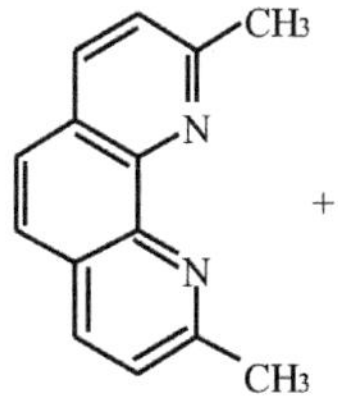 + $Hg(CH_3COO)_2$ + 2KK	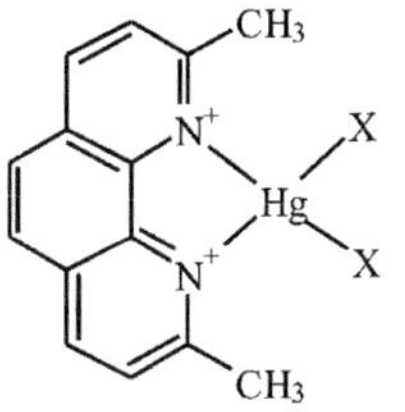	Hashemi et al. (2012)

It was one of the pioneer works in sonication, which formed, when the reaction was carried out with Zn powder, CF_3I dimethylformamide solutions of ketones, and aldehydes alcohols in good yields.

Low-intensity ultrasound was used in this closely related Reformatsky reaction.

$$RR'C = O + Zn + BrCH_2CO_2R'' \xrightarrow{\bullet))))} RR'C(OH)CH_2CO_2R''$$

Ultrasonically generated organozinc complexes for perfluoroalkylation of allyl, vinyl, and aryl halides with Palladium-catalysed cross-coupling reactions between allyl, vinyl, or aryl halide and perfluoroalkyl iodide with zinc were also carried out by Kitazume and Ishikawa (1982) using ultrasonic irradiation catalyst Pd(0).

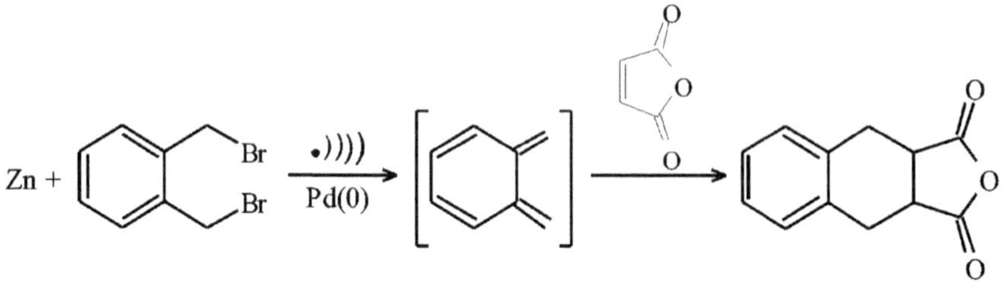

Han and Boudjouk (1982) by ultrasonic irradiation of zinc with α,α′-dibromo-*o*-xylene-generated o-xylene-like species. This σ-xylene species was reported to be trapped by dienophiles.

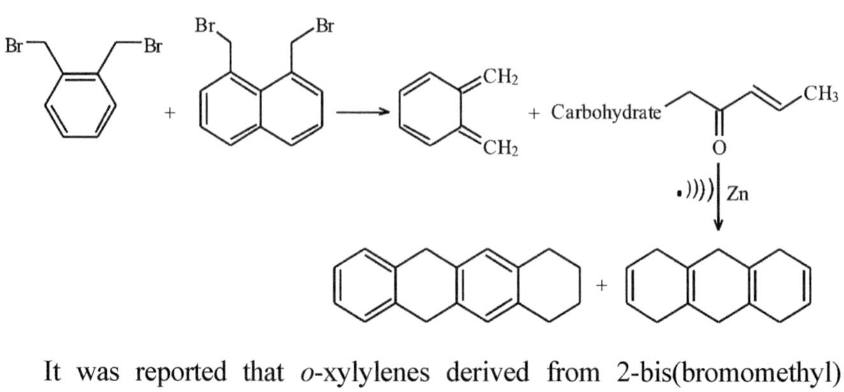

It was reported that *o*-xylylenes derived from 2-bis(bromomethyl) benzene and 2,3-bis(bromomethyl)naphthalene on reaction with so a carbohydrate enone with zinc powder under ultrasound irradiation gave some tri- and tetra-cyclic products (Chew and Ferrier, 1984), which were finally converted into the hexahydro-anthracene and –naphthacene derivatives.

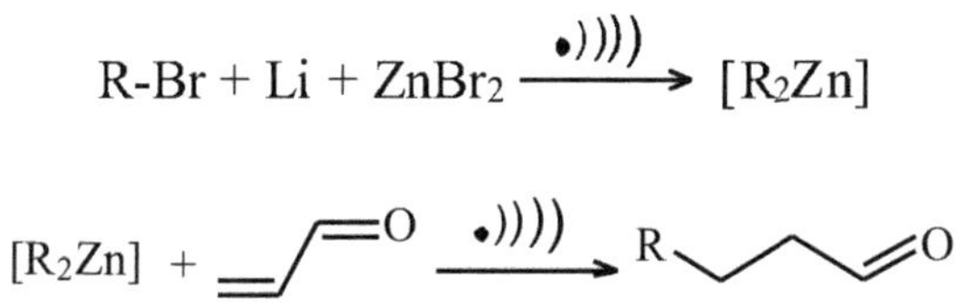

Ultrasonic irradiation of the organic halides with Li in the presence of zinc bromide results in the formation of organozinc reagents. An efficient preparation of various organozinc reagents was reported on sonication, which adds an α-enones in a conjugate fashion in the presence of $Ni(acac)_2$. These reagents undergo conjugate addition (Luche et al., 1983).

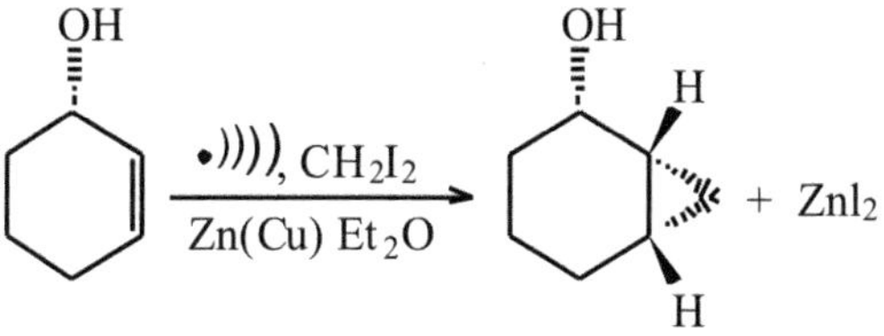

Repic and Vogt (1982) reported that Simmons–Smith reaction was facilitated by the activation of zinc with ultrasonic irradiation.

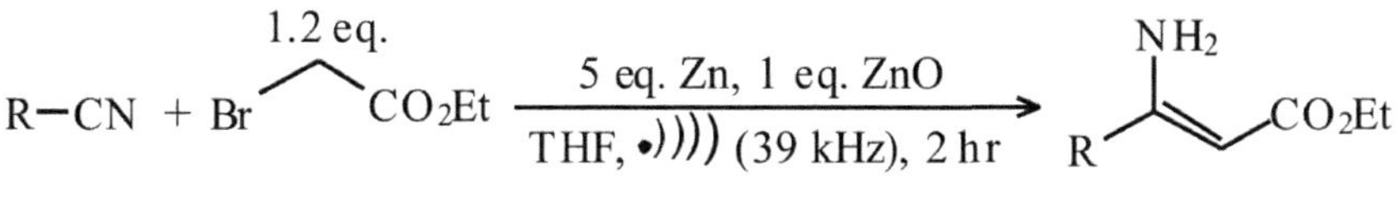

where R = Alkyl, Ar, Bn.

β-Amino-α, β-unsaturated esters were obtained by a sonochemical Blaise reaction of nitriles, zinc powder, zinc oxide, and ethyl bromoacetate in THF (Lee and Cheng, 1997). This type of sonication allows for a more convenient and one-step synthesis using zinc powders, which does not require any specific activation. This reaction also lowered the amount of self-condensation side product using only a slight excess of bromoacetate.

It has been reported by Atobe et al. (1998) that the stereoselectivity in the debromination of dl-stilbenedibromide with zinc powder depended remarkably on the frequency of ultrasounds, while this was not true for meso-form.

Ultrasonic radiation increased the rates and the yields of the reactions of some α-dicarbonyls towards zinc and trimethylchlorosilane (Boudjouk and So, 1986). It gave bis(trimethylsiloxy)alkenes as the product.

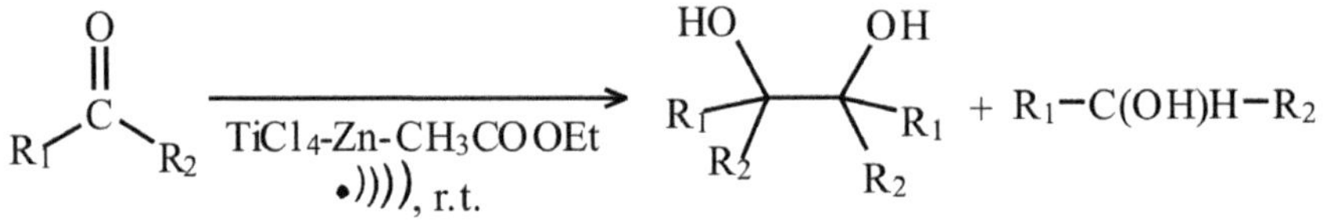

It was observed that zinc powder reduces titanium tetrachloride to its corresponding low-valent titanium complexes, which can promote coupling of some aromatic aldehydes and ketones to pinacols. These corresponding pinacols were obtained with 35–99 % yield in ethyl acetate at 20–25 °C in 20–45 min under ultrasonic radiation (Li et al., 2007).

5.5.2.3 *LITHIUM*

It was observed that the preparation of organometallic compounds of lithium and with various halide derivatives can be improved with ultrasonic waves.

$$\text{R-Br} + \text{Li} \longrightarrow \text{R-Li} + \text{Br}$$

where R = *n*-Pr, *n*-Bu, Ph.

Excellent yields of organolithium compounds were reported for *n*-propyl-, *n*-butyl-, and phenyl lithium even at room temperature in wet solvents.

A facile sonochemical preparation of organolithiums and their Barbier-type coupling with carbonyls has been observed.

$$\text{R-Li} + \text{R'R''CO} \longrightarrow \text{RR'R''COCOH}$$

The coupling of organic halides (bromides including *t*-butyl and benzyl) with carbonyl compounds (aldehydes and ketones) through lithium intermediates occurs in one step in Barbier reaction (Luche & Damiano, 1980). Such reactions are fast and also provide excellent yields (76–100 %). Trost and Coppola (1982) used intramolecular Barbier reaction to synthesize complex cyclopentanones.

S(+)2-octyl halides react spontaneously or under sonication with lithium and cyclohexanone to give an optically active alcohol (Souza-Barboza, Luche, & Petrier, 1987). Its enantiomeric excess and absolute configuration, strongly suggest the existence of two reactive intermediates following different stereochemical pathways to the condensation alcohol.

Danhui et al. (1994) reported that sonication of a *n*-butyl halide, lithium benzoate, and lithium metal in THF at room temperature provides the expected ketone and/or benzil in good yields; however, it depends on the nature of the halogen.

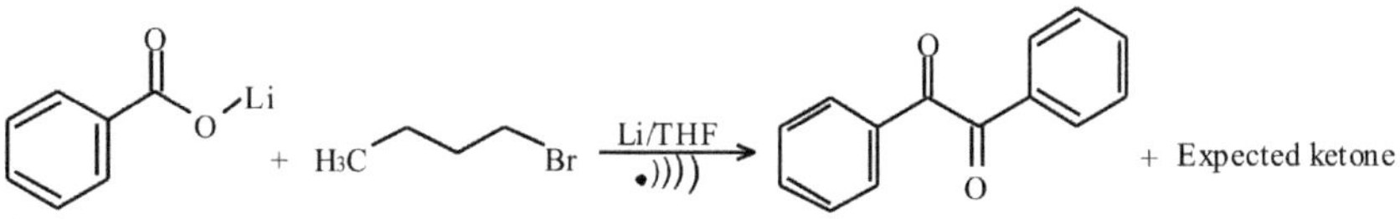

Alkyl and aryl halides were used by Luche et al. (1982) to form organocopper reagents in the presence of lithium under ultrasonic irradiation, when these were treated in situ with enones to give high yields of β-alkylated ketones.

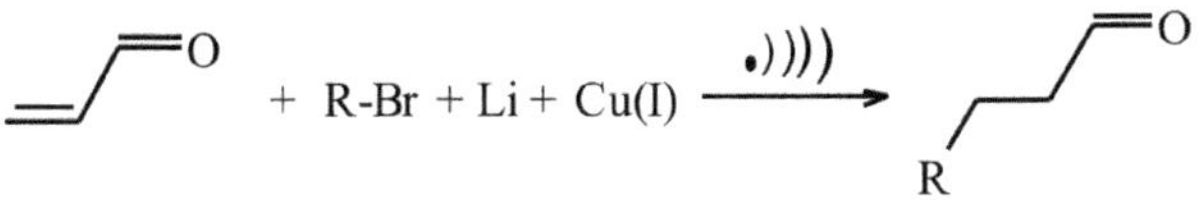

This reaction represents organocopper conjugate alkylations of oxones (Luche et al., 1982).

In a cleaning bath, Petrier et al. (1982) observed that sonication of the mixture of organic halide, dimethyl formamide and lithium sand in tetrahydrofuran provides good yields of aldehydes.

$$\text{R-Li} + (\text{H}_3\text{C})_2\text{NCHO} \longrightarrow \text{RCH(OLi)(NCH}_3)_2 \longrightarrow \text{RCHO} + (\text{CH}_3)_2\text{NH}$$

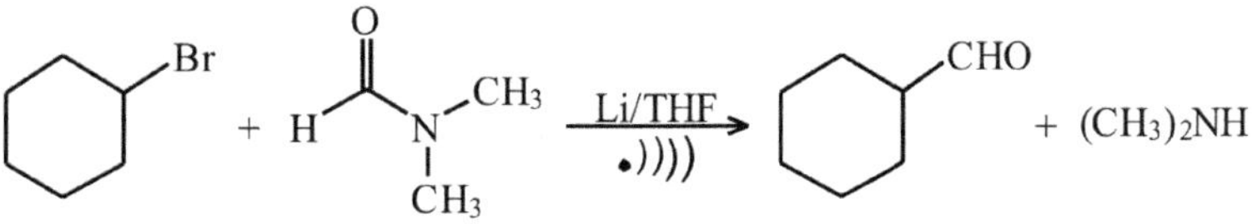

Wurtz-type coupling reactions have been reported of lithium or lithium borohydride in the presence of organic halides and chlorosilanes/chlorostannanes under sonication.

$$R_3M\text{-}Cl + Li \xrightarrow{\bullet))))} R3MMR_3$$

where M = C, Si, Sn and R = alkyl or aryl.

$$R_3M\text{-}Cl + LiAlH_4 \xrightarrow{\bullet))))} R_3M\text{-}H$$

where M = Si, Ge, Sn; X = Cl, NR_2, OR.

The preparation of some main group hydrides from corresponding chlorides and lithium aluminum hydride using low-intensity ultrasound has also been reported (Lukevics et al., 1984).

$$[2,4,6\text{-}(H_3C)_3C_6H_2]_2SiCl_2 + Li \xrightarrow{\bullet))))} R_2Si = SiR_2$$

Synthesis of West's novel disiline (West et al., 1981) upon sonication of Li with bis(mesityl)dichlorosilane was reported by Boudjouk (Boudjouk et al., 1982).

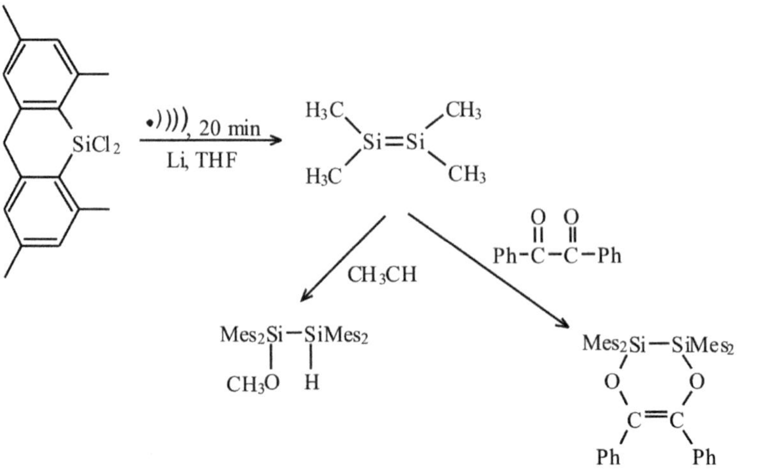

...(x)

Hexamesityl-cyclotrisilane has also been observed in this type of reaction.

5.5.2.4 OTHER METALS

5.5.2.4.1 Sodium

Ultrasound was used to produce colloidal Na, which produces radical anion salt of 5,6-benzo-quinoline (Slough & Ubbelohde, 1957). Ultrasound also facilitates the synthesis of phenylsodium (Pratt & Helsby, 1959).

$$\text{Arenes} + \text{Na} \xrightarrow{\bullet))))} \text{Na (arene}^{-}\text{)}$$

Aromatic compounds such as naphthalene and biphenyl are also reduced by alkali metals, giving aromatic anion radicals which are useful as strong bases in organic synthesis. The use of ultrasound to form such aromatic radical anions from Na in solvents has also been reported by Azuma et al. (1982).

5.5.2.4.2 Potassium

Colloidal potassium can be easily obtained by ultrasonic irradiation solution alkyl of dihalides (α, ω) in toluene (Luche et al., 1984). The ultrasonic dispersion of potassium in toluene or xylene was reported and has been used for the cyclization of α,ω-difunctionalized alkanes and other reactions.

$$XH_2C\text{-}(CH_2)_n\text{-}CH_2X + K \xrightarrow{\bullet))))} \text{Cycloalkanes}$$

5.5.2.4.3 Magnesium

Ultrasound has been used in initiating and enhancing synthetic reactions involving metals as a reagent or catalyst, for example, in the preparation of Grignard reagent. In the synthesis of Grignard reagent, oxide-free magnesium and periodic crushing of metal are essential in conventional methods. To overcome this problem, Yamaguchi et al. (1982) developed a ultrasound-assisted method for the synthesis of Grignard reagent, where any grade of magnesium can be used, and no crushing is required.

Ultrasound-accelerated coupling reaction of Grignard reagents with 1,3-dioxolanes of α, β-unsaturated aldehydes has been studied by Lu et al. (1998).

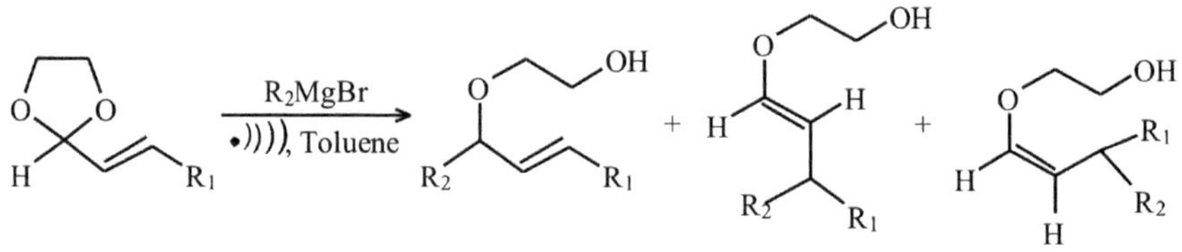

1 2 3 4 5

where $R_1 = C_6H_5$, C_3H_7, CH_3; $R_2 = CH_3$, C_2H_5, n-C_4H_9, $CH_2CH = CH_2$.

Tuulmets et al. (1995) observed the effect of ultrasonic irradiation on the kinetics of the reaction of alkyl bromides with magnesium in toluene in the presence of catalytic amounts of organic bases such as diethyl ether, tetrahydrofuran. It was observed that sonication exerts an accelerating effect on all the steps of this reaction.

It has been reported that the pinacol coupling of aromatic aldehydes and ketones by Mg powder in ammonium bromide aqueous solution ($0.1\ mol \cdot L^{-1}$) leads to the corresponding pinacols in 22–90% yields within 1–3 h at room temperature, but under ultrasound irradiation (Li et al., 2005).

They also conducted pinacol coupling reaction of aromatic aldehydes and ketones with 8–95% yield, when magnesium in 0.1 M aqueous NH_4Cl was used under ultrasound irradiation at room temperature for 3 h (Li et al., 2006).

Sano et al. (2015) reported a magnesium (II)-mediated Horner–Wadsworth–Emmons reaction of bis(2,2,2-trifluoroethyl) phosphonoacetate derivatives with disubstituted ketenes. Tri- and tetra-substituted allenes including α-fluorinated allenylcarboxamides were observed as products.

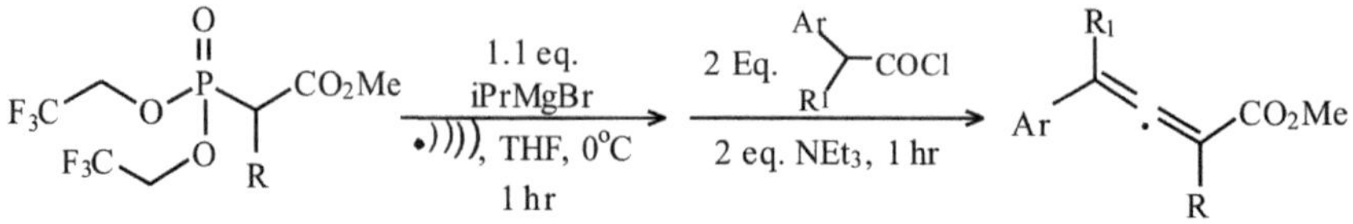

However, the sonochemistry of other alkali metals is less explored.

5.5.2.5 TRANSITION METALS

Transition metals salts have also been used in sonochemical reactions such as Fe, Mn, Ni, Cr, V, Ta, Mo, and W. Some of these reactions are reported in the following section.

A direct immersion ultrasonic horn was used for the reduction of transition metal (V, Nb, Ta, Cr, Mo, W, Mn, Fe, and Ni) halides soluble in THF or diglyme with Na to produce carbonyl anions under 1–4 atm. CO at 10 °C (Suslick and Johnson, 1984).

$$FeCl_3 + Na + CO \xrightarrow{\bullet))))} Fe(CO)_4^{2-} + Fe_2(CO)_8^{2-}$$

$$MnCl_3 + Na + CO \xrightarrow{\bullet))))} Mn(CO)_5^{-}$$

$$NiCl_2 + Na + CO \xrightarrow{\bullet))))} Ni_6(CO)_{12}^{2-}$$

$$MCl_5 + Na + CO \xrightarrow{\bullet))))} M(CO)_6^{-} \quad \text{(where M = V, Nb, Ta)}$$

$$MCl_6 + Na + CO \xrightarrow{\bullet))))} M_2(CO)_{10}^{2-} \quad \text{(where M = Cr, Mo, W)}$$

Bonnemann et al. (1983) reported the activation of transition metals, where Mg species was used to reduce metal salts in the presence of cyclopentadiene, 1,5-cyclooctadiene, and other ligands to form their metal complexes.

$$Mg + C_{14}H_{10} \xrightarrow[THF]{\bullet))))} Mg(THF)_3(\eta^2\text{-} C_{14}H_{10})$$

$$2Co(acac)_3 + 3Mg(THF)_3(\eta^2 - C_{14}H_{10}) \longrightarrow 2Co^* + 3Mg^{2+}$$

$$2Co^* + 2C_5H_6 + 31,5\text{-}C_8H_{12} \longrightarrow 2Co(Cp)(COD) + C_8H_{14}$$

The use of ultrasound in the heterogeneous reactions of some metals of high reactivity has also been reported.

Wang et al. (2006) carried out pinacol coupling of aromatic aldehydes by aqueous vanadium(II) solution under ultrasound irradiation at room temperature, which leads to the corresponding pinacols in 54–93 % yields within 15–30 min.

It was been observed that titanium tetrachloride is reduced by lanthanum filings to the corresponding low-valent titanium complex, which can induce some aromatic aldehydes to the corresponding pinacols in 28–97% yields within 10–50 min in ethyl acetate at room temperature under ultrasound irradiation (Li et al., 2006).

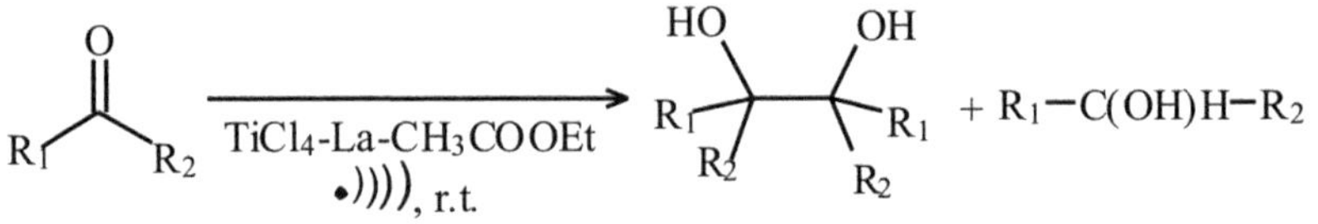

Ultrasound plays a significant role in the synthesis of metal chalcogenides. Sonication is used in the synthesis of oxides and sulfides followed by calcinations. It has also proved its worth in the preparation of some coordination polymers. Synthesis of organometallic chemical with Zn, Hg, Li and Mg like metals was also performed using sound waves and it has proved to be beneficial. Both the rate of synthesis and the amount of product were found to increase by the use of sound waves. Ultrasound is a better option to carry out such processes due to its time-saving and easy mode of operation.

REFERENCES

Aboutorabi, L.; Morsali, A. *Inorg. Chim. Acta* **2010,** *363*, 2506–2511.

Al-Juaid, S. S.; Eaborn, C.; Lickiss, P. D.; Smith, J. D.; Tavakkoli, K.; Webb, A. D. *J. Organomet. Chem.* **1996,** *510*, 143–151.

Amiri, O.; Salavati-Niasari, M.; Sabet, M.; Ghanbari, D. *Comb. Chem. High Throughput Screening* **2014,** *17*, 183–189.

Antonelli, D. M.; Ying, J. Y. *Angew. Chem., Int. Ed.* **1995,** *34*, 2014–2017.

Arenas, M. C.; Rodríguez-Núnez, L. F.; Rangel, D.; Martínez-Álvarez, O.; Martínez-Alonso, C.; Castaño, V. M. *Ultrason. Sonochem.* **2013,** *20*, 777–784.

Atobe, M.; Kado, Y.; Nonaka, T. *Chem. Lett.* **1998,** 699–700.

Avivi (Lev i), S.; Palchik, O.; Palchik, V.; Slifkin, M. A.; Weiss, A. M.; Gedanken, A. *Chem. Mater.* **2001,** *13*, 2195–2200.

Azuma, T.; Yanagida, S.; Sakurai, H.; Sasa, S.; Yoshimo, K. *Synth. Commun.* **1982,** *12,* 137–140.

Batovska, D.; Parushev, S.; Slavova, A.; Bankova, V.; Tsvetkova, I.; Ninova, M.; Najdenski, H. *Eur. J. Med. Chem.* **2007,** *42,* 87–92.

Bokhimi, X.; Zanella, R.; Morales, A. *J. Phys. Chem.* **2008,** *112,* 12463–12467.

Bonnemann, H.; Bogdanovi, B.; Brinkman, C. R.; He, D. W.; Spliethoff, B. *Angew. Chem. Int. Ed.* **1983,** *22,* 728.

Boudjouk, P.; So, J. H. *Synth. Commun.* **1986,** *16*(7), 775–778.

Boudjouk, P.; Han, B. H.; Anderson, K. R. J. *Am. Chem. Soc.* **1982,** *104,* 4992–4993.

Chew, S.; Ferrier, R. J. *J. Chem. Soc., Chem. Commun.* **1984,** 911–912.

Chong, M. N.; Jin, B.; Chow, C. W. K.; Saint, C. *Water Res.* **2010,** *44,* 2997–3027.

Cogkuner, B.; Figen, A. K.; Piskin, S. *J. Chem.* **2014,** *2014.* DOI: org/10.1155/2014/185957.

Danhui, Y.; Einhorn, J.; Einhorn, C.; Aurell, M. J.; Luche, J. L. *J. Chem. Soc., Chem. Commun.* **1994,** 1815–1816.

Dharmarathna, S.; King'ondu, C. K.; Pedrick, W.; Pahalagedara, L.; Suib, S. L. *Chem. Mater.* **2012,** *24,* 705–712.

Dhas, N. A.; Suslick, K. S. *J. Am. Chem. Soc.* **2005,** *127,* 2368–2369.

Dhas, N. A.; Ekhtiarzadeh, A.; Suslick, K. S. *J. Am. Chem. Soc.* **2001,** *123,* 8310–8316.

Dhas, N. A.; Koltypin, Y.; Gedanken, A. *Chem. Mater.* **1997,** *9,* 3159–3163.

Dhas, N. A.; Zaban, A.; Gedanken, A. *Chem. Mater.* **1999,** *11,* 806–813.

Fard-Jahromi, M. J. S.; Morsali, A. *Ultrason. Sonochem.* **2010,** *17,* 435–440.

Fry, A. J.; Bujanauskas, J. P. *J. Org. Chem.* **1978,** *43,* 3157–3163.

Fry, A. J.; Ginsburg, G. S. *J. Am. Chem. Soc.* **1979,** *101*(14), 3927–3932.

Fry, A. J.; Herr, D. *Tehrahedron Lett.* **1979,** *19*(20), 1721–1724.

Fry, A. J.; Lefor, A. T. *J. Org. Chem.* **1979,** *44,* 1270–1273.

Fry, A. J.; Ankner, K.; Hana, V. *Tetrahedron Lett.* **1981,** *22,* 1791–1794.

Fry, A. J.; Donaldson, W. A.; Ginsburg, G. S. *J. Org. Chem.* **1979,** *44,* 349–352.

Fry, A. J.; Ginsburg, G. S.; Parente, R. A. *J. Chem. Soc. Chem. Commun.* **1978,** 1040–1041.

Fuentes, A.; Marinas, J. M.; Sinisterra, J. V. *Tetrahedron Lett.* **1987,** *28,* 4541–4544.

Geoffroy, G. L.; Wrighton, M. S. *Organometallic Photochemistry;* Academic Press: New York, 1979.

Goswami, P. P.; Choudhury, H. A.; Chakma, S.; Moholkar, V. S. *Ind. Eng. Chem. Res.* **2013,** *52,* 17848–17855.

Green, M. L. H.; Thompson, M. E. *Ultrasonics* **1987,** *25,* 56–59.

Han, B. H.; Boudjouk, P. *J. Org. Chem.* **1982,** *47*(4), 751–752.

Hanifehpour, Y.; Mirtamizdoust, B.; Joo, S.-W. *J. Inorg. Organomet. Polym. Mater.* **2012,** *22,* 916–922.

Hashemi, L.; Tahmasian, A.; Morsali, A.; Abedini, J. *J. Nanostruct.* **2012,** *2,* 163–168.

Huang, W. P.; Tang, X. H.; Wang, Y. Q.; Koltypin, Y.; Gedanken, A. *Chem. Commun.* **2000,** *15,* 1415–1416.

Jugović, D.; Mitrić, M.; Cvjetićanin, N.; Jančar, B.; Mentus, S.; Uskoković, D. *Solid State Ionics* **2008,** *179,* 415–419.

Kang, Y. S.; Risbud, S.; Rabolt, J. F.; Stroeve, P. *Chem. Mater.* **1996,** *8,* 2209–2211.

Khanpour, M.; Morsali, A.; Retailleau, P. *Polyhedron* **2010,** *29,* 1520–1524.

Kim, K-H.; Kim, K-B. *Ultrason. Sonochem.* **2008,** *15,* 1019–1025.

Kim, J-H.; Son, B-R.; Yoon, D-H.; Hwang, K-T.; Noh, H-G.; Cho, W-S.; Kim, U-S. *Ceram. Int.* **2012,** *38*, 5707–5712.
Kitazume, T.; Ishikawa, N. *Chem. Lett.* **1981,** 1679–1680.
Kitazume, T.; Ishikawa, N. *Chem. Lett.* **1982,** 137–140.
Kluson, P.; Kacer, P.; Cajthaml, T.; Kalaji, M. *J. Mater. Chem.* **2000,** *11*, 644–651.
Krishna, B. M. V.; Ranjit, M.; Karunasagar, D.; Arunachalam, J. *Talanta* **2005,** *67*, 70–80.
Lee, A. S.; Cheng, R. Y. *Tetrahedron Lett.* **1997,** *38*, 443–446.
Li, N.; Gaillard, F. *Appl. Catal. B: Environ.* **2009,** *88*, 152–159.
Li, J. T.; Chen, Y. X.; Shuang, T. *Synth. Commun.* **2005,** *35*, 2831–2837.
Li, J. T.; Bian, Y. J.; Zang, H. J.; Li, T. S. *Synth. Commun.* **2006a,** *34*(4), 547–551.
Li, J. T.; Sun, X. L.; Lin, Z. P.; Li, T. S. *E-J. Chem.* **2006b,** *3*(4), 230–235.
Li, J. T.; Sun, X. L.; Lin, Z. P.; Chen, Y. X.; Li, T. S. *Indian J. Chem.* **2007,** *46B*, 1303–1307.
Li, H.; Ni, Y.; Cai, Y.; Zhang, L.; Zhou, J.; Hong, J.; Wei, X. *J. Mater. Chem.* **2009,** *19*, 594–597.
Li, Y.; Wang, J.; Li, Z.; Liu, Q.; Liu, J.; Liu, L.; Zhang, X.; Yu, J. *Chem. Eng. J.* **2013,** *218*, 295–302.
Liu, W.; Zhang, J.; Cheng, C.; Tian, G.; Zhang, C. *Chem. Eng. J.* **2011,** *175*, 24–32.
Lo, S. S.; Huang, D. *Langmuir* **2010,** *26*, 6762–6770.
Lu, T-J.; Cheng, S-M.; Sheu, L-J. *J. Org. Chem.* **1998,** *63*(8), 2738–2741.
Luche, J. L.; Damiano, J. C. *J. Am. Chem. Soc.* **1980,** *102*, 7926–7927.
Luche, J. L.; Petrier, C.; Gemal, A. L.; Zikra, N. *J. Org. Chem.* **1982,** *47*, 3805–3806.
Luche, J. L.; Petrier, C.; Lansard, J. P.; Greene, E. A. *J. Org. Chem.* **1983,** *48*, 3837–3839.
Luche, J. L.; Petrier, C.; Dupuy, C. *Tetrahedron Lett.* **1984,** *25*(7), 753–756.
Lukevics, E.; Gevorgyan, V. N.; Goldberg, Y. S. *Tetrahedron Lett.* **1984,** *25*(13), 1415–1416.
Lupan, O.; Pauport, T.; Bahers, T. L.; Ciofini, I.; Viana, B. *J. Phys. Chem. C* **2011,** *115*, 14548–14558.
Malta, M.; Silva, L. H.; Galembeck, A.; Korn, M. *Macromol. Rapid Commun.* **2008,** *29*, 1221–1225.
Miquel, P.; Granger, P.; Jagtap, N.; Umbarkarc, S.; Dongare, M.; Dujardin, C. *J. Mol. Catal. A: Chem.* **2010,** *322*, 90–97.
Mizukoshi, Y.; Oshima, R.; Maeda, Y.; Nagata, Y. *Langmuir* **1999,** *15*, 2733–2737.
Mohseni A.; Maleknia, L.; Fazaeli, R.; Ahmadi, E. *Nanocon* **2013,** *8*, 125–130.
Moosavi, R.; Abbasi, A. R.; Yousefi, M.; Ramazani, A.; Morsali, A. *Ultrason. Sonochem.* **2012,** *19*, 1221–1226.
Nagarajan, R.; Tomar, N. *J. Solid State Chem.* **2009,** *182*, 1283–1290.
Nikitenko, S. I.; Koltypin, Y.; Mastai, Y.; Koltypin M.; Gedanken, A. *J. Mater. Chem.* **2002,** *12*, 1450–1452.
Okitsu, K.; Ashokkumar, M.; Grieser, F. *J. Phys. Chem.* B **2005,** *109*, 20673–20675.
Olaya, A.; Moreno, S.; Molina, R. *Appl. Catal. A: Gen.* **2009,** *370*, 7–15.
Omidi, A.; Habibi-Yangjeh, A.; Pirhashemi, M. *Appl. Surf. Sci.* **2013,** *276*, 468–475.
Patil, V. S.; Gogate, P. R. *Chem. Eng. J.* **2016,** *289*, 513–524.
Paulusse, J. M. J.; Van Beek, D. J. M.; Sijbesma, R. P. *J. Am. Chem. Soc.* **2007,** *129*, 2392–2397.
Petrier, C.; Gemal, A.; Luche, J. L. *Tetrahedron Lett.* **1982,** *23*, 3361–3364.
Pinto, D. C. G. A.; Silva, A. M. S.; Levai, A.; Cavaleiro, J. A. S.; Patonay, T.; Elguero, J. *Eur. J. Org. Chem.* **2000,** *14*, 2593–2599.
Pratt, M. W.; Helsby, R. *Nature* **1959,** *184*, 1694–1695.
Ranjbar, Z. R.; Morsali, A. *Polyhedron* **2011,** *30*, 929–934.

Ranjbar, M.; Malakooti, E.; Sheshmani, S. *J. Chem.* **2013**. DOI: org/10.1155/2013/560983.
Renaud, P. *Bull. Soc. Chim. France* **1950,** *S17*, 1044–1445.
Repic, O.; Vogt, S. *Tetrahedron Lett.* **1982,** *23*, 2729-2732.
Reza, R-K.; Alireza, B-S.; Aslani, A.; Kaviani, K. *Phys. B: Phys. Condens. Matter* **2010,** *405*, 3099–3101.
Roger C.; Tudoret M-J.; Guerchais V.; Lapinte C. *J. Organomet. Chem.* **1989,** *365*, 347–350.
Salinas-Estevané, P.; Sánchez, E. M. *Cryst. Growth Des.* **2010,** *10*(9), 3917–3924.
Sano, S.; Matsumoto, Yano, T.; Toguchi, M.; Nakao, M. *Synlett* **2015,** *26*, 2135–2138.
Slough, W.; Ubbelohde, A. R. *J. Chem. Soc.* **1957,** 918–919.
Soltanzadeh, N.; Morsali, A. *Ultrason. Sonochem.* **2010,** *17*, 139–144.
Souza-Barboza, J. D.; Luche, J. L.; Petrier, C. *Mechanistic Consequences. Tetrahedron Lett.* **1987,** *28*(18), 2013-2016.
Suslick, K. S.; Grinstaff, M. W. *J. Am. Chem. Soc.* **1990,** *112*, 7807–7809.
Suslick, K. S.; Johnson, R. E. *J. Am. Chem. Soc.* **1984,** *106*, 6856–6858.
Suslick, K. S.; Schubert, P. F.; Goodale, J. W. *J. Am. Chem. Soc.* **1981,** *103*, 7342–7344.
Suslick, K. S.; Goodale, J. W.; Schubert, P. F.; Wang, H. H. *J. Am. Chem. Soc.* **1984,** *105*, 5781–5785.
Suwanboon, S.; Klubnuan, S.; Jantha, N.; Amornpitoksuk, P.; Bangrak, P. *Mater. Lett.* **2014,** *115*, 275–278.
Trost, B. M.; Coppola, B. P. *J. Am. Chem. Soc.* **1982,** *104*(24), 6879–6881.
Tuulmets, A.; Kaubi, K.; Heinoja, K. *Ultrason. Sonochem.* **1995,** *2*, S75–78.
Uzcanga, I.; Bezverkhyy, I.; Afanasiev, P.; Scott, C.; Vrinat, M. *Chem. Mater.* **2005,** *17*, 3575–3802.
Wang, Y. Q.; Tang, X. H.; Yin, L. X.; Huang, W. P.; Hacohen, Y. R.; Gedanken, A. *Adv. Mater.* **2000,** *12*, 1183–1186.
Wang, H.; Zhu, J-J.; Zhu, J-M.; Chen, H-Y. *J. Phys. Chem. B* **2002,** *106*, 3848–3854.
Wang, H.; Lu, Y-N.; Zhu, J-J.; Chen, H-Y. *Inorg. Chem.* **2003,** *42*, 6404–6411.
Wang, S. F.; Gu, F.; Lu M. K. *Langmuir* **2006a,** *22*, 398–401.
Wang, S. X.; Wang, K.; Li, J. T. *Synth. Commun.* **2006b,** *35*(18), 2387–2394.
Wang, P.; Zhao, Y-J.; Wen, L-X.; Chen, J-F.; Lei, Z-G. *Ind. Eng. Chem. Res.* **2014,** *53*, 20116–20123.
Weber, T.; Muijsers, J. C.; Niemantsverdriet, J. W. J. *Phys. Chem.* **1995,** *99*, 9194–9200.
West, R.; Fink, M. J.; Michl, J. *Science* **1981,** *214*(4527), 1343–1344.
Wongpisutpaisana, N.; PiyanutCharoonsuka, B.; NaratipVittayakorna, C. D.; Wisanu Pecharapaa, B. *Energy Procedia* **2011,** *9*, 404–409.
Xie, D.; Chang, L.; Wang, F.; Du, G.; Xu, B. *J. Alloys Compd.* **2012,** *545*, 176–181.
Xin, Y.; Zang, Z-H.; Chen, F-L. *Synth. Commun.* **2009,** *39*, 4062–4068.
Yahyavi, S. R.; Haghighi, M.; Shafiei, S.; Abdollahifar, M.; Rahmani, F. *Energy Convers. Manage.* **2015,** *97*, 273–281.
Yamaguchi, R.; Kawasaki, H.; Kawanisi, M. *Synth. Commun.* **1982,** *12*, 1027–1037.
Yin, J.; Qian, X.; Yin, J.; Shi, M.; Zhou, G. *Mater. Lett.* **2003,** *57*, 3859–3863.
Yu, J. C.; Zhang, L.; Yu, J. *New J. Chem.* **2002,** *26*, 416–420.
Zhang, D.; Fu, H.; Shi, L.; Pan, C.; Li, Q.; Chu, Y.; Yu, W. *Inorg. Chem.* **2007,** *46*, 2446–2451.
Zhang, X.; Jiang, W.; Gong, X.; Zhang, Z. *J. Alloys Compd.* **2010,** *508*, 400–405.

CHAPTER 6

NANOMATERIALS

MEENAKSHI SINGH SOLANKI[1], SURBHI BENJAMIN[2], SURESH C. AMETA[3]

[1]*Department of Chemistry, B. N. University, Udaipur, India*
E-mail: meenakshisingh001989@gmail.com

[2]*Department of Chemistry, PAHER University, Udaipur, India*
E-mail: surbhi.singh1@yahoo.com

[3]*Department of Chemistry, PAHER University, Udaipur, India*
E-mail: ameta_sc@yahoo.com

CONTENTS

6.1 INTRODUCTION

For many ears, scientists are trying to settle on a precise definition of nanomaterials. In all these attempts, they came to a conclusion that nanoparticles can be partially characterized by their tiny size, measured in nanometers. A nanometer is one-millionth of a millimeter—approximately 100,000 times smaller than the diameter of a human hair. More precisely, nanomaterials can be defined as a physical substance with size less than approximately 100 nm. Thus, synthesis of nanomaterials is a kind of approach that is being widely followed in the field of miniaturization of materials.

Some nanomaterials are already present in nature such as blood-borne proteins essential for life and lipids found in the blood and body fat or carbon. Nanomaterials could be synthesized from different materials. But researchers are particularly interested in engineered nanomaterials (ENMs), which are designed for use in many commercial materials, electronic device manufacture, chemistry, engineering, environmental remediation or clean-up to bind with and neutralize toxins. Already, thousands of common products manufactured using ENMs are as follows:

- **Day-to-day products:** Sunscreens, cosmetics, sporting goods, stain-resistant clothing, tires, and electronics.
- **Medical applications:** Diagnosis, imaging and drug delivery and also in environmental remediation. Nanomaterials can target specific organs or cells in the body such as cancer cells, and enhance the effectiveness of any therapy.
- **Others:** Nanomaterials can also be added to cement, cloth, and other materials to make them stronger and yet lighter.

Ultrasound provides a new method for the synthesis of nanomaterials or nanoparticles as it helps us in controlling the size and morphology of nanoparticles. The formation of nanoparticles (Fig. 6.1) includes the following steps:

- Acoustic cavitation, which involves the formation, growth, and implosive collapse of bubbles in a liquid creating extreme conditions inside the collapsing bubble, which serves as the origin of most of the sonochemical phenomena in liquids or liquid–solid slurries.

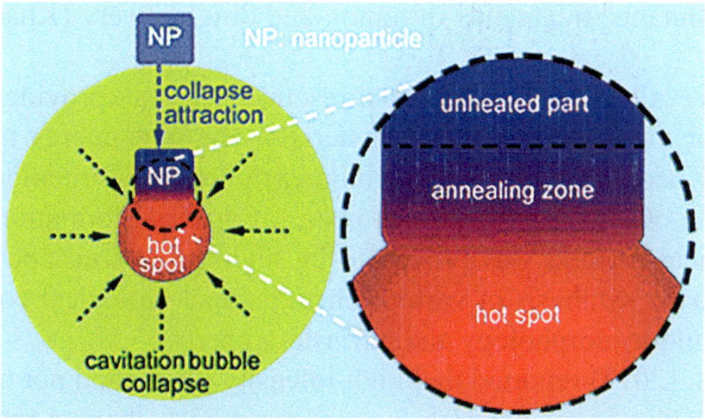

FIGURE 6.1 Formation of nanoparticles.
Source: Wang et al., 2013, with permission.

- Nebulization, which is the creation of mist from ultrasound passing through a liquid and impinging on a liquid–gas interface and it forms the basis for ultrasonic spray pyrolysis (USP) with subsequent reactions occurring in the heated droplets of the mist.

Cavitation-induced sonochemistry provides a unique interaction between energy and matter, with hot spots inside the bubbles with temperature ~5000 K, pressures ~1000 bar, and heating and cooling rates of $>10^{10}$ K·s^{-1}. Such extraordinary conditions allow a wide range of chemical reactions possible, which is normally impossible otherwise under ordinary conditions. Thus, it permits the synthesis of different types of unusual nanostructured materials. In addition to cavitational chemistry, the microdroplet reactors created by USP also facilitate the formation of a wide range of nanocomposites (Bang and Suslick, 2010).

Sonication treatment also improves the effectiveness of nanoparticles. It increases the capability of same photoactive nanoparticles, which can be widely used in photoelectrochemical (PEC), water splitting, dye-sensitized solar cells, and photocatalysis (Wang et al., 2013).

Ultrasound has various applications in the preparation of nanocomposites such as elemental metals, alloys, salts, metal oxides, carbon nanoforms, coordination compounds, macrocycles, and polymers, as well as nanoemulsions and nanogels. Not only nanocatalysts can be synthesized by ultrasound but also can function more actively in its presence. Nanocatalysts prepared by sonication could also be used in the degradation of organic pollutants, treatments of tumor, and drug delivery (Kharissova et al., 2011).

Pokhrel et al. (2016) opined that ultrasound technique provides a simple pathway for the synthesis of nanomaterials. The high-energy ultrasound creates tiny acoustic bubbles and within each period—formation, growth and collapse of bubbles, a coherent phase of material formation is there. Sonochemical method was proved to be highly localized and effective method for the synthesis of various chemical and biological compounds featuring unique morphology and intrinsic property.

Xu et al. (2013) reported that high-intensity ultrasound not only helps in the synthesis of nanomaterials but also can modify them. As sonochemistry caused acoustic cavitation, that is the formation, growth, and implosive collapse of bubbles and therefore, the effect of sonochemistry could be basically categorized into three effects. These are:

- **Primary sonochemistry:** It involves gas-phase chemistry occurring inside collapsing bubbles.
- **Secondary sonochemistry:** It involves solution-phase chemistry occurring outside the bubbles.
- **Physical modifications:** These are caused by high-speed jets or shock waves derived from bubble collapse.

Various physical, chemical, biological, and hybrid processes have been used for the synthesis of nanomaterials in different forms such as nanopowders, nanoclusters, nanocollides, nanotubes, nanorods, nanowires, nanofilms, nanoplatelets, nanopins, nanoneedles, nanopillars, nanoflowers, nanocones, nanopencils, nanoflakes, nanosheets, nanoribbons, nanobelts, nanoplates, nanodots, nanoalloys, nanodiscs, nanorings, nanotripods,

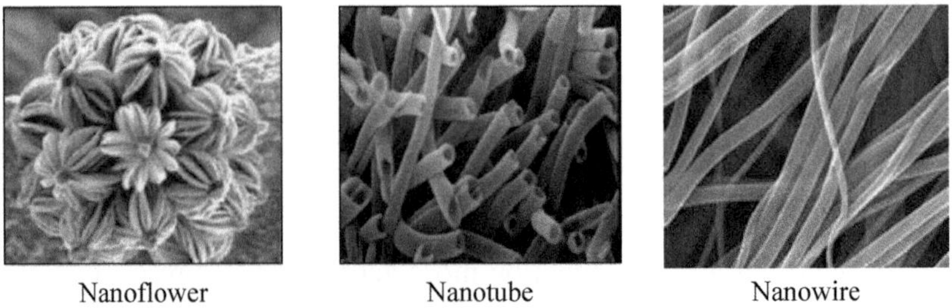

FIGURE 6.2 Nanomaterials in different shapes.

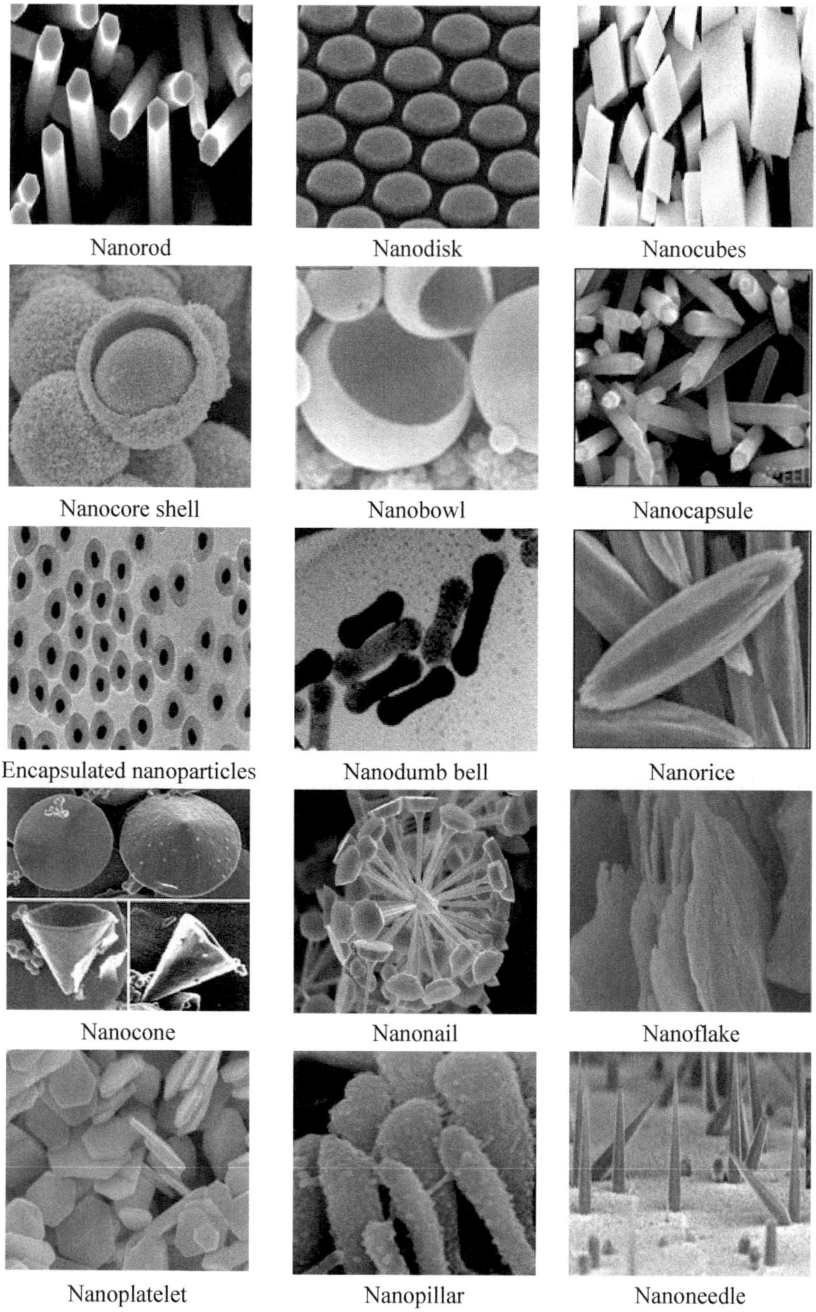
Nanorod
Nanodisk
Nanocubes
Nanocore shell
Nanobowl
Nanocapsule
Encapsulated nanoparticles
Nanodumb bell
Nanorice
Nanocone
Nanonail
Nanoflake
Nanoplatelet
Nanopillar
Nanoneedle

FIGURE 6.2 *(Continued)*

nanobricks, nanonails, nanoarrows, nanodumb-bells, nanospheres, and so on (Fig. 6.2).

Preparation methods depend on the material of interest, type of nanomaterials (zero dimension, 1D or 2D), their size and quantity (Kulkarni, 2014). Some of the methods of synthesis are vacuum synthesis, gas-phase synthesis, condensed-phase synthesis, high-speed deposition, deposition method, sol-gel method, high-energy milling, chemical vapour synthesis, precipitation method, sonochemical method, and so on (Gleiter, 1989; Solanki et al., 2015).

Among these methods, sonochemical method is the most effective technique to fabricate nanomaterial as it involves nonhazardous acoustic radiation. Ultrasound does not directly interact with molecular species but affects the physical and chemical properties because of cavitational collapse. It generates drastic conditions of temperature and pressure locally. Ultrasound irradiation helps to carry out chemical reaction effectively along with high yield and selectivity. This method is beneficial than other methods as it involves less hazardous starting material, reagent, and solvent (Gedanken, 2004).

6.2 NANOFLOWERS

Tai and Guo (2008) used ultrasound and microwave simultaneously for the preparation of single crystalline CdS nanoflowers. The synergistic effect of both these processes was responsible for formation of well-defined flower-like CdS nanostructures. It consists of hexagonal nanopyramids and/or nanoplates depending on different sources of sulphur. CdS nanoflowers exhibited large blue-shift up to 100 nm as compared to simple low-dimensional CdS nanostructures. Such nanostructure could be widely used in optoelectronics devices, catalysis, and solar cells, because of this shift in optical properties.

Flower-shaped ZnO nanostructures were also synthesized by Singh et al. (2008) by sol-gel method under exposure to ultrasound. The root size of ZnO nanoflower structure was 1 μ with a tip of 150 nm solvent. They observed that the morphology of nanoparticles changes depends on the nature of solvent. It means that the shape and size of a sample changes on changing the solvent. ZnO nanorods with diameter <60 nm and length >1 μ were synthesized using methanol as solvent. As-prepared ZnO were highly pure and crystalline in nature. They also studied the effect of vapour pressure of solvent on this sonication process. The presence of vapours affected the intensity of sonication, which affects various chemical reactions taking place during the formation of ZnO. They also analysed effect of ethylenediamine on the chemical reactions leading to the formation of these ZnO nanostructures.

8-Hydroxyquinoline aluminum nanoflowers were prepared via sonochemical method (Mao et al., 2011). It was made up of nanobelts with thickness about 50 nm, average width of 200 nm, and length up to 10 μm. As-prepared sample nanoflowers showed good electro-generated chemiluminescence behaviour.

Nanoflower Pb(II) coordinated compound, $[Pb(phen)_2(4\text{-}abs)_2]_n$, was prepared with the ligands 1,10-phenanthroline (phen) and 4-aminobenzene sulphonic acid (4-abs) in the presence of ultrasound (Mirtamizdoust et al. 2012). In this coordinated nanoflower compound, metal ion has six coordination numbers, that is, (PbN_4O_2) and it has a stereochemically active lone pair of electrons and a hemi-directed coordination sphere. The chains interact with each other through π–π stacking to create a 3D framework.

Zak et al. (2013) used Zn salt, sodium hydroxide, and ammonia solution as precursors without any other structure directing agent or surfactant for preparation of different-shaped ZnO nanostructures (ZnO–NS). They used a simple, feasible, and green sonochemical method for this purpose as it does not require high temperature and/or highly toxic chemicals. They observed that the shape of the ZnO–NS can be changed by adjusting the ultrasound energy dissipated via variation of the exposure time of ultrasound from 5 to 60 min. Following different shapes were observed on varying the ultrasonication time.

- Plate-like, irregular-shaped nanoparticles were observed on 5 min ultrasonication.
- Uniform ZnO nanorods with diameter around 50 nm were formed on 15 min of ultrasonication.
- Flower-like ZnO–NS was formed, when time was increased to 30 min ultrasonication.
- Well-defined and magnified flower-shaped ZnO–NS was observed after 60 min ultrasonication.

Thus, this method provided high-quality ZnO–NS with controllable shapes, uniformity, and purity.

Flower-like-shaped nanostructures (nanoflowers) of a protein-inorganic hybrid consisting of laccase as a model protein and copper phosphate were synthesized by Batule et al. (2015). It was synthesized within 5 min in the presence of ultrasound. This sample exhibited interesting good features such as:

- Increased activity
- Enhanced stability
- Reusability

These properties were demonstrated by its application in the colorimetric detection of epinephrine. They also found application of this sample in biosensor and enzyme catalysis indicating further utilization of nanoflowers in diverse fields of biotechnology.

Hanifehpour et al. (2015) synthesized two neutral nanoflower polymeric Pb(II) coordination compounds under sonication. These are $[Pb(tmph)(l\text{-}SCN)_2]_n$, $[Pb(tmph)(l\text{-}NO_3)_2]_n$, [where, tmph = 3,4,7,8-tetramethyl-1,10-phenanthroline]. Flower-shaped nanostructures of both these complexes were observed by SEM. Single-crystal X-ray data indicated 1D double-chain net-like coordination polymers.

Qi et al. (2016) reported synthesis of hydroxyapatite nanoflowers (HAFs) by using creatine phosphate disodium salt as an organic phosphorus source in single step via a sonochemical method. It was observed that the HAFs (diameters 300 nm) are formed by self-assembly of hydroxyapatite nanosheets with thicknesses of <10 nm. As-prepared sample showed excellent cytocompability, which was proved by its essentially inappreciable toxicity to MC-3T3 osteoblast cells. Not only this but these HAFs also exhibited good protein adsorption ability on using hemoglobin (Hb) as a model protein, which suggests that HAFs may have some promising applications in different biomedical fields such as protein and drug delivery (Fig. 6.3).

Shahverdizadeh and Hossein (2016) prepared 3D Cd(II) coordination polymer nanoflowers, $[Cd_2(pcih)(N_3)_3(CH_3OH)]_n$, (where pcih=2-pyridinecarbaldehyde isonicotinoylhydrazonate), under ultrasonic irradiation. These individual particles were assembled by strong π–π-staking interaction to form a compressed supramolecular 3D network. Later on, pure CdO nanoparticles were also prepared by thermolysis of $[Cd_2(pcih)(N_3)_3(CH_3OH)]_n$ precursor in oleic acid at 180 °C in the presence of air.

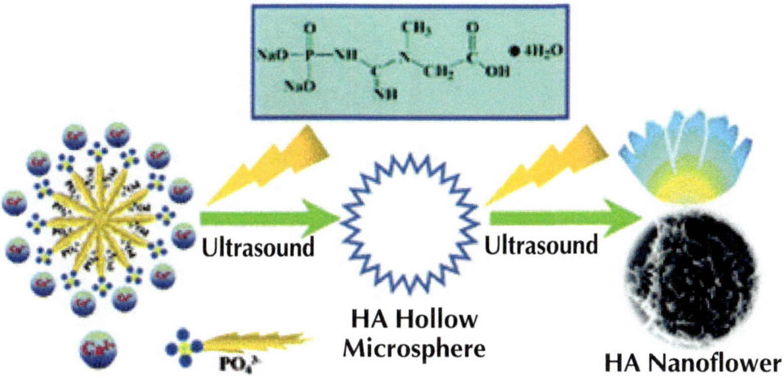

FIGURE 6.3 Synthesis of hydroxyapatite nanoflowers (HAFs).
Source: Qi et al., 2016, with permission.

Chung et al. (2014) reported sonochemical synthesis of discrete Pb(II) coordination compound. The lead ion was coordinated by six ligands, that is, four nitrogen and two oxygen atoms from two pcih ligands. It was observed that the overall structure of 1D supramolecular chain and with other directional intermolecular interactions, it is further extended into a 3D supramolecular structure via a strong π–π interaction. This complex was used to obtain PbO nanoparticles by thermolysis at 180 °C with oleic acid as a surfactant.

6.3 NANORODS

Single-crystalline needle-shaped ZnO nanorods were prepared using zinc acetate dihydrate and sodium hydroxide as starting materials at room temperature in the presence of high-energy sound (Wahab et al., 2007). The needle-shaped sample was made up of hexagonal surfaces along the length with typical diameter and length in the range of 120–160 nm and 3–5 μm, respectively. The nanorods were found to be single crystalline with wurtzite hexagonal phase. They also found that the sonication exposure time affects the shape of nanoparticles. As-synthesized material has good optical properties as evident from ultraviolet-visible and room temperature photoluminescence spectroscopy.

Zhu et al. (2007) synthesized $CeVO_4$ nanorods using $Ce(NO_3)_3$ and NH_4VO_3 in aqueous solution without any surfactant or template under ultrasound irradiation. They also prepared high specific surface area

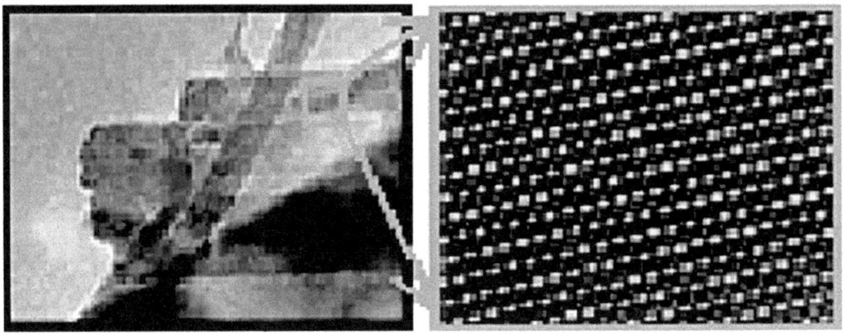

FIGURE 6.4 SEM image of plate-like hydroxyapatite (HAp) nanoparticles.
Source: Jevtic et al., 2008, with permission.

mesoporous $CeVO_4$ by the same method by using $Ce(NO_3)_3$, V_2O_5, and NaOH. Mesoporous $CeVO_4$ has a specific surface area of 122 $m^2 \cdot g^{-1}$ and an average pore size of 5.2 nm, while $CeVO_4$ nanorods have a diameter and a length of 5 nm and 100–150 nm, respectively. The rod-like shape formation takes place due to two key factors and these are ultrasound irradiation and the presence of ammonia in the reactive solution.

Plate-like hydroxyapatite (HAp) nanoparticles were synthesized in an ultrasound field by homogeneous precipitation method (Jevtic et al., 2008). This plate-like structure consists of specifically oriented and laterally connected nanorods with a length of about 500 nm and a diameter of 100 nm (Fig. 6.4). A single nanorod showed a highly regular and defect-free lattice with unique crystallographic plane orientations. They also noticed the influence of the ultrasound on the growth mechanism.

Warule et al. (2009) prepared various morphologies of the ZnO nanostructures such as nanoneedles, tetrapods, nanowires, nanopetals, self-assembled hexagonal rods, and nanocups (Fig. 6.5). All these shapes were achieved by varying some parameters such as:

- Concentration of the precursors
- Ultrasonic irradiation period
- Duty cycle

The samples were prepared by a sonochemical as well as a sonochemical-assisted hydrothermal method. The physicochemical properties were determined by different methods.

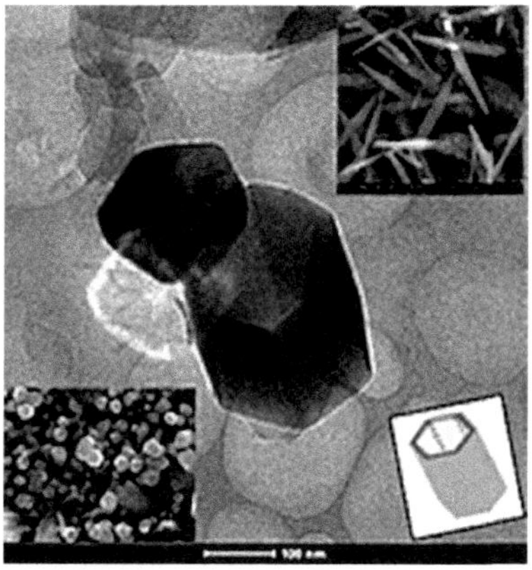

FIGURE 6.5 SEM image ZnO nanostructures.
Source: Warule et al., 2009, with permission.

- SEM showed the surface morphology of the ZnO nanostructures. The nanoneedles originate from the hexagonal tube.
- TEM analysis demonstrated the nanocrystalline nature of the ZnO structures with unique morphologies like hexagonal nanocups.

These ZnO nanostructures (nanocups and nanoneedles) had an excellent photocatalytic activity. These unique nanostructures can find application in ZnO-based dye-sensitized solar cells as a prospective semiconductor, because they have a wide band gap.

Li et al. (2009) used a mixture of ZnO nanorods, $[Ag(NH_3)_2]^+$, and formaldehyde in an aqueous medium to form Ag nanoparticles on ZnO nanorods sonochemically. X-ray diffraction of the ZnO/Ag composites showed that this sample has the face-center-cubic-structured crystalline Ag. ZnO nanorods were coated with Ag nanoparticles with a mean size of several tens of nanometer in ZnO/Ag composites. The absorption band of these composite is distinctly broadened and red-shifted, which indicates a strong interfacial interaction between ZnO nanorods and Ag nanoparticles. This sonochemical method is simple, mild, and can be readily scaled up, affording a simple way for the synthesis of other composites also.

Ultrasound-assisted synthesis of two coordination polymers of Hg(II) was reported by Yaser et al. (2010) with polyimine ligands, that is, 4-bpdh and 3-bpdh, as $[Hg(4\text{-bpdh})Br_2]_n$ and $[Hg(3\text{-bpdh})Br_2]_n$ (where 4-bpdh = 2,5-bis(4-pyridyl)-3,4-diaza-2,4-hexadiene and

3-bpdh=2,5-bis(3-pyridyl)-3,4-diaza-2,4-hexadiene). Thermal gravimetric and differential thermal analyses revealed that these nanostructures were relatively thermally less stable than their bulk materials.

Jung et al. (2010) synthesized zinc oxide nanorod arrays on a quartz wafer without using any metal catalysts via a sonochemical route. The average diameter of ZnO nanorods was 156 nm and length was 933 nm. This sonication method constitutes a practical technique for the design of state-of-the-art nanodevices (based on laterally grown ZnO nanorod arrays on planar substrates) as compared to conventional standard semiconductor microfabrication technologies.

Cho et al. (2012) also fabricated ZnO nanorods by a sonochemical method using zinc acetate dihydrate as a precursor. A thin film of ZnO was formed on fluorine-doped tin oxide (FTO) glass by decomposition of zinc acetate dehydrate. It was deposited on FTO glass by sputtering method. They could achieve different thicknesses (20, 40, and 60 nm) by adjusting the sputtering time. This was observed that the ZnO nanorods grown on the catalytic layer of 40 nm thickness represented the best crystal and spatial orientation leading to the appreciable optical properties. As-prepared ZnO thin film (40 nm thickness) yields well aligned and high-quality ZnO nanorods, due to small surface roughness and structural strain.

Kianpour et al. (2013) prepared $NiMoO_4$ nanorods by a reaction between $Ni(CH_3COO)_2{\cdot}4H_2O$ and $(NH_4)_6Mo_7O_{24}{\cdot}4H_2O$ under ultrasound irradiation. They also studied the effect of various parameters such as ultrasonic power, ultrasonic irradiation time, stirring, solvent, and surfactant. This method provides a general, simple, and an effective way to control the composition and morphology of $NiMoO_4$ in aqueous solution, which can prove to be important for inorganic synthesis methodology.

Zhang et al. (2005) synthesized ZnO nanorods and trigonal-shaped ZnO ultrafine particles by decomposition of zinc acetate dihydrate in paraffin oil on irrradating with ultrasound. As-prepared were characterized by different techniques such as XRD, TEM, selected area electron diffraction (SAED), SEM, and UV-visible techniques.

Single-phase antlerite, $Cu_3(OH)_4SO_4$, nanorods were synthesized by the sonochemical method from an aqueous solution of $CuSO_4$, without any additives (Segal et al., 2015). This reaction is a single-step reaction process with short irradiation time of <30 min. Here, protonation of sulphate ions resulted in bisulphate and hydroxyl ions. The extreme local conditions developed inside the cavity during the collapse of bubble, that is, higher pressure and temperature, lead to the formation of the crystalline mineral. On implosive collapse of the bubbles, the distances between the oppositely charged cupper and sulphate ions were decreased and as a result, the crystallization process was initiated.

6.4 NANOCUBES

Wu et al. (2006) dissociated $K_4Fe(CN)_6$ in acidic media under ultrasonic condition to obtain $Fe_4[Fe(CN)_6]_3$ nanocubic particles. These nanocubic particles were found to be regular, single-crystalline, and Prussian blue in colour. They observed that the size and size distribution of particles depend on reaction temperature, the concentration of $K_4Fe(CN)_6$, and ultrasonic condition.

Chen et al. (2012) prepared plasmonic photocatalyst AgCl with a small amount of Ag metal species. These nanocubes of Ag/AgCl hybrid were formed by sonication method. It was observed that the concentration of Ag ions and polyvinylpyrrolidone molecules in precursor solutions affects the size of Ag/AgCl photocatalysts. The compositions, microstructures, influencing factors, and possible growth mechanism of the Ag/AgCl hybrid nanocubes were also studied. They also used this hybrid for degradation of various dye molecules under visible light.

Abbas et al. (2013) carried out sonochemical reaction using inexpensive and nontoxic metal salt ($FeSO_4 \cdot 7H_2O$) as reactant for the preparation of monodisperse magnetite nanocubes in aqueous medium. The nanoparticles were found to be of uniform size of approximately 80 nm. These particles showed enhancement in their crystallinity by annealing treatment up to 600 °C with uniform shape and size distribution. When annealing temperature was 700 °C, then a distortion in shape and a broad size distribution was observed.

$CoSn(OH)_6$ nanocubes were synthesized using $CoCl_2$, $SnCl_4$, and NaOH as precursors without any surfactant or template by a simple sonochemical route at room temperature for 30 min (Cheng et al., 2013). It was further calcined at 500 °C for 120 min in air to obtain SnO_2–Co_3O_4 hybrids. The SnO_2–Co_3O_4 hybrids also had cubic morphology. They observed that SnO_2–Co_3O_4 hybrids exhibited more photocatalytic activity than $CoSn(OH)_6$ nanocubes for the degradation of methylene blue.

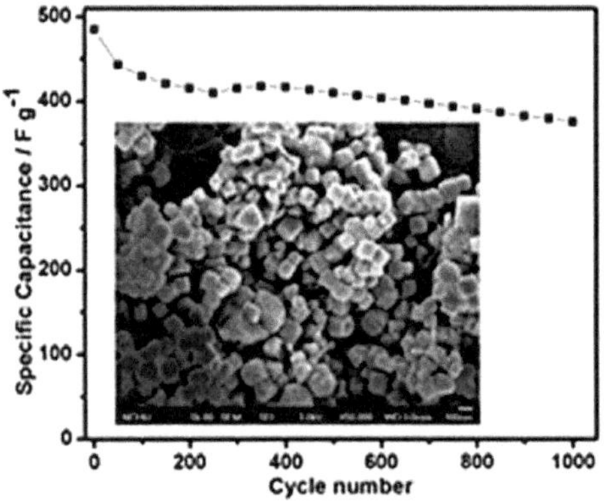

FIGURE 6.6 SEM image of SnO_2–Co_3O_4 hybrid nanocubes.
Source: Raj et al., 2016, with permission.

Raj et al. (2016) also prepared SnO_2–Co_3O_4 hybrid nanocubes from $CoSn(OH)_6$ nanocubes using sonication method. They avoided the use of any surfactant or template in this reaction. The sample was prepared at room temperature in the presence of stannous and cobalt precursors. The electrochemical performance of the hybrid SnO_2–Co_3O_4 nanocubes exhibited enhanced supercapacitor performance (Fig. 6.6). This enhanced electrochemical property of the electrode material was due to the synergetic effect between nanostructured SnO_2 and Co_3O_4. It was observed that about 77% of its initial capacitance was retained even after 1000 cycles. It showed its high-capacitance retention and electrochemical stability. These results confirmed that this material is a promising candidate for supercapacitor applications.

Cu_2O nanocubes were synthesized in a single step by reacting copper sulphate in the presence of polyvinylpyrrolidone and ascorbic acid at pH 11 under ultrasound irradiation (Kaviyarasan et al., 2016). Ultrasound acoustic cavitation-generated radicals reacted with $CuSO_4$ and resulted in $Cu(OH)_2$ intermediate, which finally reacts with ascorbic acid to generate Cu_2O nanocubes. As-synthesized Cu_2O nanocubes were found to be very effective for enhancing chemiluminescence in the presence of luminol-H_2O_2 system.

6.5 NANOTUBES

Katoh et al. (1999) applied high-intensity sound in homogeneous liquid consisting of chlorobenzene with $ZnCl_2$ particles and *o*-dichlorobenzene

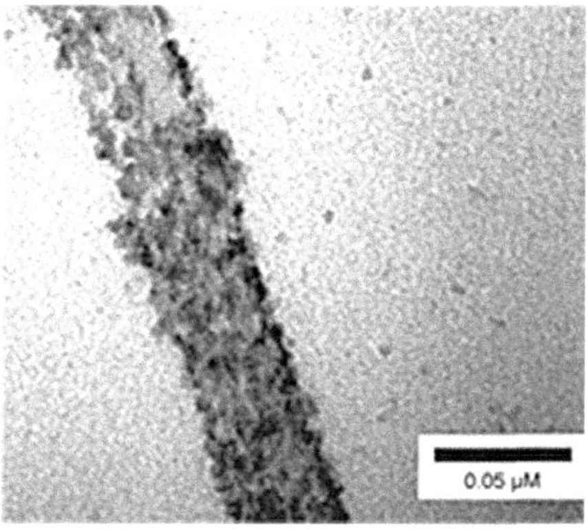

FIGURE 6.7 TEM image of tin nanoparticle-coated MWCNTs.
Source: Qui et al., 2004, with permission.

with $ZnCl_2$ and Zn particles to form carbon nanotubes (CNTs). This high-intensity sound helps in the formation of polymer and the disordered carbon by cavitational collapse in homogeneous liquid. These were annealed by the interparticle collision induced by the turbulent flow and shockwaves.

Sonication time has a major influence on reduction in amorphous carbon. Raja and Ryu (2009) prepared single and multiwalled carbon nanotubes (MWCNTs) using dichlorobenzene and $ZnCl_2$ and characterized it. The electronic band structural characterization confirmed the existence of metallic and semiconducting CNTs.

MWCNTs surface was also uniformly decorated with crystalline tin nanoparticles (<5 nm) via a sonication route (Qui et al., 2004). MWCNTs were coated by time on treating it in a solution of $SnCl_2$ in ethylene glycol under an N_2 atmosphere and ultrasound irradiation (Fig. 6.7). Formation of as-prepared sample supported that this technique can also be used for coating different metals on CNTs.

The oxidation of CNTs in the presence of ultrasound helps in increasing the density of surface functional group. Xing et al. (2005) reported that functionalization of CNTs enhanced deposition of metal nanoparticles in the preparation of supported catalyst. Thus, sonication method is very effective in functionalization of CNTs. Most of the CNT surface oxidation takes place between one to two hours.

Sonication method promotes the deposition of metal nanoparticles on the surface of carboxylate functionalized MWCNTs. Pan and Wai (2009) deposited rhodium nanoparticles with an average diameter of 2.5 ± 0.7 nm on the surface of carboxylate functionalized MWCNTs. The sample was prepared in ethanol solution containing $RhCl_3$ and an organic-reducing agent, borane morpholine complex ($C_4H_{12}BNO$). CNT-supported Rh nanoparticles showed highly active and reusable catalytic activity for hydrogenation of benzene and its derivatives at room temperature. It was found that complete ring saturation of polycyclic aromatic hydrocarbons (PAHs) can be obtained by mild hydrogenation conditions using the Rh/MWCNT catalyst, which is not possible by commercially available Rh nanocatalysts otherwise. This one-pot synthesis technique gave a simple and rapid way of preparing highly active and recyclable CNT-supported nanocatalysts for hydrogenation reactions at low temperatures.

Later, they also studied the effect of deposition of bimetallic nanoparticles (Pt-Rh) on the surface of carboxylate functionalized MWCNTs (Pan and Wai, 2011). These metals show a synergistic effect and enhance catalytic hydrogenation of PAHs, neat benzene and alkyl benzenes. Bimetallic nanoparticles-coated MWCNTs promote complete ring saturation of PAHs even at room temperature. CNT-supported monometallic and bimetallic nanocatalysts were prepared in single step and these could be recycled or reused for further hydrogenation reaction and that too at low temperatures.

Chuong et al. (2011) fabricated TiO_2 and TiO_2-ZrO_2 composite nanotubes using ultrasound. These nanomaterials were used for synthesis of lead zirconate titanate (PZT) ceramic, ($Pb[Zr_{(x)}Ti_{(1-x)}]O_3$), at low sintering temperature. It is one of the widely used piezoelectric ceramic materials. PZT has a perovskite crystal structure, each unit of which consists of a small tetravalent metal ion in a lattice of large divalent metal ions.

The surface modification of amine-grafted or epoxy-grafted MWCNTs composites in polyethylene polyamine or in liquid epoxy resin, respectively, was done by Kotsilkova et al. (2014). The amine epoxy grafted MWCNT composites produced an extra phase on the CNT walls, which was confirmed by the presence of a thick polymer layer. It positively affects the electrical conductivity and radiofrequency response properties. The epoxy-grafted composites could be applied for producing effective antistatic and electrostatic dissipation coatings.

Colloidal silver nanoparticles were prepared using starch by sonochemical reduction process (Kumar et al., 2014). Starch was used as a reducing as well as stabilizing agent. As-prepared silver nanoparticles were spherical, polydispersed, and amorphous with diameter ranging from

23 to 97 nm with mean particle size of 45.6 nm. These nanoparticles could be used for efficient catalyst in the synthesis of 2-aryl substituted benzimidazoles, which have numerous biomedical applications. The optimized conditions for this reaction are 10 mL of 1 mM $AgNO_3$, 25 mg starch, pH 11, and sonication for 20 min at room temperature.

Zhang et al. (2007) decomposed iron pentacarbonyl in *cis-trans* decalin by irradiating it with ultrasound for very short time. The high-energy ultrasonic irradiation promoted the process of crystallization, which further extended to monodisperse iron oxide nanoparticles at low temperature. They also studied the effect of concentration of the surfactant and the refluxing time on the size as well as the size distribution of iron oxide nanoparticles.

Hassanjani-Roshan et al. (2011) prepared iron oxide (α-Fe_2O_3) nanoparticles (smaller than 19 nm) via a sonochemical method. They also noticed that temperature, power of ultrasonication, and sonication time had a major effect on size and morphology of the iron oxide.

The sensing properties of $FeSbO_4$ nanoparticles were tested toward 1000 ppm and 1, 2, 4, and 8% hydrogen at 300 °C and 450 °C. It was observed that sample prepared at 300 °C exhibited a response of 76% toward 4% H_2 gas, whereas sample calcined at 450 °C showed a higher response of 91%.

Safaei-Ghomi et al. (2016) prepared benzo[*g*]chromenes using Fe_3O_4/polyethylene glycol (PEG) core/shell nanoparticle via ultrasonic irradiation. These nanoparticles have advantages such as mild reaction conditions, high-yield products, short reaction times, reusability of the catalyst and requirement of little catalyst loading as compared to conventional method.

6.6 NANORIBBONS

Wu et al. (2010) used sonochemical method for preparation of narrow and long graphene nanoribbons (GNRs) by chemically derived graphene sheets (GSs). The yield of GNRs was approximately 5 wt.% of the starting GSs (Fig. 6.8). The final GNRs have several micrometers in length, with ~75% being single-layer, and ~40% being narrower than 20 nm in width.

Graphene nanoribbons have properties of electronic and spin transport and if it has width <10 nm, then it has a band gap that can be utilized to fabricate field effect transition. But today also, preparing nanoribbons of

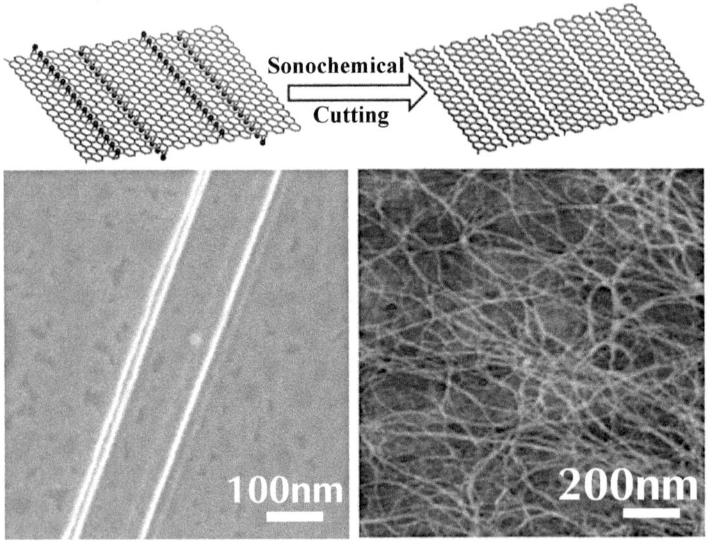

FIGURE 6.8 SEM image of GNRs.
Source: Wu et al., 2010, with permission.

excellent quality and of high volume is a major challenge. Jiao et al. (2010) made pristine few-layer nanoribbons from unzipping mildly gas-phase-oxidized MWCNTs using mechanical sonication in an organic solvent. As-prepared nanoribbons had some novel properties such as:

- High quality
- Smooth edge
- Low ratios of disorder to graphitic Raman bands
- The highest electrical conductance and mobility up to 5 e^2/h and 1500 $cm^2 \cdot V^{-1} \cdot s^{-1}$, respectively, for ribbons 10–20 nm in width.

Temperature also affects types of nanoribbons. At low temperatures, the nanoribbons exhibited phase-coherent transport and Fabry–Perot interference. It shows minimal defects and edge roughness. The yield of nanoribbons was approximately 2% of the starting raw nanotube soot material. It had even higher quality and narrow ribbons than nanoribbons prepared from the earlier methods. The relatively high-yield synthesis of pristine GNRs will make these materials easily accessible for a wide range of fundamental and practical applications.

Nanoribbons graphene of narrow, straight-edged striped and single-layer graphite proved to be useful in electronic devices. Nanroribbons graphene (less than 10 nm) prepared from 2-D starting material graphene showed semimetallic behavior, quantum confinement, and edge effects. As stated earlier, preparation of GNRs of smaller size than 10 nm with chemical precision remains still a significant challenge, whether it is synthesized by chemical, sonochemical, and lithographic methods as well as through the unzipping of CNTs (Cai et al., 2010).

Xie et al. (2011) prepared GNRs of widths ~ 10–30 nm from sonochemical unzipping of MWCNTs. The layer–layer stacking angles ranged from 0° to 30°. It includes average chiral angles near 30° of armchair orientation or 0° of zigzag orientation. A large fraction of GNRs is with bent and smooth edges, whereas the rest portion exhibited flat and less smooth edges with roughness of ≤ 1 nm.

When the width of a graphene nanoribbon is only a few nanometers, it possesses semiconducting properties that enable various high-end electronic applications. Yoon et al. (2015) used the dense and stable dispersion of a natural graphite made up of flavin mononucleotide (FMN) as a surfactant to form GNRs as small as 10 nm in width. In this sonochemical method, they unzipped graphene with a template of 1-D FMN supramolecular to form nanoribbon structure. Thermal annealing showed increase in the optical contrast and van der Waals interactions of the graphene film, which led to the enhancement of conductivity compared to the as-prepared graphene film, which is also better than that of reduced graphene oxide (rGO).

6.7 NANOSHEETS

By the reduction of Rh^{3+} salt on poly(ethylene oxide)/poly(propylene oxide)/poly(ethylene oxide) (PEO/PPO/PEO) triblock copolymer or pluronic-stabilized graphene oxide (GO) nanosheets with borohydride, highly water-dispersible rhodium–graphene nanocomposite was prepared by sonochemical process (Chandra et al., 2011). Rhodium nanoparticles of size 1–3 nm were homogeneously distributed throughout the GSs. They also observed some porous structures of GSs after the reduction of pluronic-stabilized GO in the presence of metal ions. This material has proved to be very effective for hydrogenation of arenes, especially for benzene as the substrate material at room temperature and 5 atm. pressure of hydrogen.

Iqbal et al. (2016) prepared gallium oxynitride (GaON) nanosheets at a reaction time of 24 h at 180 °C via direct sono-assisted solvothermal approach. Its band gap energy was observed to be ~1.9 eV. It was calculated from both absorption and diffused reflectance spectroscopy, which indicated stronger *p-d* repulsions in the Ga (3d) and N (2p) orbitals. This effect and chemical nitridation led to an upward shift of valence band and band gap reduction. The GaON nanosheets were found to have application in the field of photoelectrochemical cells (PECs) (Fig. 6.9). It was studied with a standard three-electrode system under 1 Sun irradiation in 0.5 M Na_2SO_4. These GaON nanosheets can be used as potential photocatalysts for solar water splitting also.

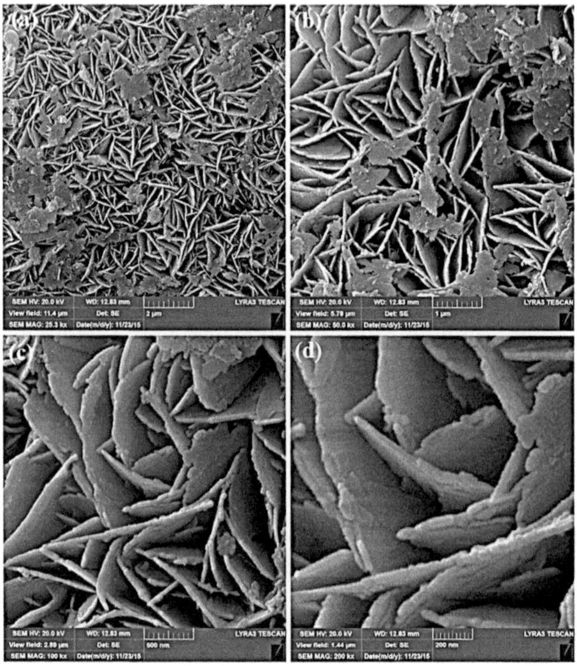

FIGURE 6.9 SEM images of GaON nanosheets (a, b) low-resolution micrographs at 2 and 1 μm showing growth patterns of nanosheets (c, d) higher-magnification micrographs show internal texture and defects attributing toward availability of high surface area. *Source*: Iqbal et al., 2016, with permission.

6.8 NANOSPHERES

Qi et al. (2015) used the adenosine-5′-triphosphate (ATP) disodium salt for the preparation of amorphous calcium phosphate (ACP) vesicle-like nanospheres by sonochemical method (Fig. 6.10). Here, ATP disodium salt was used as a biocompatible phosphorus source and stabilizer. The sample was prepared in mixed solvents of water and ethylene glycol. The ACP vesicle-like nanospheres showed essentially inappreciable toxicity to the cells in vitro. These could be used as anticancer drug nanocarriers. It exhibited a pH-responsive drug-release behaviour using doxorubicin (Dox) as a model drug. The ACP vesicle-like nanosphere drug delivery system showed a high ability to damage cancer cells. It shows that it has promising application in pH-responsive drug delivery.

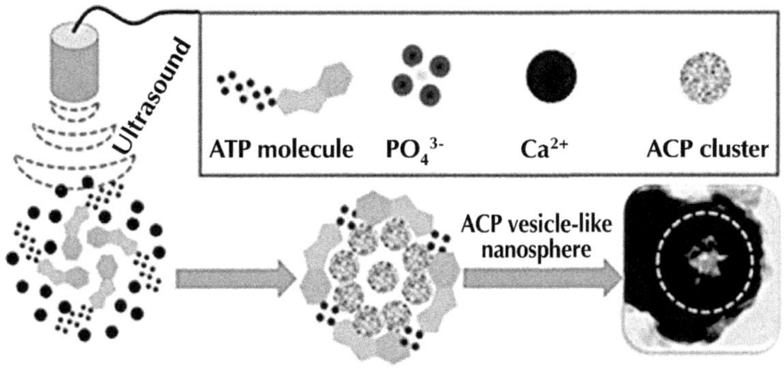

FIGURE 6.10 Sonochemical synthesis of ACP.
Source: Qi et al., 2015, with permission.

Dhas and Suslick (2005) prepared ceramic hollow spheres of MoS_2 and MoO_3. These ceramic samples were synthesized using MoS_2 and MoO_3 template on silica nanoparticles (diameters 50 to 500 nm), which were prepared under irradiation of ultrasound. MoS_2 and MoO_3 template on silica nanoparticles was further treated with acid to remove the silica core. These hollow sphere nanoparticles were found to be active catalysts for the hydrodesulphurization (HDS) of thiophene.

PbS hollow nanospheres were synthesized by a surfactant-assisted sonochemical pathway (Wang et al., 2006). The nanospheres were of 80–250 nm. Structural study showed that shells of the hollow spheres were made up of PbS nanoparticles with diameters of about 12 nm. This formation of the hollow nanostructure was explained by a vesicle-template mechanism, in which sonication and surfactant played important roles. They also coated uniform silica layers onto the hollow spheres via a modified Stöber method to enhance their performance for some promising applications.

Magnetic fluorescent nanospheres (MFNs) were prepared by Wu et al. (2011). MNFs were synthesized by sonochemical method in the presence of hydrophobic Fe_3O_4 nanopaticles. Further, the nanospheres were modified with rhodamine B through an electrostatic interaction. These nanosphere had 160 nm particle size and showed supermagnetic property. MNFs have their wide applications in medical imaging, drug targeting, and catalysis.

Zhu et al. (2007) doped 5.0 mol.% of Eu^{3+} dopant in BaF_2 nanocrystals via ultrasonic solution route. In this method, reactions between $Ba(NO_3)_2$, $Eu(NO_3)_3$ and KBF_4 under ambient conditions was carried out under ultrasound. They observed the effect of sonication time on the morphology of the $FBaF_2$:Eu^{3+} particles. The caddice-sphere-like particles with an average diameter of 250 nm were obtained with ultrasonic irradiation, whereas only olive-like particles were produced without ultrasonic irradiation. It was found that BaF_2:Eu^{3+} nanospheres crystallized well with a cubic structure.

Dai et al. (2013) synthesized pectin-coated Fe_3O_4 magnetic nanospheres (PCMNs) by sonochemical method. The pectin-coated nanosphere had magnetite content up to 63 %, with the saturation magnetization of about 32.69 emu/g. PCMNs could be applied in some biomedical applications.

6.9 NANOFLAKES

Ghosh et al. (2014) prepared ZnO nanoflakes via sonochemical method by using cetyltrimethylammonium bromide (a cationic surfactant) as a structure-directing agent. It was observed that surfactant concentration modified the morphology of ZnO from starlet-like 3D structures to 2D flakes. As-prepared sample of ZnO flakes was 200–400 nm wide and a few nanometres thick. Its band gap was found to be 3.37 eV.

Vabbina et al. (2015) prepared 2D ZnO nanoflakes in one step under ultrasound irradiation. It was fabricated on Au-coated substrates. As-prepared sample was label-free, highly sensitive and selective electrochemical immunosensors and it was used in the detection of cortisol. It was also found useful in sensing advantages as compared to bulk materials. Nanoflakes possess large surface area. ZnO-NSs also exhibited higher chemical stability, high catalytic activity, and biocompatibility. ZnO nanoflakes were having the lowest detection limit of 1 pM, which was almost 100 times better than conventional enzyme-linked immunosorbant assay (ELISA). It was also used to examine human saliva samples, and the results were compared with the performance of conventional (ELISA) method. Sensors could be integrated with microfluidic system and miniaturized potentiostat for point-of-care cortisol detection and such developed protocol can be used in personalized health monitoring/diagnostics.

Liu et al. (2012) prepared $CdWO_4$ nanoflakes without using mild template or any surfactant through ultrasonication method. Ultrasound plays a major role in the formation of nanocrystals. First nanoflakes are formed, which further change to nanorods on increasing ultrasonic period from 0.5 to 1.5 h. Zhou et al. (2007) also prepared Bi_2Te_3 hexagonal nanoflakes by an ultrasonic-assisted disproportionation route. It edge length was in the range of 150 nm to as small as 10 nm. The mechanochemical effects of the ultrasonic irradiation increased the rate of reaction and were also helpful in obtaining relatively small and uniform nanocrystals. They also studied electro-generated chemiluminescence of the as-prepared sample.

6.10 NANOWIRES

Chiniforoshan and Tabrizi (2015) synthesized Au nanowires of gold-containing polymers of the formula [(Au(I)-NCN-R-NCN-Au(III)-NCN-R-NCN-)]$_n$; (where R = biphenyl or NCN-R-NCN, 4,4′-dicyanamidobiphenyl, bpH_2). Au nanowires were of 20–100 nm, length up to several microns and showed thermal stability. As-prepared sample was used as gas sensor. It was highly sensitive toward CO with the detection limit (1 ppm) at room temperature.

β-$AgVO_3$ nanowires of 30–60 nm diameter and length 1.5–3 μ were synthesized by sonochemical reaction (Mao et al., 2008). As-prepared sample has electrochemical properties and was used as cathode materials for lithium-ion batteries. In the initial discharge and charge process, the as-prepared β-$AgVO_3$ nanowires showed the initial charge and discharge capacities of 69 and 102 (mAh)·g^{-1}, respectively.

Chen et al. (2006) synthesized ceria (CeO_2) nanowires in ambient air and alkali aqueous solution from CeO_2 nanoparticles without using any templates in the presence of ultrasound irradiation. In this reaction involving the fabrication of nanowires, alkali concentration and ultrasonic irradiation both played a critical role. The catalytic activity of Au/CeO_2 using CeO_2 nanowires as support for CO conversion showed better performance than bulk CeO_2.

Ultrasound was used as a driving force for both nucleation and dispersion for the preparation of selenium nanowires (Mayers et al., 2003). They observed that the trigonal Se seeds were formed during sonication and grew in hours at the expense of the amorphous Se colloids in a process very similar to Ostwald ripening. As-prepared nanowires were typically single crystals of trigonal Se. The morphology of these wires could be tuned by adjusting the reaction conditions.

6.11 NANOPOROUS MATERIALS

Talebi et al. (2010) studied addition of silver ions into the nanoporous (1.2 nm) zeolite lattice by ion-exchange route. Silver ions were introduced by 24 kHz ultrasonic waves. This range of ultrasound reduced silver ions to silver nanoparticles. After the reduction process, silver nanoparticles were placed in the cavities having size about 1 nm and also on the external surfaces of the zeolite, with sizes about less than 10 nm. Many methods have been used for the reduction of silver ions but the use of ultrasonic waves was found to be a new, simple, and size-controllable method with a high practical value, which does not need any complicated facilities. As a huge amount of energy is provided by the collapse of bubbles, this released energy causes the formation of reducing radicals that consequently reduce the silver ions. It was also observed that increasing the irradiation time and ultrasonic power do not affect the silver crystal growth significantly, but the extent of silver ion reduction was found to increase as the power of ultrasonic waves was increased.

Nanoporous amorphous-MnO_2 was fabricated by Hasan et al. (2015) by using solid manganese (II) acetate tetrahydrate in 0.1 M $KMnO_4$. Absorption study of as-prepared sample was done by remazol reactive dye or red 3BS dyes. In batch experiment, 10 ppm of remazol reactive dye was used, and the experiment was carried out at room temperature. Adsorption of remazol dye on 0.2 g synthesized nanoporous amorphous-MnO_2 showed almost 100% decolorization.

Zhan et al. (2014) prepared solid-state dodecyl perylene diimide (DDPDI) fluorescent materials (DDPDI/MCM-41) by introducing the DDPDI molecules into the nanopores of MCM-41 in toluene solution under ultrasonic irradiation. The DDPDI molecules were incorporated into the mesopores of MCM-41 in the monomeric or dimeric state after being treated with ultrasound (Fig. 6.11).

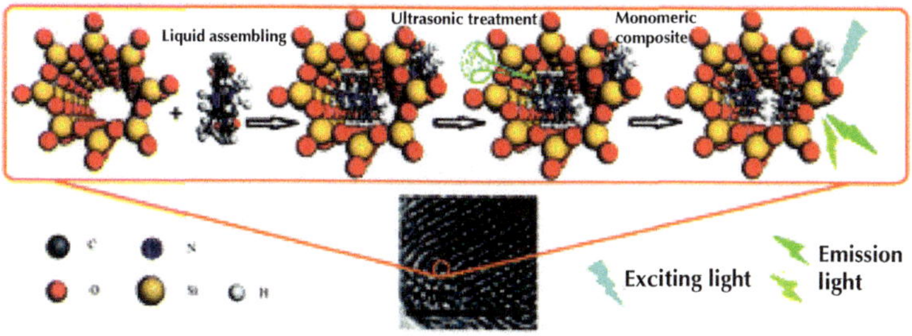

FIGURE 6.11 Incorporation of DDPDI into nanopores of HCM-41.
Source: Zhan et al., 2014, with permission.

6.12 GRAPHENE AND RELATED MATERIALS

6.12.1 GRAPHENE

Graphene consists of an atomic-scale honeycomb crystal lattice with sp^2-hybridized carbon. It possesses unique mechanical, electrical, chemical, and physical properties. But its zero band gap puts hurdle on its utilization in field-effect transistors (Fig. 6.12). Graphite is 2D sheet of sp^2-hybridized six-member carbon atoms ring (the graphene), which are regularly arranged. Graphene has excellent strength and firmness along its basal levels that reaches with approximately 1020 GPa. This is almost close to the strength value of diamond. It has varied methods of its synthesis and finds numerous applications as its electrical, physical and mechanical properties are increasing dramatically. Even the barrier properties of its polymer composites change a lot at extremely low loadings. Due to these characteristics, it is considered a material of interest and, therefore, it is a promising material for various industries

FIGURE 6.12 Graphene sheet.

producing composites, coatings, biosensors, drug delivery, microelectronics and so on.

Widenkvist et al. (2009) prepared suspensions of GSs by a combination of solution-based bromine intercalation with mild sonochemical exfoliation. Graphite formed suspensions of graphite flakes in water through ultrasound treatment. The delamination showed improvement by intercalation of bromine into the graphite before sonication. SEM and TEM showed a significant content of few-layer graphene with sizes up to 30 μm, corresponding to the grain size of the starting material.

Chia et al. (2014) prepared graphene (1 $mg \cdot ML^{-1}$) by exfoliating raw graphite sonochemistry in ethanol–water mixtures. The effective surface area of the graphene was found around 1000 $m^2 \cdot g^{-1}$ and the average thickness of GSs was <2 nm. The cyclic voltammetry test was used to study the electrochemical characteristics of the graphene and amperometric detections of hydrogen peroxide (H_2O_2), a by-product of most oxidase based enzymatic reactions. The graphene was deposited onto the surface of screen-printed carbon electrodes (SPCEs). This modified electrode showed increased electrocatalytic response of more than two-fold as compared to unmodified electrodes. The detection of hydrogen peroxide exhibited good stability with a more than 20 times improvement in sensitivity by optimizing the number of deposition layers on the electrode surface.

Yang et al. (2014) studied that graphene nanomesh (a new graphene nanostructure) with tunable band gap has outstanding performance. It can be used in electronic or photonic devices such as highly sensitive biosensors, new generation of spintronics and energy materials. The use of graphene nanomesh will add a step further in nanoscience and technology (Fig. 6.13).

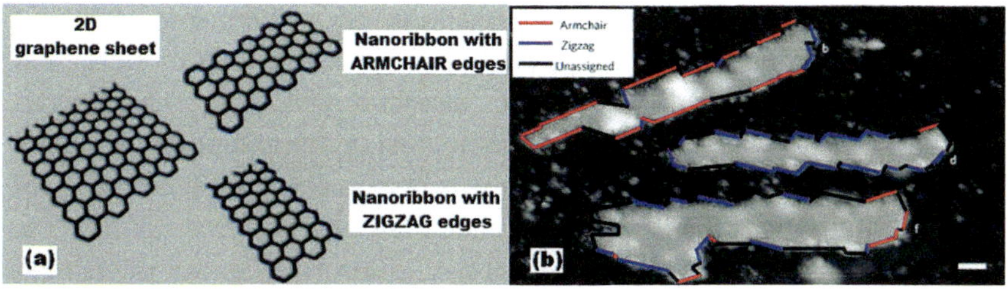

FIGURE 6.13 Graphene nanomesh.
Source: Yang et al., 2014, with permission.

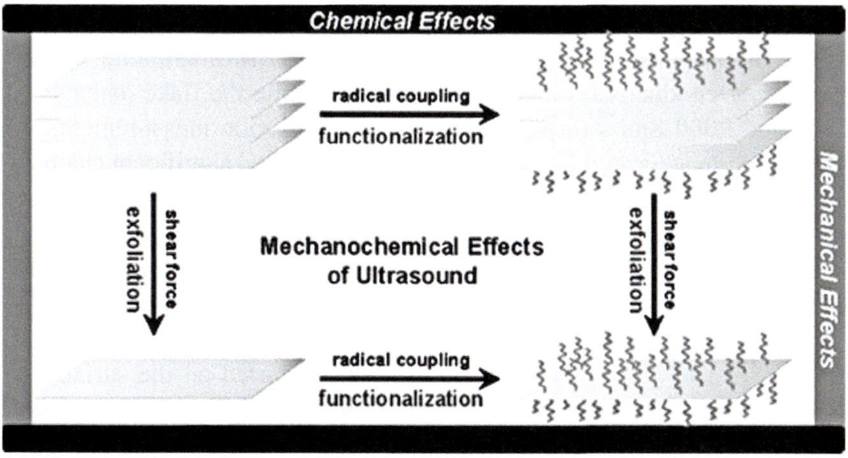

FIGURE 6.14 Mechanochemical effect.
Source: Xu and Suslick, 2011, with permission.

The reaction of graphite with styrene was studied under ultrasonic irradiation by Xu and Suslick (2011), which resulted in the mechanochemical exfoliation of graphite flakes to single-layer and few-layer GSs (Fig. 6.14). These sheets were combined with functionalization of the graphene with polystyrene chains. The polystyrene chain was formed by the radical polymerization of styrene initiated by this sonochemical process. The functionalized graphene was formed up to 18% weight. Such a one-step protocol can be used for preparation of graphene-based composite materials by using functionalization of graphenes with other vinyl monomers.

Sahoo et al. (2013) modified graphene as graphene/polypyrrole and graphene/polyaniline. They also observed the effect of this modification on thermal and electrical properties of nanocomposites. It was prepared by in situ-oxidative polymerization method using ammonium persulphate as an oxidant. The modified graphene was responsible for enhanced electrical conductivity of the as-prepared nanocomposites.

Graphitic precursors in the presence of chloroform under ultrasound resulted in the formation of exfoliated graphitic materials as mesographite and mesographene. The properties of material depend on the number of layers and exfoliation conditions. Mesographene was made up to nanoscaled two-dimensional graphene layers, and three-dimensional carbon nanostructures sandwiched between these layers, similar to those found in ball-milled and intercalated graphites. Mesographite exhibited higher conductivity than mesographene despite the flake damage as 2700 and 2000 Sm^{-1}, respectively. Optical absorption measurements of mesographite sonicated in different solvents exhibited significant changes in dispersion characteristics, and also indicated significant changes to mesoscopic colloidal behaviour (Srivastava et al., 2014).

Geetha et al. (2016) carried out sonochemical synthesis of highly crystalline graphene-gold (G-Au) nanocomposite. Ultrasound was used for exfoliation of graphite as well as the reduction of gold chloride to give the desired product. In situ growth of gold nanoparticles occurred on the surface of exfoliated few layer GSs. This nanocomposite was used to fabricate glassy carbon electrode (GCE) for an electrochemical sensor in the selective detection of nitric oxide (NO), which is prominently a cancer biomarker. G-Au modified GCE presented an enhanced electrocatalytic response for the oxidation of NO compared to other control electrodes. The combination of graphene and Au NPs provided exceptional electron transfer processes between the electrolyte and the GCE. It provides excellent sensing performance of the fabricated G-Au-modified electrode with stable and reproducible responses. They found that this nanocomposite will prove to be an electrode material in the sensitive and selective detection of NO.

6.12.2 GRAPHENE OXIDE

Park et al. (2012) used potassium permanganate to oxidize the carbon atoms in the graphene oxide (GO) support. Reduced manganese oxide and

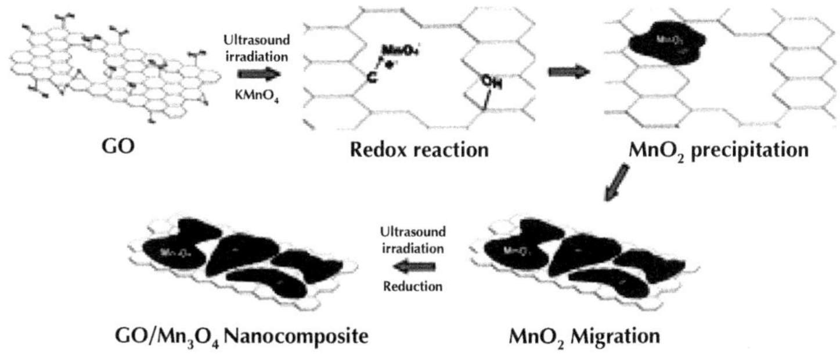

FIGURE 6.15 Graphene oxide–manganese oxide nanocomposite.
Source: Park et al., 2012, with permission.

graphene oxide (GO) have formed graphene oxide–manganese oxide nanocomposite (GO–Mn_3O_4) under ultrasound irradiation (Fig. 6.15). This ultrasound-assisted synthesis is a green chemical pathway as the number of steps is reduced along with temperature, and the rate of reaction is enhanced. It also helped in obtaining Mn_3O_4 phase. The effect of change in the ratio of permanganate to GO dispersion on coverage and crystallinity of Mn_3O_4 was also studied. It was used as a catalyst for poly(ethylene terephthalate) (PET) depolymerization into its monomer, bis(2-hydroxylethyl) terephthalate (BHET). The highest monomer yield was more than 95% with the nanocomposite containing the lowest amount of Mn_3O_4. On the other hand, PET glycolysis with the Mn_3O_4 without GO yielded 82.7% BHET.

Sonochemical approach was used for in situ reduction and direct functionalization of graphene oxide by Maktedar et al. (2016). They functionalized graphene oxide directly with tryptamine (TA) within 20 min. This was a rapid, robust, scalable, and nonhazardous sonochemical approach without using any hazardous acylating and coupling reagents. The direct covalent functionalization and formation of f-(TA) GO was observed. An enhanced thermal stability of f-(TA) GO was also observed. It was suggested that f-(TA) GO is emerging as an advanced functional biomaterial for thermal and biomedical applications.

The graphene oxide nanosheets (GOS) and hydrazine hydrate were used as template and reductant, respectively, to form highly dispersed 2D copper/rGO nanosheets (Cu/RGOS) nanocomposites under ultrasound irradiation. Here, uniform layers of Cu of 60 nm thickness have single-crystalline with (111) preferred crystalline direction and have tight binding with RGOS. In this electroless Cu plating, ultrasound helps in promoting interfacial bonding and avoids the chance of aggregation of 2D Cu/RGOS nanocomposing (Peng et al. 2014).

One-pot sonochemical synthesis of graphene oxide-wrapped gold nanoparticles (Au NPs) hybrid materials was carried out by Cui et al. (2015) using ethylene glycol as the reducing agent. This reaction took nearly 60 min to complete. The hybrid materials exist as spheres wrapped with gauze-like GO sheets via ionic interaction-based self-assembly. Graphene oxide sheet helped in the dispersion of the Au NPs. As-prepared sample showed enhanced photocatalytic activity under visible light irradiation due to synergistic effect of these two components in the hybrid materials.

6.12.3 REDUCED GRAPHENE OXIDE

Pt and Sn precursors were used for simultaneous loading of Pt and Sn (monometallic) and Pt–Sn (bimetallic) nanoparticles on the surface of reduced GO (rGO) (Anandan et al., 2012). They used low-frequency ultrasound of 20 kHz for this purpose. These reduced monometallic and bimetallic nanoparticles were spherical in shape with diameters of about 2–6 nm and also uniformly embedded on rGO sheets of few layers thickness. The electrocatalytic performance of the prepared nanomaterials was tested.

Shi et al. (2014) used sonochemical method for fabricating CuS nanoparticle-decorated rGO (CuS/rGO) composites. CuS nanoparticles were of 10–25 nm size and these were well distributed on the rGO nanosheets. Its efficiency was evaluated for photocatalytic degradation of methylene blue solution under natural light. As-prepared nanocomposites showed enhanced photocatalytic activity compared with pure CuS. This enhancement may be due to the enhanced light adsorption, strong absorption of the dye, and efficient charge transport after the introduction of rGO.

An aqueous solution containing a silver ammonia complex, $[Ag(NH_3)_2OH]$, and graphene oxide was used to prepare rGO uniformly decorated with silver nanoparticles (Ag NPs) via ultrasonic method. This involves simultaneous formation of cubic-phase AgNPs and the reduction of GO. The proportion of the precursors and the ultrasonic irradiation period were the deciding factors for the size of nanoparticles. A uniform distribution of ultrafine spherical Ag NPs with a narrow size distribution on the rGO sheets was obtained by silver ammonia complex and not by silver nitrate precursor. The average particle size of the silver with the narrowest size distribution was 4.57 nm. As-prepared modified GCE was found to have electrocatalytic activity toward the non-enzymatic detection of H_2O_2 with a wide linear range of 0.1–70 mM and a detection limit of 4.3 μM (Golsheikh et al., 2014).

TEM images and size distribution diagrams of AgNPs–rGO prepared by using solution with GO (1.0 $mg \cdot mL^{-1}$) to $Ag(NH_3)_2OH$ (0.04 M) volume ratio of 4 at different ultrasonic irradiation times of 15 min (a and b) and 30 min (c and d), and HRTEM image of AgNPs anchored on surface of GS (e).

rGO decorated with zinc sulphide nanospheres (ZnS NSs) was also fabricated under ultrasound by Golsheikh et al. (2015). Here, aqueous solution containing zinc acetate dehydrate, thioacetamide, and graphene oxide was irradiated with ultrasound. They reported that this involves simultaneous formation of cubic-phase ZnS NSs along with reduction of GO. Incorporation of rGO sheets with ZnS NSs reduced the electron-hole recombination and increased the surface area of the composite. Therefore, a significant enhancement in the photocatalytic degradation of methylene blue was observed with the ZnS NSs–rGO nanocomposite, as compared to the bare ZnS particles.

Fe_3O_4 nanoparticles were prepared by sonochemical method, where these nanoparticles were uniformly dispersed on the rGO sheets (Fe_3O_4/rGO) (Zhu et al., 2013). Hemoglobin, a biosensor, was prepared using modified GCE with the combination of Fe_3O_4/rGO. This biosensor showed an excellent electrocatalytic reduction toward H_2O_2 at a wide, linear range from 4×10^{-6} to 1×10^{-3} M, and with a detection limit of 2×10^{-6} M. The high performance of H_2O_2 detection was attributed to the synergistic effect of the combination of Fe_3O_4 nanoparticles and rGO, promoting the electron transfer between the peroxide and electrode surface.

6.13 MISCELLANEOUS

Dharmarathna et al. (2012) used water/acetone co-solvent for the synthesis of cryptomelane-type manganese octahedral molecular sieve (OMS-2) materials under nonthermal condition on ultrasonic irradiation. Co-solvent was used to decrease the reaction time and it was observed that it reduced reaction time by 50%. Calcination was not needed in this process to obtain the pure cryptomelane phase. OMS-2 material possesses greater amount of defect, surface area of 288 ± 1 m^2g^{-1} and small particle sizes in the range of 1–7 nm. Because of greater amounts of such defects, OMS-2 showed excellent catalytic efficiency for the oxidation of benzyl alcohol as compared to OMS-2 prepared by reflux methods and commercial MnO_2.

Yayapao et al. (2013) synthesized hexagonal wurtzite ZnO nanoneedles via ultrasonic solution method. It was doped with 0–3 % Ce, which resulted into a direct energy gap (3.00 eV). As-prepared nanoneedles were used for photocatalytic degradation of methylene blue.

Khanjani and Morsali (2013) carried out consecutive dipping of silk yarn in alternating bath of magnesium nitrate and potassium hydroxide to develop magnesium hydroxide nanostructures under ultrasound irradiation. It was observed that different factors such as ultrasound irradiation, concentration, pH, and sequential dipping steps affect the growth of the $Mg(OH)_2$ nanostructures. These nanoparticles were transformed to nanoneedles on decreasing pH from 13 to 8.

Sonochemical reaction of iron pentacarbonyl in water with protein bovine serum albumen (BSA) was carried out to prepared nanofiber and nanoneedle. These nanomaterials consisted of mixtures of goethite (α-FeOOH), lepidocrocite (γ-FeOOH), and hematite (α-Fe_2O_3). The reaction proceeds with thermal decomposition mechanism for iron pentacarbonyl with BSA acting as a templating agent (Bunker et al. 2007).

Water-dispersed two-dimensional materials in the form of platelets have wide applications in the field of inkjet printing on hard and flexible substrates. Water is an ideal dispersion medium as it is readily available in abundance and it is also of low cost. But, the hydrophobicity of platelet surfaces restricts its widespread use. Kim et al. (2015) reported that these 2D materials could be exfoliated and dispersed in water by increasing the temperature of the sonication bath. They also investigated inkjet printing on hard and flexible substrates as a potential application of water-dispersed 2D materials. Palanisamy et al. (2015) prepared metastable cadmium sulphide nanoplatelets (10 nm thickness) in a continuous flow sonochemical reactor. This reactor helps in micromixing of reagents. The unique shape and crystal structure of the obtained nanoplatelets were due to high localized temperatures within the sonochemical process. The particle size was more uniform in case of continuous sonochemical process in comparison to the batch sonochemical process and conventional synthesis processes.

In situ template approach for the synthesis of metal chalcogenides in the hollow form of spherical assemblies was also carried out in the presence of ultrasound (Xu et al., 2004). They observed self-assembly of metallic hydroxide particles into spherical templates generated in situ and subsequent surface crystal growth resulting in hollow spherical structures. In addition, ultrasound played a major role in the formation of the intermediate templates and also crystal growth process. This approach provides a simple and an efficient one-step procedure to the large-scale synthesis of hollow spherical nanostructures. Zhu et al. (2003) prepared CdSe hollow spherical assemblies by sonochemical method. In situ template amorphous $Cd(OH)_2$ helps in the development of primary CdSe nanoparticles on its surface and their assembly into hollow spherical structures.

One-pot synthesis of water-soluble Ag nanoclusters (NCs) was carried out by Liu et al. (2013) by using bovine serum albumin as a stabilizing agent and reducing agent in aqueous solution via a sonication process. This sample showed intensive electrogenerated chemiluminescence (ECL). It was used for ECL detection of dopamine with high sensitivity and a wide detection range. Dopamine concentration was found in the range of 8.3×10^{-9} to 8.3×10^{-7} mol·L^{-1} without the obvious interference of uric acid, ascorbic acid, and some other neurotransmitters, such as serotonin, epinephrine, and norepinephrine, and the detection limit was 9.2×10^{-10} mol·L^{-1} at a signal/noise ratio of 3. Xu and Suslick (2010) used polyelectrolyte, polymethylacrylic acid, as a capping agent for synthesis of highly fluorescent, stable, water-soluble Ag nanoclusters via sonochemical method. Sonication period, stoichiometry of the carboxylate groups to Ag^+, and polymer molecular weight were found to control the optical and fluorescence properties of the Ag nanoclusters.

PbS nanobelt was fabricated via a sonochemical method by using a solution of $PbCl_2$ and $Na_2S_2O_3$. As-prepared nanobelt possesses a width of 80 nm, length up to several millimeters, and width-to-thickness ratio of nearly 5 (Zhou et al., 2005).

Nanomaterials have a wide range of applications in widely different areas such as electronic device manufacture, catalyst, chemical sensors, hydrogen production, environmental remediation, or clean-up to bind with and neutralize toxins due to their optical, magnetic, electrical, and other properties, and so on like degradation of organic dye (Chen et al. 2012, Cheng et al. 2013, Hasan et al. 2015, Guo et al., 2016), solar-driven photoelectrochemical water-splitting (Iqbal et al. 2016), hydrogenation of arenes (Chandra et al. 2011), pH-responsive drug delivery (Qui et al. (2015), supercapacitor (Raj et al. 2016), photoelectrochemical devices and sensing (Tan et al. 2013), better chemiluminescence (Kaviyarasan et al. 2016), chemical and biological assay (Ruan et al. 2007), and so on. Nanomaterials have been also used as electrode-building material in rechargeable Li batteries, and also in the preparation of rare-earth-doped optical fibers (Gedanken, 2003).

REFERENCES

Abbas, M.; Takahashi, M.; Kim. C. J. *J. Nanopart. Res.* **2013,** *15*, 1354–1366.
Anandan, S.; Manivel, A.; Kumar, M. A. *Fuel Cells* **2012,** *12*(6), 956–962.
Bang, J. H.; Suslick, K. S. *Adv. Mater.* **2010,** *22*, 1039–1059.
Batule, B. S.; Park, K. S.; Kim, M. I.; Park, H. G. *Int. J. Nanomed.* **2015**, *10*, 137–142.
Bunker, C. E.; Novak, K. C.; Guliants, E. A.; Harruf, B. A.; Meziani, M. J.; Lin, Y.; Sun, Y. P. *Langmuir* **2007,** *23*(20), 10342–10347.
Cai. J.; Ruffieux, P.; Jaafar, R.; Bieri, M.; Blankenberg, S.; Muoth, M.; Seitsonen, A. P.; Saleh, M.; Feng, X.; Mullen, K.; Fasel, R. *Nature* **2010,** *466*(7305), 470–473.
Chandra, S.; Bag, S.; Bhar, R.; Pramanik, P. *J. Nanopart. Res.* **2011,** *13*(7), 2769–2777.
Chen, H. L.; Zhu, H. Y.; Wang, H.; Dong, L.; Zhu, J. J. *J. Nanosci. Nanotechnol.* **2006,** *6*(1), 157–161.
Chen, D.; Yoo, S. H.; Huang, Q.; Ali, G. *Chemistry (Weinheim Bergstrasse)* **2012**. DOI :10.1002/chem.201103787.
Cheng, P.; Ni, Y.; Yuan, K.; Hong, J. *Mater. Lett.* **2013,** *90*, 19–22.
Chia, J. S. Y.; Tan, M. T. T.; Khiew, P. S.; Chin, J. K.; Lee, H.; Bien, D. C. S. et al. *Chem. Eng. J.* **2014,** *249*, 270–278.
Chiniforoshan, L.; Tabrizi, H. *Dalton Trans.* **2015,** *44*(5), 2488–2895.
Cho, S. C.; Lee, H. S.; Sohn, S. H. *J. Nanosci. Nanotechnol.* **2012,** *12*(7), 6080–6084.
Chung, J. H.; Min, B. K.; Kim, Y. K.; Kim, K. H.; Kwon, T. Y. *J. Mol. Struct.* **2014,** *1076*, 698–703.
Chuong, T. V.; Dung, L. Q. T.; Luan, N. D. T.; Huy, T. T. *Int. J. Nanotechnol.* **2011,** *8*(3–5), 291–299.
Cui, Y.; Zhou, D.; Sui, Z.; Han, B. *Chin. J. Chem.* **2015,** *33*(1), 119–124.
Dai, J.; Wu, S.; Jiang, W.; Li, P.; Chen, X.; Liu, L. et al. *J. Magn. Magn. Mater.* **2013,** *331*, 62–66.
Dharmarathna, S.; King'ondu, C. K.; Perdrick, W.; Pahalagedara, L.; Suib, S. L. *Chem. Mater.* **2012,** *24*(4), 705–712.
Dhas, N. A.; Suslick, K. S. *J. Am. Chem. Soc.* **2005,** *127*(8), 2368–2369.
Gedanken, A. *Curr. Sci.* **2003,** *85*(12), 1720–1722.
Gedanken, A. *Ultrason. Sonochem.* **2004,** *11*(2), 47–55.
Geetha, B. R.; Muthoosamy, K.; Zhou, M.; Kumar, A. M.; Huang, N. M.; Manickam, S. *Biosens. Bioelectron.* **2016,** *87*, 622–629.
Ghosh, S.; Majumdar, D.; Sen, A.; Roy, S. *Mater. Lett.* **2014,** *130*, 215–217.

Gleiter, H. *Prog. Sci. Mater. Sci.* **1989,** *33*, 223–315.
Golsheikh, A. M.; Huang, N. M.; Lim, H. N.; Zakaria, R. *RSC Adv.* **2014,** *4*, 36401–36411.
Golsheikh, A. M.; Lim, H. N.; Zakaria, R.; Huang, N. M. *RSC Adv.* **2015,** *5*, 12726–12735.
Guo, X.; Fu, Y.; Hong, D.; Yu, B.; He, H.; Wang, Q.; Xing, L.; Xue, X. *Nanotechnology* **2016,** *27*(37). DOI: 10.1088/0957-4484/27/37/375704.
Hanifehpour, Y.; Safarifard, V.; Morsali, A.; Mirtamizdoust, B.; Joo, S. W. *Ultrason. Sonochem.* **2015,** *23*, 282–288.
Hasan, S. Z.; Yusop, M. R.; Othman, M. R. *AIP Conf. Proc.* **2015,** *1678*, 050006, http://dx.doi.org/10.1063/1.4931285.
Hassanjani-Roshan, A.; Vaezi, M. R.; Shokuhfar, A.; Rajabali, Z. *Particuology* **2011,** *9*(1), 95–99.
Iqbal, N.; Khan, L.; Yamani, Z. H.; Qurashi, A. *Sci. Rep.* **2016,** *6*. DOI: 10.1038/srep32319.
Jevtic, M.; Mitric, M.; Skapin, S.; Jancar, B.; Lgnjatovic, N.; Uskpkovic, D. *Cryst. Growth Des.* **2008,** *8*(7), 2217–2222.
Jiao, L.; Wang, X.; Diankov, G.; Wang, H.; Dai, H. *Nat. Nanotechnol.* **2010,** *5*, 321–325.
Jung, S. H.; Shin, N.; Kim, N. H.; Lee, K. H.; Jeong, S. H. *IEEE Trans. Nanotechnol.* **2010,** *10*(2), 319–324.
Katoh, R.; Tasaka, Y.; Sekreta, E.; Yumura, M.; Ikazaki, F.; Kakudate, Y.; Fujiwara, S. *Ultrason. Sonochem.* **1999,** *6*(4), 185–187.
Kaviyarasan, K.; Anandan, S.; Mangalaraja, R. V.; Sivasankar, T.; Ashokkumar, M. *Ultrason. Sonochem.* **2016,** *29*, 388–393.
Khanjani, S.; Morsoli, A. *Ultrason. Sonochem.* **2013,** *20*(2), 729–733.
Kharissova, O. V.; Kharisov, B. I.; Valdes, J. J. R.; Mendez, U. O. *Synth. React. Inorg., Met.-Org. Nano-Met. Chem.* **2011,** *41*(5), 429–448.
Kianpour, G.; Salavati-Niasari, M.; Emadi, H. *Ultrason. Sonochem.* **2013,** *20*(1), 418–424.
Kim, J.; Kwon, S.; Cho, D. H.; Kang, B.; Kwon, H.; Kim, Y.; Park, S. O. et al. *Nat. Commun.* **2015,** *6*. DOI: 10.1038/ncomms9294.
Kotsilkova, R.; Ivanov, E.; Bychanok, D.; Paddubskaya, A.; Demidenko, M.; Macutkevic, J.; Maksimenko, S. A.; Kuzhir, P. *Compos. Sci. Technol.* **2014.** DOI: 10.1016/j.compscitech.2014.11.004.
Kulkarni, S. K. Synthesis of Nanomaterial-I (Physical method). In *Nanotechnology: Principles and Practices*; Kulkarni S. K., Ed.; Springer: Basel, **2014**; pp. 55–76.
Kumar, B.; Smita, K.; Cumbal, L.; Debut, A.; Pathak, R. N. *Bioinorg. Chem. Appl.* **2014,** DOI: org/10.1155/2014/784268.
Li, F.; Liu, X.; Qin, Q.; Wu, J.; Li, Z.; Huang, X. *Cryst. Res. Technol.* **2009,** *44*(11), 1249–1254.
Liu, J.; Yang, L. L.; Wang, Y. J.; Wang, X. F. *Appl. Mech. Mater.* **2012,** 174–177, 413–416.
Liu, T.; Zhang, L.; Song, H.; Wang, Z.; Lv, Y. *Luminescence* **2013,** *28*(4), 530–535.
Maktedar, S. S.; Mehetre, S. S.; Avashthi, G.; Singh, M. *Ultrason. Sonochem.* **2016,** *34*, 67–77.
Mao, C.; Wu, X.; Zhu, J. J. *J. Nanosci. Nanotechnol.* **2008,** *8*(6), 3203–3207.
Mao, C. J.; Wang, D. C.; Pan, H. C.; Zhu, J. J. *Ultrason. Sonochem.* **2011,** *18*(2), 473–476.
Mayers, B. T.; Liu, K.; Sunderland, D.; Xia, Y. *Chem. Mater.* **2003,** *15*(20), 3852–3858.
Mirtamizdoust, B.; Shaabani, B.; Joo, S. W.; Viterbo, D.; Croce, G.; Hanifehpour, Y. *J. Inorg. Organomet. Polym. Mater.* **2012,** *22*(6), 1397–1403.
Palanisamy, B.; Paul, B.; Chang, C. H. *Ultrason. Sonochem.* **2015,** *26*, 452–460.
Pan, H. B.; Wai, C. M. *J. Phys. Chem. C* **2009,** *113*(46), 19782–19788.
Pan, H. B.; Wai, C. M. *New J. Chem.* **2011,** *35*, 1649–1660.

Park, G.; Bartolome, L.; Lee, K. G.; Lee, S. J.; Kim, D. H.; Park, T. J. *Nanoscale* **2012,** *4,* 3879–3885.

Peng, Y.; Hu, Y.; Han, L.; Ren, C. *Compos. Part B: Eng.* **2014,** *58,* 473–477.

Pokhrel, N.; Vabbina, P. K.; Pala, N. *Ultrason. Sonochem.* **2016,** *29,* 104–128.

Qi, C.; Zhu, Y. J.; Zhang, Y. G.; Jiang, Y. Y.; Wu, J.; Chen, F. *J. Mater. Chem. B* **2015,** *3,* 7347–7354.

Qi, C.; Zhu, Y. J.; Wu, Z. T.; Sun, T. W.; Jiang, Y. Y.; Zhang, Y. G.; Wu, J.; Chen, F. *RSC Adv.* **2016,** *6,* 9686–9692.

Qui, L.; Pol, V. G.; Wei, Y.; Gedanken, A. *New J. Chem.* **2004,** *28,* 1056–1059.

Raj, B. G. S.; Wu, J. J.; Asiri, A. M.; Anandan, S. *RSC Adv.* **2016,** *6,* 33361–33368.

Raja, M.; Ryu, S. H. *Nanoscale* **2009,** *9*(10), 5940–5945.

Ruan, C.; Eres, G., Wang, W.; Zhang, Z.; Gu, B. *Langmuir* **2007,** *23*(10), 5757–5760.

Safaei-Ghomi, J.; Eshteghal, F.; Shahbazi-Alavi, H. *Ultrason. Sonochem.* **2016,** *33,* 99–105.

Sahoo, S.; Bhattacharya, P.; Hatui, G.; Ghosh, D.; Das, C. K. *J. Appl. Polym. Sci.* **2013,** *128*(3), 1476–1483.

Segal, E.; Perelshtein, I.; Gedanken, A. *Ultrason. Sonochem.* **2015,** *22,* 30–34.

Shahverdizadeh, Hossein, G. *Main Group Chem.* **2016,** *15*(2), 179–189.

Shi, J.; Zhou, X.; Liu, Y.; Su, Q.; Zhang, J.; Du, G. *Mater. Lett.* **2014,** *126,* 220–223.

Singh, P. K.; Mittal, A.; Agarwal, V. International Conference on Nanomaterials and Devices, Processing and Applications, Roorkee, India, **2008.**

Solanki, M. S.; Ameta, R.; Benjamin, S. *Int. J. Adv. Chem. Sci. Appl.* **2015,** *3,* 24–30.

Srivastava, V. K.; Quinlan, R. A.; Agapov, A. L.; Kisliuk, A.; Bhat, G. S.; Mays, J. W. *Ind. Eng. Chem. Res.* **2014,** *53*(23), 9781–9791.

Tai, G.; Guo, W. *Ultrason. Sonochem.* **2008,** *15*(4), 350–356.

Talebi, J.; Halladj, R.; Askari, S. *J. Mater. Sci.* **2010,** *45*(12), 3318–3324.

Tan, S. T.; Umar, A. A.; Balouch, A.; Yahava, M.; Yap, C. C.; Salleh, M. M.; Oyama, M. *Ultrason. Sonochem.* **2013,** *21*(2), 754–760.

Vabbina, P. K.; Kaushik, A.; Pokhrel, N.; Bhansali, S.; Pala, N. *Biosens. Bioelectron.* **2015,** *63,* 124–130.

Wahab, R.; Ansari, S. G.; Kim, Y. S.; Seo, H. K.; Shin, H. S. *Appl. Surf. Sci.* **2007,** *253*(18), 7622–7626.

Wang, S. F.; Gu, F.; Lu, M. K. *Langmuir* **2006,** *22*(1), 398–401.

Wang, H.; Jia, L.; Bogdanoff, P.; Firchter, S.; Mohwald, H.; Shchukin, D. *Energy Environ. Sci.* **2013,** *6,* 799–804.

Warule, S. S.; Chaudhari, N. S.; Kale, B. B.; More, M. A. *Cryst. Eng. Commun.* **2009**, *11,* 2776–2783.

Widenkvist, E.; Boukhvalov, D. W.; Rubino, S.; Akhtar, S.; Lu, J.; Quinlan, R. A. et al. *J. Phys. D: Appl. Phys.* **2009,** *42*(11), 112003.

Wu, X.; Cao, M.; Hu, C.; He, X. *Cryst. Growth Des.* **2006,** *6*(1), 26–28.

Wu, Z. S.; Ren, W.; Gao, L.; Liu, B.; Zhao, J.; Cheng, H. M. *Nano Res.* **2010,** *3*(1), 16–22.

Wu, S.; Jiang, W.; Sun, Z.; Dai, J.; Liu, L.; Li, F. *J. Magn. Magn. Mater.* **2011,** *323*(16), 2170–2173.

Xie, L.; Wang, H.; Jin, C.; Wang, X.; Jiao, L.; Suenaga, K.; Dai, H. *J. Am. Chem. Soc.* **2011,** *133*(27), 10394–10397.

Xing, Y.; Li, L.; Chusuei, C. C.; Hull, R. V. *Langmuir* **2005,** *21*(9), 4185–4190.

Xu, H.; Suslick, K. S. *ACS Nano* **2010,** *4*(6), 3209–3214.

Xu, H.; Suslick, K. S. *J. Am. Chem. Soc.* **2011,** *133*(24), 9148–9151.

Xu, S.; Wan, H.; Zhu, J. J.; Xin, X. Q.; Chen, H. Y. *Eur. J. Inorg. Chem.* **2004,** *2004*(23), 4653–4659.

Xu, H.; Zeiger, B. W.; Suslick, K. S. *Chem. Soc. Rev.* **2013,** *42*, 2555–2567.

Yang, J.; Ma, M.; Li, L.; Zhang, Y.; Huang, W.; Dong, X. *Nanoscale* **2014,** *6*, 13301–13313.

Yaser, M. M.; Ghodrat, M.; Ali, M. *J. Coord. Chem.* **2010,** *63*(7), 1186–1193.

Yayapao, O.; Thongtem, S.; Phuruangral, A.; Thongtem, T. *Sonochemical Synthesis, Photocatalysis and Photonic Properties of 3% Ce-Doped ZnO Nanoneedles*. The 8th Asian Meeting on Electroceramics (AMEC-8), **2013**, 39, S563–S568.

Yoon, W.; Lee, Y.; Jang, H.; Jang, M.; Kim, J. S.; Lee, H. S.; Im, S.; Boo, D. W.; par, J.; Ju, S. Y. *Carbon* **2015,** *81*, 629–638.

Zak, K.; Majid, W. H.; Wang, H. Z.; Yousefi, R.; Golsheikh, M.; Ren, Z. F. *Ultrason. Sonochem.* **2013,** *20*(1), 395–400.

Zhan. X.; Cui, K.; Dou, M.; Jin, S.; Yang, X.; Guan, H. *RCS Adv.* **2014,** *4*, 47081–47086.

Zhang, X.; Zhao, H.; Tao, X.; Zhao, Y.; Zhang, Z. *Mater. Lett.* **2005,** *59*(14–15), 1745–1747.

Zhang, G. Q.; Wu, H. P.; Ge, M. Y.; Jiang, Q. K.; Chen, L. Y.; Yao, J. M. *Mater. Lett.* **2007,** *61*(11–12), 2204–2207.

Zhou, S. M.; Zhang, X. H.; Meng, X. M.; Fan, X.; Lee, S. T.; Wu, S. K. *J. Solid State Chem.* **2005,** *178*(1), 399–403.

Zhou, B.; Liu, B.; Jiang, L. P.; Zhu, J. J. *Ultrason. Sonochem.* **2007,** *14*, 229–234.

Zhu, J. J.; Xu, S.; Wang, H.; Zhu, J. M.; Chen, H. Y. *Adv. Mater.* **2003,** *15*(2), 156–159.

Zhu, L.; Li, Q.; Li, J.; Liu, X.; Meng, J.; Cao, X. *J. Nanopart. Res.* **2007,** *9*(2), 261–268.

Zhu, S.; Guo, J.; Dong, J.; Cui, Z.; Lu, T.; Zhu, C. et al. *Ultrason. Sonochem.* **2013,** *20*(2), 872–880.

CHAPTER 7

POLYMERS

KIRAN MEGHWAL[1], GUNJAN KASHYAP[2], RAKSHIT AMETA[3]

[1]*Department of Chemistry, M. L. Sukhadia University, Udaipur, India*
E-mail: meghwal.kiran1506@gmail.com

[2]*Department of Chemistry, M. L. Sukhadia University, Udaipur, India*
E-mail: gunjankashyap0202@yahoo.in

[3]*Department of Chemistry, J.R.N. Rajasthan Vidyapeeth (Deemed to be University), Udaipur, India, E-mail: rakshit_ameta@yahoo.in*

CONTENTS

7.1 INTRODUCTION

Polymer is a large molecule or macromolecule, composed of many repeating subunits. Because of their broad range of properties, both synthetic and natural polymers play an essential and ubiquitous role in day-to-day life. Polymers range from familiar synthetic plastics to natural biopolymers (DNA and proteins) that are fundamental to the biological structure and function. Polymers (natural and synthetic) are formed via polymerization of many small molecules, known as monomers. Their consequently large molecular mass relative to small molecules produces unique physical properties, including toughness, viscoelasticity and a tendency to form glasses and semicrystalline structures rather than crystals. Polymerization is the process of combining many small molecules (monomers) into a covalently bonded chain or network. During the polymerization process, some chemical groups may be lost from each monomer. Synthetic methods are generally divided into two categories: step-growth polymerization and chain-growth polymerization. The essential difference between the two is that in chain-growth polymerization, monomers are added to the chain one at a time only, as in polyethylene, whereas in step-growth polymerization chains of monomers may combine with one another directly, such as in polyester. Synthetic polymerization reactions may be carried out with or without a catalyst.

Polymers are among the most important materials produced commercially and these are used in a wide range of applications. Their synthetic methods are generally straightforward, but there are often several processing steps and these steps depend critically on the bulk properties such as melting temperature and viscosity. At this point, there is an extra consideration of the molecular weight also. Any polymer will contain a distribution of chain lengths and this plays an important role in determining the properties of that material.

Most effects in sonochemistry arise from cavitation. As a result of cavitation radicals are produced, which are used in making polymers. However, the exact origin of cavitation effects is relatively unimportant to a polymer chemist, whether from thermal "hot spots" or electrical or corona discharges. Most of the effects involved in controlling molecular weight can be attributed to the large shear gradients and shock waves generated around collapsing cavitation bubbles (Basedow and Ebert, 1977; Price et al., 1990).

Organic polymers are prepared under ultrasound exposure. These are polymerized by a variety of mechanisms, but the most widely used method is initiation by radicals undergoing chain growth or addition reactions. A majority of organic polymers are prepared from monomers containing a reactive double bond (e.g. α-olefins and vinyl monomers).

The mechanistic work of Price et al. (1992) suggested high-molecular weight polymers are formed under sonochemical polymerization early in the reaction, but the average chain length shortens at longer times. This is caused by the onset of the degradation process, once sufficient long chains are formed. Miyata and Nakashio (1975) and Price et al. (1990) found that ultrasound affects the propagation, and hence, differences in the sequences of the two monomers along the chain were observed.

Ultrasound irradiation is used in polymer chemistry to initiate radical-mediated polymerizations and also to cleave polymer chains. Ultrasound (high-intensity energy) provides a facile and multipurpose synthetic tool to generate novel materials with unusual properties. Ultrasound irradiation has been utilized in the synthesis of polymers and composites. As compared to traditional energy sources, ultrasonic irradiation offers rather some uncommon reaction conditions (mainly enormously high temperatures and pressures in liquids, but for short periods) that cannot be reached at, otherwise using other methods. It also provides some control over the molecular weight, tacticity and polydispersity (Bhavase et al., 2012). Polymerization can be enhanced by the extreme conditions produced by ultrasound and these are known to act as a special initiator to allow chemical bonds to break.

Ultrasound, the high-intensity energy provides a facile and multipurpose synthetic tool to generate some novel materials with unusual properties. Ultrasound irradiation has been proved to be an effective method for the synthesis of polymers and composites (Bradley and Griser, 2002; Xia and Wang, 2002). Polymerization of styrene has been proposed by Kojima et al. (2001), Kobayashi et al. (2008), acrylonitrile, acrylamide, and acrylic acid by (Xiuyuan et al. (2001). Ultrasound polymerization of methyl methacrylate (MMA) (Price et al., 1992), styrene (Biggs and Griaser, 1995), and *n*-butyl acrylate (Xia et al., 2002) was also studied. Ultrasonic irradiation in copolymerization has also been widely studied since it does not only allow shorter reaction times, but it is also easy to handle. Particularly, this method is very useful for those copolymers, which are difficult to prepare by conventional methods (Fujiwara et al., 1992, 1999).

The susceptibility of a polymer to degradation depends on its structure. Polymer degradation means a change in the properties, that is, tensile strength, colour, shape or molecular weight of a polymer or polymer-based product under the influence of one or more environmental factors such as heat, light, chemicals and, in some cases, galvanic action. It is due to the scission of polymer chain bonds via hydrolysis, thereby leading to a decrease in the molecular mass of the polymer. The degradation of polymers via ultrasound is best known in solutions.

Sonochemistry opens up a new frontier of chemistry as ultrasound leads to some new chemical reactions as well as the enhanced rate of a chemical reaction (Xu et al., 2013). The ultrasound is high-intensity radiation, which can accelerate and break the aggregation of the heterogeneous liquid–liquid chemical reactions. The particle size can be reduced by using the techniques such as dispersion, crushing, emulsifying and activation effect, and, thus, it has a better control on the morphology of particles, especially on the hard solid particles.

7.2 POLYANILINE

Polyaniline (PANI) is a polymer of aniline. Electrical properties of PANI are reversibly controlled by charge-transfer doping and protonation. It is environmentally stable and inert and has electrical, electrochemical, and optical applications. It finds use in cell phones, calculators, LCD technology, and so on.

Advanced conducting polymeric materials can be prepared by ultrasonic-assisted polymerization. PANI is an easily aggregated substance and, therefore, this method is especially useful for the preparation of PANI–EB because in strong acidic medium, conductive PANI can be easily prepared. Many researchers have adopted the strategy to co-dope PANI with both inorganic and organic acids so as to synthesize semiconducting PANI-EB so as to accomplish good conductivity and thermal stability (Bhadra et al., 2007; Trovati et al., 2010) efficiently.

Xia and Wang (2003) prepared PANI/nano–SiO_2 particle composites through polymerization of aniline by ultrasonic irradiation in the presence of two types of nano–SiO_2 (porous nanosilica and spherical nanosilica).

A sonochemical synthesis of PANI nanotubes containing Fe_3O_4 nanoparticles has been carried out by Lu et al. (2006), starting from aqueous solutions of aniline, ammonium peroxydisulphate (APS), phosphoric acid (H_3PO_4), and 10 or 20 wt.% of Fe_3O_4. It was shown that Fe_3O_4 nanoparticles were embedded in PANI nanotubes. The mechanism of the formation of PANI/Fe_3O_4 nanotubes could be attributed to the ultrasonic irradiation and the H_3PO_4–aniline salt template. On polymerizing the solution of aniline, APS, phosphoric acid, and 10 or 20 wt. % Fe_3O_4 nanoparticles under magnetic stirring, no PANI/Fe_3O_4 nanotubes could be obtained, but only PANI/Fe_3O_4 powder was produced. However, PANI/Fe_3O_4 composite nanotubes were obtained under the ultrasonic irradiation.

Ho et al. (2009) synthesized nanotubular, nanofibrouspolyaniline (PANINT). These were prepared by an emulsion polymerization in the presence of *n*-dodecylbenzene sulphonic (DBSA) and hydrochloric acids. The building of the tubular structure was based on the tubular micelles from the accumulation of micelles at high surfactant concentration. It leads to early stage of centipede-like and eventual tubular morphologies. The solid rods were filled with helical emeradine base (nano-EB) molecules associated with intermolecular H-bonding. These solid rods of nano-EB went on cross-linking and carbonization after 200–300 °C by opening the quinoid rings and the inter-molecular H-bondings were destroyed with nano-EB molecules cross-linking into ladder-like structure.

Kowsari and Faraghi (2010) applied ultrasound for the synthesis of PANI–Y_2O_3 nanocomposites. These nanocomposites were prepared under controlled conductivity with the assistance of an ionic liquid (IL). Ultrasound energy and the IL replace conventional oxidants and metal complexes in promoting the polymerization of aniline monomer, respectively. This nanocomposite consists of regular solid microspheres of average diameter 3–5 nm, which were covered with some 40 nm nanoparticles. Under certain different polymerization conditions, PANI nanofibres and nanosheet were also obtained. This method opens a new pathway for the preparation of nanoscale-conducting polymer nanocomposites with the help of ILs. It was observed that the conductivity of the product varied with the mass ratio of aniline monomer to Y_2O_3 and IL. This composite was more thermally stable than pure PANI. The reaction conditions have been optimized by varying parameters such as the aniline/Y_2O_3 ratio and also the type and amount of IL used. The effect of the ultrasonic irritation time and frequency on the morphology, conductivity, and yield was also observed.

Silver wire/PANI composites were synthesized by De Barros and Azevedo (2010) via a sonochemical process using aniline monomer, silver nitrate, and anilinium nitrate aqueous solution in the presence of isopropyl alcohol. Isopropyl alcohol has a remarkable effect on the growth and the morphological structure of reduced silver. The reaction time is also an important parameter, which allowed to obtain silver wires with a mean diameter of 120 nm and a medium length of 4 μm instead of the spherical particles obtained by its synthesis without alcohol. Bigger silver wires with a mean diameter of 1.5 μm and length of 85 μm were obtained on keeping reaction medium in a dark place after the completion of sonication process. The reduced silver was highly crystalline, and the polymer obtained was PANI in the emeraldine salt form.

Ai and Jiang (2011) applied ultrasound for the synthesis of PANI nanosticks. They successfully synthesized PANI nanosticks by a facile template-free method in aqueous solution using ultrasonic irradiation. These synthesized PANI nanosticks found potential application in water treatment for the removal of Cu(II) ions by adsorption. In the beginning, the adsorption rate is relatively fast as the Cu(II) is adsorbed on the exterior surface of PANI nanosticks, but after some time, the adsorption capacity becomes constant because the free exterior surface of adsorption sites of the adsorbent gets diminished.

Barkade et al. (2011) also synthesized PANI–Ag (polyaniline/Ag) nanocomposite using ultrasound radiation. These nanocomposites were prepared under in situ miniemulsion polymerization of aniline along with different loading of silver nanoparticles. PANI–silver nanocomposites have potential application as sensor for ethanol vapour. Colloidal silver nanoparticles were synthesized by reduction of silver nitrate with sodium borohydride using sodium dodecyl sulfate. The sonochemically prepared PANI–silver (PANI/Ag) nanocomposites and the average particle size of Ag nanoparticles consists of open spherical aggregates of 50 nm and ~5–10 nm in diameter, respectively.

Manuel et al. (2014) used ultrasonic irradiation for synthesis of PANI nanofibres involving oxidative polymerization of aniline monomer, dilute hydrochloric acid, and doping with ammonium persulphate. As-prepared PANI was evaluated for its suitability in lithium cells after doping with lithium perchlorate salt. DC conductivity of doped PANI nanofibres was measured at different temperatures and compared with the conductivity of emeraldine base form of PANI, which was found to be 5.0×10^{-11} S·cm^{-1} after doping with ammonium persulphate, but it was dramatically improved to 1.0×10^{-2} S·cm^{-1} at room temperature after doping with lithium salt. Electrochemical properties of nanofibres were evaluated, where a remarkable increase in cycle stability was achieved on comparing with that to PANI prepared by simple oxidative polymerization of aniline. This cell delivers a stable and higher discharge capacity even at 2 C—rate compared to the cell prepared with bulk PANI doped with $LiClO_4$.

PANI nanocomposites (PANI/$NiCoFe_2O_4$) were synthesized with and without ultrasonic irradiation (Chitra et al., 2014). They observed that PANI nanocomposites (PANI/$NiCoFe_2O_4$) were obtained by in situ chemical polymerization of aniline in the presence of $NiCoFe_2O_4$ nanoparticles (20%, 10% w/w of fine powders). Ferromagnetic nature of PANI/$NiCoFe_2O_4$ nanocomposites was confirmed by vibrating sample magnetometer.

Jangid et al. (2014) reported ultrasound-assisted synthesis of conducting functional polymers such as PANI and ring-substituted PANIs, which was conventionally prepared by the stoichiometric oxidative polymerization of aniline (ANI) and its derivatives with ammonium persulphate. Here, PANIs were synthesized by oxidative polymerization of aniline in the presence of different dopants such as inorganic (orthophosphoric acid, PA) and organic acids (hippuric, HA and tosic acids, TA). It was observed that the synthesized PANIs form of emeraldine base show reasonably good room temperature conductivity and better thermal stability. *N,N'*-Dimethylformamide (DMF)was used as a solvent in the UV-Visible spectra of the PANI-PA, PANI-TA and PANI-HA in UV-vis absorption spectroscopy.

Bhanvase et al. (2015) synthesized PANI and PANI/$ZnMoO_4$ nanocomposite with different loading of $ZnMoO_4$ (ZM) nanoparticles using ultrasound-assisted in situ semibatch emulsion. ZM nanoparticles were functionalized for better compatibility with PANI by using myristic acid (MA). During this process, the cavitational effects induced due to ultrasonic irradiations show significant enhancement in the dispersion of functionalized ZM nanoparticles into the PANI. The presence of ZM nanoparticles in PANI/ZM nanocomposite showed significant improvement in the different types of physical properties like mechanical (cross-cut adhesion), thermal, anticorrosion and sensing properties of PANI/ZM nanocomposite/alkyd coatings over PANI/alkyd and neat alkyd resin coating. PANI (*p*-type)/ZM (*n*-type) heterojunction has the improved LPG sensing ability with minimized response time to sense LPG significantly as compared to neat PANI.

7.3 POLYPYRROLE

Polypyrrole (PPy) is a type of organic polymer formed by the polymerization of pyrrole. These conductive polymers are used in electronic devices and chemical sensors. It is a potential vehicle for drug delivery. Together with other conjugated polymers such as PANI, poly (ethylenedioxythiophene) and so forth, polypyrrole may also find its use as a material for artificial muscles.

Very fine and stable uniform monomer droplets were created in the nanometer range in the ultrasound-assisted miniemulsion process, due to the cavitational activity quite close to the interface of the immiscible organic liquid phase. This emulsion polymerization has improved polymerization rate, narrow molecular weight, and particle size distribution, as well as higher monomer conversion and several other advantages are these over the conventional emulsion polymerization process (Hu et al., 2011; Teo et al., 2008). There is a rapid dissociation of the initiator, which generates an enhanced quantum of radicals during the miniemulsion polymerization process. It is all due to the mechanical effects of ultrasound (intense turbulence associated with shear and liquid circulation) (Bang and Suslick, 2010; Sivakumar and Gedanken, 2005; Bhanvase et al., 2011).

Qiu et al. (2006) prepared polypyrrole (PPy)/Fe_3O_4 magnetic nanocomposites using ultrasonic irradiation. These nanocomposites were prepared by chemical oxidative polymerization of pyrrole in the presence of Fe_3O_4 nanoparticles. PPy was found deposited on the surface of Fe_3O_4 nanoparticles whereas Fe_3O_4 nanoparticles were dispersed at the nanoscale, which leads to the formation of polypyrrole-encapsulated Fe_3O_4 composite particles. The doping level of PPy in PPy/Fe_3O_4nanocomposite was found to be higher than that of neat PPy. These composites possess good electrical and magnetic properties. It was observed that the magnetization increases and the conductivity first increases and then decreases with an increase in the Fe_3O_4 content. The conductivity reached its maximum value of 11.26 S cm^{-1} with Fe_3O_4 content 40 wt.%, which is about nine times higher than that of neat PPy. The introduction of Fe_3O_4 nanoparticles enhanced the thermal stability of PPy/Fe_3O_4 composite.

Deshpande et al. (2009) synthesized core/shell structure containing organic–inorganic phases using ultrasound-assisted in situ miniemulsion polymerization process, but they obtained a range of grain sizes rather than monodispersed particles. The sensing performance of polypyrrole was improved by incorporating, zinc oxide nanoparticles into polypyrrole.

Ansari and Mohammad (2011) synthesized (pTSA/Pani:TiO_2) an electrically conductive polymer, *p*-toluene sulphonic acid-doped PANI/titanium dioxide nanocomposites using in situ polymerization of aniline along with TiO_2 nanoparticles. It was observed that this polymer showed superior ammonia-sensing capacity as compared to the pure PANI. Tang et al. (2012) synthesized PANI/ZnO organic–inorganic hybrids by chemical oxidative polymerization, and used these for detecting VOCs at low temperature.

Roy et al. (2012) developed a rapid and direct sonochemical method to synthesize polypyrrole salt. As-synthesized particles were found to be irregular in shape and the particle morphology with diameter in the range 200–500 nm was observed. It was found that polyprrole salt has effective chromium(VI) sorption–desorption properties and the sorption of chromate did not change the morphology significantly.

The sonochemical synthesis of hybrid nanocomposites titania (anatase):polypyrrole (TiO_2:PPy) wasreported by Arenas et al. (2013). Theses hybrid nanocomposites were prepared from a simple, one-step, ultrasonic-assisted synthesis. The properties of these crystalline nanocomposites were compared with others, which were prepared under without any ultrasound assistance, but it required a hydrothermal treatment to yield crystalline titania(anatase) in the nanocomposite (TiO_2:PPy) at low temperature (130 °C) and in a short time (3 h). Strong interaction between the secondary amine groups (N-H) of PPy and the oxygen from $TiO_2$is detected in crystalline nanocomposites. Absorption in the visible region (λ_{max}=670 nm) by this composite indicates that these are good candidates for harvesting energy in solar cells. An increase in the photocurrent was observed for the devices with such nanocomposites prepared from the ultrasound-assisted synthesis.

An ultrasound-assisted in situ miniemulsion polymerization of pyrrole in the presence of ZnO nanoparticles was reported (Barkade et al., 2013). Synthesized polypyrrole–zinc oxide (PPy/ZnO) hybrid nanocomposites were of uniform size around 100 nm and these can be used as a sensor for the detection of liquefied petroleum gas (LPG). (Fig. 7.1).

ZnO nanoparticles were encapsulated by the polymerized polypyrrole as evident for Fourier Transform Infrared Spectroscopy (FTIR). They indicated that the PPy/ZnO composite particles were of the crystalline

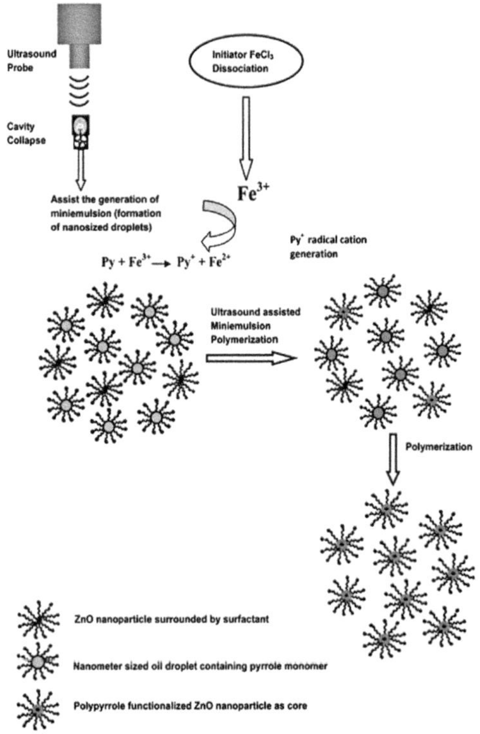

FIGURE 7.1 Mechanism of functionalization of ZnO nanoparticles using polypyrrole in the presence of ultrasound during miniemulsion polymerization.
Source: Barkade et al., 2013, with permission.

nature of ZnO (Wurtzite) giving a well-defined hybrid nanocomposite structure suitable as an LPG sensor. The controlled size of the hybrid particles obtained using this ultrasound synthesis strategy minimizes its response time to sense the LPG significantly (2.2 min for PPy/ZnO). A LPG sensing mechanism for the PPy(ptype)/ZnO(n-type) heterojunction has been presented through a change in the height of the barrier potential.

7.4 POLYTHIOPHENE

Polythiophenes (PTP) are polymerized thiophenes, which is a sulphur heterocycle. These polymers can become conducting on oxidation. Optical properties (colour changes) and conductivity of polythiophenes change, by twisting the polymer backbone, disrupting conjugation making conjugated polymers, and so on. Polythiophenes have found their wide applications in field-effect transistors, solar cells, batteries, chemical sensors, and so on.

Barkade et al. (2013) synthesized polythiophene (PTP)-coated SnO_2 nano-hybrid particles using an ultrasound assisted-in situ oxidative polymerization of thiophene monomers. Control experiments have also been performed in the absence of ultrasound to know the effect of ultrasonic irradiation. Conjugation and chemical interactions between PTP and SnO_2 particles have been studied using FTIR spectroscopy. This hybrid nanocomposite has the potential application as chemical sensors due to the strong synergetic interaction between the SnO_2 nanoparticles and polythiophene. LPG having high sensitivity at room temperature could be detected by PTP/SnO_2 hybrid sensors. Hybrid composite containing 20 wt.% SnO_2 showed the maximum sensitivity at room temperature. The sensing mechanism of PTP/SnO_2 hybrid nanocomposites to LPG was mainly attributed to the effects of p–n heterojunction between PTP and SnO_2.

7.5 POLY (METHYL METHACRYLATE)

Poly (methyl methacrylate) (PMMA) is also known as acrylic or acrylic glass. It is a transparent thermoplastic, which is often used in sheet form as a strong and lightweight material. PMMA has been used in a wide variety of fields as rear lights, instrument clusters for vehicles, appliances and lenses for glasses, LCD screens, medical, dental, and so on.

Choi et al. (2007) polymerized MMA in the presence of polycarbonate by ultrasonic irradiation. The homopolymerization of MMA formed PMMA, while copolymerization with macroradicals of PC formed PC and PC-PMMA copolymer; thereby the resulting products consisted of a mixture of homopolymers of PMMA, PC and PC-PMMA copolymer.

The PMMA synthesis by emulsion polymerization under ultrasonic radiation was also carried out by Albano et al. (2010). These products were obtained by varying the concentration of the anionic surfactant and monomer concentration. At low concentrations of the surfactant (0.5% p/v), the latex particles presented a rough surface but at high surfactant concentration, nanometric spherical particles were obtained with a smooth surface.

Patra et al. (2012) studied ultrasound-assisted emulsifier-free emulsion polymerization and synthesized PMMA/clay nanocomposites. The dispersion of the clay layers with polymer matrix can be enhanced by using different power and frequencies of ultrasound waves. The structural information of the PMMA/clay nanocomposites by XRD revealed that the interlayer spacing increased with clay loading. The magnitude of dispersion of the clay in the polymer matrix was detected by transmission electron microscopy (TEM). Various physical properties such as Young's modulus, breaking stress, elongation at break, toughness, yield stress and yield strain of the nanocomposites as a function of different clay concentrations and ultrasonic power were also measured. Oxygen permeability of the samples was studied, and the oxygen flow rate was found to be reduced by the combined effect of clay loading and ultrasound. The flame-retardant property of the nanocomposites due to clay dispersion was also investigated by measuring limiting oxygen index.

Prasad et al. (2013) observed polymerization of MMA and MMA–$CaCO_3$ systems that possessed the combined effects of sonochemical and conventional chemical initiation. Ultrasound and conventional initiation by potassium persulphate (KPS) were combined to increase the final conversion for the MMA and MMA–$CaCO_3$ systems. An increase of 15% for the MMA only system (from 72 to 87%) and of 10% for the MMA–$CaCO_3$ system (from 76 to 86%) was observed as compared to the initiation by KPS alone. Also, an increase of 18% (from 69 to 87%) and 20% (from 66 to 86%) for the MMA only and MMA–$CaCO_3$ systems, respectively, was observed for the combined initiation as compared to that of ultrasound alone. Although the size of all particles synthesized varied in between 60 and 130 nm, for both PMMA and PMMA–$CaCO_3$ systems, an excellent dispersion ability of ultrasound helped us to obtain narrow size distribution and smaller average sizes.

Vargas-Salazar et al. (2015) synthesized polymer of MMA with the reactive surfactant polyoxyethylene (10) alkylphenyl ether ammonium sulphate (HBC10) in the presence of ultrasound at 50 °C by using semi-continuous heterophase polymerization technique. The effects of HBC10 concentrations and the monomer addition rate (R_a) on kinetics, colloidal behaviour and molar masses were studied. Polymer content between 20 and 24% was obtained by latexes.

When HBC10 concentration was increased, it was observed to decrease very high-average molecular weights ($1.63 \times 10^6 \leq Mw \leq 2.34 \times 10^6$ g/mol), while it increases with increased R_a. The corresponding polydispersity indexes $\left(\overline{M}w/\overline{M}n\right)$ were in the range of 1.55–2.30, showing relatively wide molar masses distributions. Polymers with HBC10 showed two T_gs (123 and 178 °C).

The influence of ultrasound and cavitation in the synthesis of PMMA and its nanocomposite using the organo-modified Cloisite 30B clay was observed by Poddar et al. (2016). The nanocomposites revealed better thermal properties (maximum degradation temperature of 277 °C at T5 % and glass-transition temperature of 120 °C at 2 % clay loading) as compared to neat PMMA. The physical effect of ultrasound and cavitation on generation of intense microturbulence in the system played a major role in the process rather than the chemical effect of generation of radicals.

MMA–HNT nanocomposites (Halloysite nanotubes) were synthesized (Kumar et al., 2016). These nanocomposites were synthesized by a solution-casting method using ultrasound irradiation. The use of ultrasound energy provides is a greener, safer, faster, and facile way of synthesizing PMMA–HNT nanocomposites. HNTs are naturally occurring hollow nanotubes. The effect of clay loading and ultrasound energy on the structure, morphology, and thermal properties of nanocomposites were also investigated. As-synthesized nanocomposites utilizing ultrasound revealed complete exfoliation of filler into the polymer matrix. Sonication time of 60 min resulted in uniform distribution of particles. Increase in sonication time resulted in a decrease in glass transition temperature (T_g) of nanocomposites due to breaking of polymer chains, because of an excessive temperature and pressure generated during sonication. Successful incorporation of HNTs in the polymer matrix was confirmed by FTIR.

7.6 POLYURETHANE

Polyurethane (PU) is a polymer composed of organic units joined by carbamate (urethane) links. PUs are thermosetting polymers but now thermoplastic PUs are also available. PU products are available in the form of foams. Materials made of PUs have wide range of stiffness, hardness, and densities and find their application in bedding, automotive, seating, thermal insulation, gel pads, print rollers, footwear, and so on.

Polymerization of PU proceeds through two commonly used synthetic strategies: one- and two-shot methods. The first method, one-shot method requires the addition of all the components simultaneously (Rogers and Long, 2003). The second method is as two-shot method. It requires the synthesis of oligomer with isocyanate functionality (end-capping) first, and then adding the chain extender in a following step (Nelson and Long, 2014). Ion-containing PUs represent an industrially established class of waterborne polyurethanes (WPUs), or waterborne PU dispersions (Lee and Kim, 2009; Lee et al., 2012).

Chen et al. (2007) prepared PU/titania (PU/TiO_2) nanocomposites viaan ultrasonic process. These PU/titania nanocomposites were dispersed homogeneously in PU matrix on nanoscale as supported by TEM and SEM. It was observed that the heat stability of the composite was also improved.

Tarek et al. (2016) synthesized waterborne PU copolymer by ultrasonic irradiation via acoustic cavitation, which allows the synthesis of such new materials with unusual properties. It was observed that the acoustic power, type of solvent, catalyst, and monomers molar ratio affect the polymerization of urethane. As-prepared waterborne polyurethanes were duly characterized by FTIR and GPC. The molecular weight of ultrasonically obtained polyurethane was found to be large in magnitude as compared to that of polyurethane which is prepared thermally. The increase in ultrasonic irradiation power also had noticeable increases in the molecular weight of a polymer.

7.7 POLYVINYL ALCOHOL

Polyvinyl alcohol (PVA) is a water soluble synthetic polymer, which is commonly used in paper making, textiles, and so on. It has certain biomedical application also of contact lenses, artificial heart surgery, drug delivery system, wound dressings, and so on. PVA has properties like nontoxicity, noncarcinogenic, swelling properties, bioadhesive characteristics, and so on.

Uniformly dispersed amorphous nanoparticles of magnetite in PVA have been obtained by Kumar et al. (2000) using ultrasound radiation. The size of magnetite particles was 12 ± 20 nm in diameter, and these particles were very well dispersed in the PVA. Such a composite material was found to be superparamagnetic in nature.

Kumar et al. (2002) synthesized nanospherical Ag_2S/PVA (PVA) and nanoneedles of CuS/PVA composite. These have been prepared by 10% ethylenediamine–water solution of elemental sulfur, silver nitrate, or copper acetate in the presence of polyvinyl under using sonochemical irradiation. The resulting nanocomposites particle sizes were 25 and

225 nm for Ag_2S/PVA and CuS/PVA, respectively. A band gap of 1.05 and 2.08 eV was estimated for Ag_2S/PVA and CuS/PVA nanocomposites, respectively. Micro scale zinc oxide-PVA (ZnO–PVA) composite has also been prepared by ultrasound irradiation (Kumar et al., 2003). The properties of the as-prepared ZnO–PVA composite material were characterized by different techniques such as XRD, TGA, TEM and DRS. A band gap of 3.25 eV was estimated for this composite.

7.8 COMPOSITES OF CARBON NANOTUBES

Ultrasonication has been used for functionalization of CNTs (Xing et al., 2005; Chakraborty and Raj, 2005) and fabrication of polymer/CNTs nanocomposites (Xia et al., 2003; Park et al., 2002). Park et al. (2003) studied different multiwalled carbon nanotube (MWCNTs) contents of PMMA/MWCNT nanocomposites prepared by in situ bulk polymerization under ultrasonication (frequency of 28 kHz, power of 600 W). Initially, the acid-treated MWCNT was dispersed in a liquid MMA monomer state. Then, the presence of 2,2-azoisobutyronitrile (AIBN) initiator polymerized MMA and MWCNT-MMA suspension. Molecular weight of PMMA was found to increase as the MWCNT contents wereincreased due to the involvement of MWCNTs in polymer reaction and the consumption of AIBN (Jia et al., 1999).

MWCNT and polysytrene (PS) composites were synthesized via an in situ bulk polymerization under the application of ultrasonication without any added initiator (Kim et al., 2007). Polymerization was initiated by ultrasound and the radicals generated by decomposition of styrene monomer. The acid-treated MWCNTs were dispersed in such a monomer. The role of nanocomposites as an initiator was observed during polymerization in comparison with AIBN.

The nanocomposite MWCNTs are almost completely embedded in the PS matrix. The diameter of MWCNTs becomes larger than the pristine one after removing the ungrafted PS with excess chloroform, which indicates that the PS was successfully adhered to the MWCNTs surface making this combination quite strong. The wrapped MWNT plays an important role in dispersing the MWNTs.

7.9 COORDINATION POLYMERS

New synthetic methods such as sonochemical synthesis can provide control over size, morphology, and nano/microstructure. The utilization of high-intensity ultrasound offers a facile, environmental-friendly, and versatile synthetic tool for these nanostructured materials that are often not obtained by conventional methods.

Silicon-based inorganic polymer, poly (organosilanes) was prepared by a Wurtz-type coupling of dichlorodiorganosilanes using molten sodium in refluxing toluene to remove the halogens (Miller and Michl, 1989). Han and Boudjouk (1981) synthesized R_3SiR_3 in the presence of ultrasound using lithium to couple organochlorosilanes. It is highly desirable that such polymers should be prepared under more environmentally acceptable conditions with a controlled structure and preferably monomodal distribution. Kim and Matyjaszewski (1989) also reported sonochemical synthesis of such polymers.

Miller et al. (1991) reported some conflicting results because only sonication did not yield polymers with a monomodal distribution unless diglyme or 15-crown-5 was added to the solvent. Price (1992) and Price and Patel (1997) have reported that a number of polysilanes with diverse substituents can be prepared under a wide range of conditions, and all these reactions gave higher yields and faster reactions under ultrasound. The effect of ultrasonic intensity on polymer molecular weights and distributions established that higher intensities lead to narrower distributions and smaller amounts of low-molar-mass material. A similar reaction was used by Bianconi et al. (1989) to prepare polysilynes, $(RSi)_n$, with sonochemical activation of Na–K alloy. Here also, the use of ultrasound removed the very high-molecular-weight fractions and allowed the synthesis of polymers with molecular weights in the range of 10,000–100,000 and monomodal distributions.

A new nanosized lead(II) coordination polymer, $\{[Pb_2(\mu\text{-}3\text{-bpdh})(\mu\text{-}NO_3)_3(NO_3)]_n$; (3-bpdh = 2,5-bis(3-pyridyl)-3,4-diaza-2,4-hexadiene)$\}$, was synthesized sonochemically by Hashemi and Ali (2010). This coordination polymer of lead(II) consists of metallocyclic chains formed by bridging NO_3^- and 3-bpdh ligands, thus, making a 2D array of $Pb(NO_3)_2$ and 3-bpdh. Thermal stability of this compound was also studied. The particle size of coordination polymerprepared by the sonochemical method was about 95 nm. The PbO nanoparticles were obtained by its thermolysis at

180 °C with oleic acid as a surfactant. Scanning electron microscopy study showed that the size of such PbO particles is ~60 nm.

Hojaghani et al. (2013) synthesizednano-sized cobalt(II) coordination polymer has been prepared by a sonochemical method using ligand, 5-(4-carboxy phenyl azo) salicylic acid. These synthesized nanosized polymer have thermal stability. Nanoparticles of Co_3O_4 were obtained by calcination of these nano-sized polymer at 550 °C.

Yawer et al. (2014) synthesized nanoparticles of coordination polymer $[Co_2(pydc)_2(H_2O)_6]_n \cdot 2n\ H_2O$[$H_2$pydc = pyridine-2,5-dicarboxylic acid] by sonochemical method. The structure of single-crystalline coordination polymer was determined by X-ray crystallography, which revealed that nanoparticles of coordination polymer have the same structure as that of a bulk single crystalline polymer. This coordination polymer possesses a 1-D chain-like extended structure with binuclear cobalt(II) nodal unit. The average particle size of nanoparticles calculated was found to be ~27 nm. The nanoparticles are composed of polyhedral blocks with definite edges. These nanoparticles are thermally stable up to 352 K and thereafter decompose in well-defined steps.

Safarifard and Morsal (2015) reported the sonochemical synthesis of crystalline metal–organic coordination polymers. They discussed sonochemical synthesis of nanostructured metal complexes, including mononuclear complexes, one-, two-, and three-dimensional coordination/supramolecular polymers (CPs), and metal–organic frameworks (MOFs).

Ranjbar and Yousefi (2016) prepared nanosized Pb(II) coordination polymer, $\{[Pb(pydc)(pydc.H_2)(H_2O)_2]_2\}_n$, (where pydc.$H_2$=2,6-pyridinedicarboxylic acid) by sonochemical method. The sizes of the nanostructures were ~90 nm. Pure phase PbO nanoparticles were obtained by calcination of nanosized compound at 600 °C under argon atmosphere, and the particles size of the PbO were approximately 26 nm. The coordination polymers are suitable precursors for the simple one-pot preparation of nanoscale metal oxide materials with different and interesting morphologies.

7.10 OTHERS

Polymer surface are modified also by ultrasound and this was first reported by Urban and Salazar-Rojas (1988), who worked on a piezoelectric material, that is poly(vinylidene difluoride). It is normally an insulator, but after dehydrofluorination produced a surface with carbon–carbon double bonds (C=C). This process can be accelerated by ultrasound. Ultrasound helps in enhancing contact between the solid surface and the solution, which in turn provided excellent wetting of the surface and also better mass transport of reagents.

The conversion of solar energy into electrical energy requires semiconductor materials such as photo anodes in photoeletrochemical cells (Hillhouse and Beard, 2009; Kalambhe and Kharat, 2002). Polymer–semiconductor devices based on metal oxide materials offer cost effectiveness for the energy conversion process (Xiong et al., 2008). Two approaches were used in the preparation of thin film-based photo anodes and these are:

- Metal or semiconductor metal oxides are deposited onto a conductive glass (ITO) at higher temperature using chemical vapour deposition or electrochemical deposition and
- Semiconductors incorporating conducting polymers, such as PANI/polypyrrole/porphyrin (Wang and Jing, 2007; Brune et al., 2004; Du Pasquier et al., 2009; Claudia et al., 2002) are coated onto the nonconducting transparent matrix such as poly(methyl methacryalate) and polyethylene terphthalate.

Gumel et al. (2012) reported enzymatic synthesis of poly-E-caprolactone using ultrasound. The rate constant for chain propagation was only slightly enhanced. The acoustic effects allowed this reaction to continue longer and the viscosity remains under control as compared to the nonsonicated process, where it became almost impossible to operate due to the highly elevated viscosity (>2000 times increase from initial viscosity) of reaction mixture.

Zhao et al. (2016) investigated the ultrasound-assisted enzymatic synthesis of poly(ethylene glutarate). Diethyl glutarate and ethylene glycol diacetate were used as start-up materials, without the need of addition of any extra solvent. Lipase B catalysed the reactions by from *Candida antarctica* immobilized on glycidyl methacrylate-ter-divinylbenzene-ter-ethylene glycol dimethacrylate at 40 °C during 1 h in ultrasonic bath followed by 6 h in vacuum in both the cases for evaporation of ethyl acetate. The same degree of polymerization was obtained for the same enzyme loading in reaction but on using the ultrasound treatment required for lesser time. The ultrasound treatment enhanced initial kinetics with high degree of polymerization and it is considered demonstrated to be an effective green approach to intensify the polyesterification reaction. The effect of reaction operating parameters on the time of reaction, degree of polymerization, and monomer conversion rate were also observed.

Sonawane et al. (2010) prepared metal oxide-loaded poly(butyl methacrylate) films sonochemically. These films were used as photo anodes in photoelectrochemical cells. A small quantity of PANI was added during the latex synthesis to facilitate the transport of charge carriers. The semiconductor metal oxides used were ZnO, TiO_2 and $Bi_2O_3/ZrTiO_4$. It was found that there was a decrease in photocurrent with an increase in the thickness of the latex film, but photocurrents generated were directly proportional to the amount of metal oxide loading. $Bi_2O_3/ZrTiO_4$ latex at 0.4 g loading, showed the highest photocurrent of 0.9 $\mu A \cdot cm^{-2}$ in aqueous formic acid solution. The reaction of the three different oxide films was found as:

$$Bi_2O_3/ZrTiO_4 > ZnO > TiO_2$$

7.11 POLYMER DEGRADATION

The polymer chain degradation (an irreversible lowering of the chain length caused by cleavage, and not necessarily any chemical change) in solution was one of the first reported effects of high-intensity ultrasound on polymers (Flosdorf and Chambers, 1933). This common behaviour is a result of a physical process that is independent of the chemical nature of the polymer, but rather it depends on the polymer chain dimensions in solution. Several studies have demonstrated that when sonicated under the same conditions, the rate of degradation and M_{lim} are not sensitive to the nature of the polymer.

The degradation proceeds faster and to lower molecular weights at lower temperatures in more dilute solutions, and in solvents with low volatility. This pattern follows the effect of the parameters on cavitational bubble collapse. At higher temperatures or in volatile solvents, sonication results in more vapours entering the bubble and so cushioning the collapse, making it less violent. The polymer chains are not entangled in dilute solutions and are free to move in the flow fields around the bubbles. The degradation is more efficient at high ultrasonic intensities, owing to the greater number of bubbles with larger radii. Many other factors have been quantified including the nature of any dissolved gases, and the conformation that the polymer adopts in solution.

Hence, one can exert a great deal of control over the degradation process by suitable variation of the experimental conditions. It has been observed that degradation does not occur under conditions, which suppress cavitation. The polymer chains are subjected to extremely large forces in the rapid liquid flows near collapsing cavitation bubbles and in the shock after bubble implosion. The polymer chains have negligible vapour pressure and are unlikely to be found at the bubble interface that is why there is no evidence that in nonaqueous liquids, the extreme conditions of temperature found in cavitation bubbles contribute to polymer degradation.

The molecular weight dependence of the degradation indicates that longer chains are preferentially removed from a sample and the polydispersity of the polymer is also changed. Thus, degradation can be used as an additional processing parameter to control the molecular weight distribution. A commercial process uses a sonochemical treatment during the final processing stage to control the molecular weight distribution to give the desired processing properties of the polymer (Price, 1996).

Ultrasonic degradation is better than the thermal degradation, which produces cleavage along the chain at random points and in ultrasonic degradation, cleavage occurred preferentially near the middle of the chain (Van der Hoff and Glynn, 1974). Van der Hoff and Gall (1977) also investigated the degradation of polystyrene in tetrahydrofuran. They found that the degradation could be best modeled, when it was assumed that the probability of chain breakage was distributed in a Gaussian manner within ~ 15 % of the center of the chain. This center cleavage model is also consistent with the stretching and breakage mechanism.

Degradation of poly(vinyl acetate) was carried out in the presence of an oxidizing agent, benzoyl peroxide, using ultrasonic irradiation. It was reported that degradation rate coefficient of the polymer decreased with increasing benzoyl peroxide concentration (Madras and Chattopadhyay, 2001a). Daraboina and Madras (2009) observed that the ultrasonic degradation of different polymers is also influenced by the alkyl group substituents present in it.

The primary products of the degradation are radical species arising from homolytic bond breakage along the chain in all the carbon backbone polymers. Evidence for macromolecular radicals may be obtained from radical trapping experiments as well as using electron-spin resonance spectroscopy (Tabata et al., 1980). One of the applications of degradation is that it provides those macromolecular radicals, which can be used as initiating species in the preparation of copolymers. When mixture of two polymers dissolved in a common solvent was sonicated, then combination of the two different macromolecule fragments leads to the formation of a block co-polymer. However, it is slightly difficult to separate these copolymers from the starting materials, and there are also some problems in controlling the block structure. An alternative approach is also there, which involves sonicating a polymer dissolved in a solution containing the second monomer. The macromolecular radical generated initiates the polymerization of the second monomer.

Encina et al. (1980) reported that the peroxide linkages in the backbone increased rate of degradation of poly(vinyl pyrrolidone) about 10-folds, when the polymer was prepared with a small number, suggesting that chain cleavage can occur preferentially at weak spots in the chain.

Although heat, light, chemical reagents, or ultrasonic radiation can be used for the degradation of polymers, but in ultrasonic degradation, the chain cleavage is preferentially near the middle of the chain whereas thermal degradation occurs principally at the chain end and/or there is a random chain scission. There are several unique features of ultrasound that make it a viable option from practical viewpoints. The limiting molecular weight polymer degradation by ultrasound is unique. The rapid growth and violent collapse of bubbles is responsible for their degradation. Relative motion of the polymer segments and solvents produces stresses in polymer chains leading to scission of chain (Madras et al., 2000; Madras and Chattopadhyay, 2001a, 2001b). Passage of ultrasonic waves through the solution creates the localized shear gradient, which is responsible for tearing off the molecules leading to chain scission and also for a decrease in molecular weight (Singh and Sharma, 2008). Concentration, initial molecular weight, temperature, frequency, and intensity of ultrasound are the major factors that affect the degradation rates of polymers (Vijayalakshmi and Madras, 2004).

The application of ultrasonic energy for polymer degradation dates back to the 1930s. When natural polymers were subjected to sonication, it resulted in a reduction in viscosity (Price et al., 1994). Suslick (1988) reported in the late 1930s, that ultrasonic irradiation depolymerized starch, gum arabic, gelatine and polystyrene.

Ultrasonic degradation of PVA was carried out by Taghizadeh and Mehrdad (2003) in aqueous solution at 25 °C. The effect of solution concentration on the rate of degradation was investigated. Kinetics of degradation was studied using viscometry method. They indicated that degradation rate of PVA solutions decreases with increasing solution concentration. This behaviour in the rate of degradation was interpreted in terms of viscosity and concentration of polymer solution. With increasing solution concentration, viscosity was found to increase and it causes a reduction in the cavitation efficiency and, as a result, the rate of degradation also decreased.

Ultrasonic degradation of polybutadiene and isotactic polypropylene (PP) was studied by Chakraborty et al. (2004). Degradation of these polymers was studied at different temperatures and in different solvents. The variation of rate coefficients with vapour pressure and kinematic viscosity was also investigated.

A new method using high-intensity ultrasonic wave has been developed for the degradation of polymer melt using an intensive mixer (Kim and Lee, 2002). They observed that a significant degradation of PP melt in a mixer occurred due to the action of ultrasonic wave without any additives or solvents. The degradation efficiency of ultrasonic irradiation was compared with that of peroxide, dicumyl peroxide. They also found that the direct sonication of polymer mixture in melt state reduces the domain sizes and stabilizes the phase morphology of the blend. Ultrasound-assisted melt mixing can lead to in situ copolymer formation between the components and consequently provide an effective route to compatibilize immiscible polymer blends.

Konaganti and Madras(2010) synthesized and characterized the copolymers, PMMAMA, poly(methyl methacrylate-co-ethyl acrylate) (PMMAEA), and poly(methyl methacrylate-co-butyl acrylate) (PMMABA), of different compositions. The effect of alkyl acrylate content, alkyl group substituents, and solvents on the ultrasonic degradation of these copolymers was studied. A model based on continuous distribution kinetics was used to study the kinetics of degradation. The rate coefficients depend linearly on the logarithm of the vapour pressure of the solvent, which indicated that vapour pressure is on important parameter controlling the degradation process. The rate of degradation was found to increase with an increasing alkyl acrylate content. At a particular copolymer composition, the rate of degradation follows the order:

$$\text{PMMAMA} > \text{PMMAEA} > \text{PMMABA}$$

It was observed that the degradation rate coefficient varies linearly with the mole percentage of the alkyl acrylate in the copolymer.

Koda et al. (2011) utilized ultrasonic irradiation for the degradation of methyl cellulose, pullulan, dextran, and poly(ethylene oxide) in aqueous solutions. They investigated these degradations at the frequencies of 20 and 500 kHz. The average molecular mass and the polydispersity were observed as a function of sonication time. It was observed that the degradation under sonication at the 500 kHz frequency proceeded faster in comparison with the 20 kHz sonication for all the four polymers. The addition of a radical scavenger, *t*-BuOH, resulted in the suppression of degradation of water-soluble polymers. The degradation rate of methyl cellulose was the largest one among these polymers. They discussed differences in the degradation rates in terms of flexibility and the hydrodynamic radius of polymer chains in aqueous solutions.

Mohod and Gogate (2011) also used ultrasound for the polymer degradation. The ultrasonic degradation of two water-soluble polymers such as carboxy methyl cellulose (CMC) and PVA has been studied. The effect of different operating parameters such as time of irradiation, immersion depth of horn and solution concentration has been investigated with different additives such as air, sodium chloride, and surfactant. They showed that the viscosity of polymer solution decreased with an increase in the ultrasonic irradiation time and it approached a limiting value and the use of additives such as air, sodium chloride, and surfactant helps in increasing the extent of viscosity reduction.

Wang et al. (2015) studied that the sonocatalytic degradation of methyl orange (MO) with Fe_3O_4/polyaniline (Fe_3O_4/PANI) microspheres in near neutral solution (pH ~ 6) using ultrasound-assisted advanced oxidation process (AOP) (Fig. 7.2).These Fe_3O_4/PANI microspheres were characterized and tested for adsorption and sonocatalytic decolourization of MO in solution. The isotherms and kinetics of MO adsorption with Fe_3O_4/PANI followed Langmuir model and the pseudo-second-order model, respectively, as Fe_3O_4/PANI has a high capacity to adsorb MO in solution. The percentage of room temperature sonocatalytic degradation of MO with Fe_3O_4/PANI is about 4.8, 8.8, and 5.7 times that with Fe_3O_4, dedoped Fe_3O_4/PANI, and ultrasonication alone, respectively. The eco-friendly Fe_3O_4/PANI added with superparamagnetism and excellent reusability offers to be a promising sonocatalyst for rapid decolourization and enhanced degradation of azo dyes in effluents.

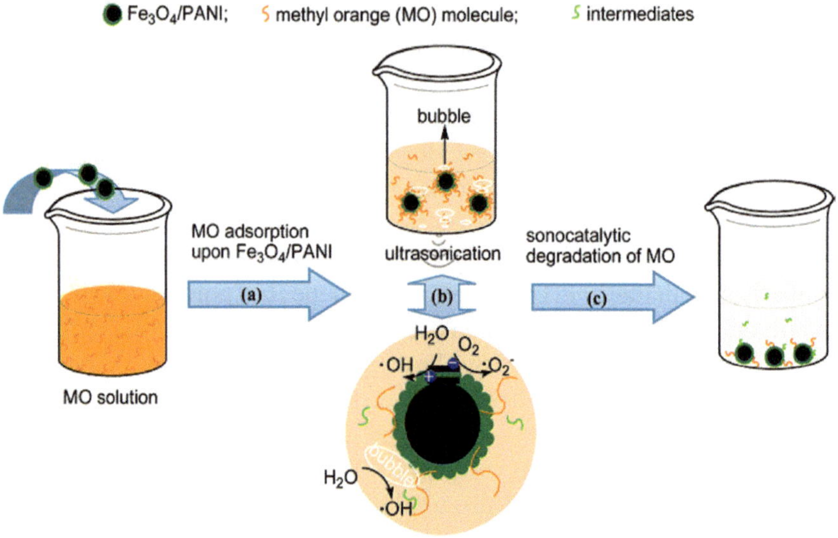

FIGURE 7.2 Proposed sonocatalytic degradation mechanism of MO upon Fe_3O_4/PANI. *Source*: Wang et al., 2015, with permission.

The use of polymer in day-to-day life has proved to be a boon for humans as it has almost replaced traditional materials such as iron, wood, paper, cloth, and so on. Polymers have been used in many things such as furniture, vehicle bodies, wrapping materials, nonstick coating of utensils, textile, and so on. The synthesis of these polymers using ultrasound has been a great development. In this polymer era, the most developed countries are extensively using polymers in different kinds for one purpose or the other and are therefore facing a new emerging face of pollution, that is, polymer pollution. The degradation and disposal of polymers is posing a serious problem nowadays. In this aspect, ultrasonic treatment of polymers for their degradation and disposal will play an important role and prove to be a technology involving green chemical routes.

REFERENCES

Ai, L.; Jiang, J. *Mater. Lett.* **2011,** *65*, 1215–1217.
Albano, C.; Gonza´lez, G.; Parra, C. *Polym. Bull.* **2010,** *65*, 893–903
Ansari, M.; Mohammad, F. *Sens. Actuators B* **2011,** *157*, 122–129.

Arenas, M. C.; Rodríguez-Núnez, L. F.; Rangel, D.; Martínez-Álvarez, O.; Martínez-Alonso,C.; Castaño, V. M. *Ultrason. Sonochem.* **2013,** *20*, 777–784.
Bang, J. H.;Suslick, K. S. *Adv. Mater.* **2010,** *22*, 1039–1059.
Barkade, S. S.; Naik, J. B.; Sonawane, S. H. *Colloids Surf. A* **2011,** *378*, 94–98.
Barkade, S. S.; Pinjari, D. V.; Nakate, U. T.; Singh, A. K.; Gogate, P. R.; Naik, J. B.; et al. *Chem. Eng. Process.: Process Intensif.* **2013a,** *74*, 115–123.
Barkade,S. S.; Pinjari,D. V.; Singh, A. K.; Gogate,P. R.; Naik, J. B.; Sonawane, M.; Ashokkumar, S. H.; Pandit A. B. *Ind. Eng. Chem. Res.* **2013b,** *52*(23), 7704–7712.
Basedow, A. M.; Ebert, K. H. *Adv. Polym. Sci.* **1977,** *22*, 83–148.
Bhadra, S.; Chattopadhyay, S.; Singha, N. K.; Khastgir, D. *J. Polym. Sci. Part B* **2007,** *45*(15), 2046–2059.
Bhanvase, B. A.; Pinjari, D. V.; Gogate, P. R.; Sonawane, S. H.; Pandit, A. B. *Chem. Eng. Process.* **2011,** *50*, 1160–1168.
Bhanvase, B.; Pinjari, D.; Sonawane, S.; Gogate, P.; Pandit, A. *Ultrason. Sonochem.* **2012,** *19*, 97–103.
Bhanvase, B. A.; Darda, N. S.; Veerkar, N. C.; Shende, A. S.; Satpute, S. R.; Sonawane, S. H. *Ultrason. Sonochem.* **2015,** *24*, 87–97.
Bianconi, P. A.; Schilling, F. C.; Weidman, T. W. *Macromolecules* **1989,** *22*(4), 1697–1704.
Biggs, S.; Grieser, F. *Macromolecules* **1995,** *28*, 4877–4882.
Bradley, M; Grieser, F. *J. Colloid Interface Sci.* **2002,** *251*, 78–84.
Brune, A.; Jeong, G.; Liddell, A.; Sotomura, T.; Moore, T.; Moore, A.; Gust, D. *Langmuir* **2004,** *20*, 8366–8371.
Chakraborty, S.; Raj, C. R. *J. Electroanal. Chem.* **2005,** *609*, 155–162.
Chakraborty, J.; Sarkar, J.; Kumar, R.; Madras, G. *Polym. Degrad. Stab.* **2004,** *85*(1), 555–558.
Chen, J.; Zhou, Y.; Nan, Q.; Sun, Y.; Ye, X.; Wang, Z. *Appl. Surf. Sci.* **2007,** *253*(23), 9154–9158.
Chitra, P.; Muthusamy, A.; Jayaprakash, R.; Kumar, E. R. *J. Magn. Magn. Mater.* **2014,** *366*, 55–63. Choi, M.; Lee, K.; Lee, J.; Kim, H. *Macromol. Symp.* **2007,** *249–250*, 350–356.
Claudia, L.; Nogueira, A.; De Paoli, M. *J. Phys. Chem. B* **2002,** *106*, 5925–5930.
Daraboina, N.; Madras, G. *Ultrason. Sonochem.* **2009,** *16*, 273–279.
De Barros, R. A.; De Azevedo, W. M. *Synth. Met.* **2010,** *160*, 1387–1391.
Deshpande, N.; Gudage, Y.; Sharma, R.; Vyas, J.; Kim, J.; Lee, Y. *Sens. Actuators, B* **2009,** *138*, 76–84.
Du Pasquier, A.; Stewart, M.; Spitler, T.; Coleman, M. *Sol. Energy Mater. Sol. Cells* **2009,** *93*, 528–535.
Encina, M. V.; Lissi, E.; Sarasusa, M.; Gargallo, L.; Radic, D. *J. Polym. Sci. C: Polym. Lett.* **1980,** *18*, 757–760.
Flosdorf, E. W.; Chambers, L. A. *J. Am. Chem. Soc.* **1933,** *55*, 3051–3052.
Fujiwara, H.; Kimura, T.; Segi, M.; Nakatuka, T.; Nakamura, H. *Polym. Bull.* **1992,** *28*, 189–196.
Fujiwara, H.; Ishida, T.; Taniguchi, N.; Wada. S. *Polym. Bull.* **1999,** *42*, 197–204.
Gumel, A. M.; Annuar, M. S. M.; Chisti, Y.; Heidelberg, T. *Ultrason. Sonochem.* **2012,** *19*, 659–667.
Han, B. H.; Boudjouk, P. *Tetrahedron Lett.* **1981,** *22*, 3813–3814.
Hashemi, L.; Ali, M. *J. Inorg. Organomet. Polym. Mater.* **2010,** *20*(4),856–861.
Hillhouse, H. W.; Beard, M. C. *J. Colloid Interface Sci.* **2009,** *14*, 245–259.

Ho, K.-S.; Han, Y.-K.; Tuan, Y.-T.; Huang, Y.-J.; Wang Y.-Z.; Ho, T.-H.; et al. *Synth. Met.* **2009,** *159*(12), 1202–1209.
Hojaghani, S.; Sadr, M. H.; Morsali, A. *J. Nanostruct.* **2013,** *3*, 109–114.
Hu, J.; Chen, M.; Wu, L. *Polym. Chem.* **2011,** *2*, 760–772.
Jangid, N. K.; Chauhan, N. P. S.; Goswami, S.; Punjabi, P. B. *Malays. Polym. J.* **2014,** *9*, 45–53.
Jia, Z. J.; Wang, Z. Y.; Xu, C. L.; Liang, J.; Wei, B. Q.; Wu, D. H.; Zhu, S. W. *Mater. Sci. Eng.* **1999,** *A271*, 395–400.
Kalambhe, B.; Kharat, B. *Prog. Cryst. Growth Charact. Mater.* **2002,** *44*, 141–146.
Kim, H. K.; Matyjaszewski, K. *J. Am. Chem. Soc.* **1989,** *110*, 3321–3323.
Kim, H.; Lee, J. W. *Polymer* **2002,** *43*(8), 2585–2589.
Kim, S. T.; Choi, H. J.; Hong, S. M. *Colloid Polym. Sci.* **2007,** *285*, 593–598.
Kobayashi, D.; Matsumoto, H.; Kuroda, C. *Chem. Eng. J.* **2008,** *135*, 43–48.
Koda, S.; Taguchi, K.; Futamura, K. *Ultrason. Sonochem.* **2011,** *18*(1), 276–281.
Kojima, Y.; Koda, S.; Nomura, H. *Ultrason. Sonochem.* **2001,** *8*, 75–79.
Konaganti, V. K.; Madras, G. *Ultrason. Sonochem.***2010,** *17*(2), 403–408.
Kowsari, E.; Faraghi, G. *Ultrason. Sonochem.* **2010,** *17*, 718–725.
Kumar, R. V.; Koltypin, Y.; Cohen, Y. S.; Yair Cohen, Aurbach, D.; Palchik, O.; Felnerb, I.; Gedanken, A. *J. Mater. Chem.* **2000,** *10*, 1125–1129.
Kumar, R. V.; Palchik, O.; Koltypin, Y.; Diamant, Y.; Gedanken, A. *Ultrason. Sonochem.* **2002,** *9*(2), 65–70.
Kumar, R. V.; Elgamiel, R.; Koltypin, Y.; Jochen, N.; Gedanken, A. *J. Cryst. Growth* **2003,** *250*(3–4), 409–417.
Kumar, M. J. K.; Kezia, B.; Jagannathan, T. K. *Indian J. Adv. Chem. Sci.* **2016,** *S1*, 213–216.
Lee, S. K.; Kim, B. *J. Colloid Interface Sci.* **2009,** *336*, 208–214.
Lee, D. K.; Tsai, H. B.; Yang, Z. D.; Tsai, R. S. *J. Appl. Polym. Sci.* **2012,** *126*(S2), 275–282.
Lu, X.; Mao, H.; Chao, D.; Zhang, W.; Wei, Y. *J. Solid State Chem.* **2006,** *179*, 2609–2615.
Madras, G.; Chattopadhyay, S. *Polym. Degrad. Stab.* **2001a,** *73*, 33–38.
Madras, G.; Chattopadhyay, S. *Polym. Degrad. Stab.* **2001b,** *71*, 273–278.
Madras, G.; Kumar, S.; Chattopadhyay, S. *Polym. Degrad. Stab.* **2000,** *69*, 73–78.
Manuel, J.; Kim, J.; Fapyane, D.; Chang, S.; Ahn, H.; Ahn, J. *Mater. Res. Bull.* **2014,** *58*, 213–217.
Miller, R. D.; Michl, J. *Chem. Rev.* **1989,** *89*, 1359–1410.
Miller, R. D.; Thompson, D.; Sooriyakumaran, R.; Fickes, G. N. *J. Polym. Sci. A: Polym. Chem.* **1991,** *29*, 813–824.
Miyata, T.; Nakashio, F. *J. Chem. Eng. Jpn.* **1975,** *8*, 463–467.
Mohod, A. V.; Gogate, P. R. *Ultrason. Sonochem.* **2011,** *18*(3), 727–734.
Nelson, A. M.; Long, T. E. *Macromol. Chem. Phys.* **2014,** *15*, 2161–2174.
Park, C.; Ounaies, Z.; Watson, K. A.; Crooks, R. E.; Smith, J.; Lowther, S. E.; Connell, J. W.; Siochi, E. J.; Harrison, J. S.; St. Clair, T. L. *Chem. Phys. Lett.* **2002,** *364*, 303–308.
Park, S. J.; Cho, M. S.; Lim, S. T.; Choi, H. J.; Jhon, M. S. *Macromol. Rapid Commun.* **2003,** *24*, 1070–1073.
Patra, S. K.; Prusty, G.; Swain, S. K. *Bull. Mater. Sci.* **2012,** *35*, 27–32.
Poddar, M. K.; Sharma, S.; Moholkar, V. S. *Macromol. Symp.* **2016,** *361*, 82–100.
Prasad, K.; Sonawane, S.; Zhou, M.; Ashokkumar M. *Chem. Eng. J.* **2013,** *219*, 254–261.
Price, G, J.; Daw, M. R.; Newcombe, N. J.; Smith, P. F. *Brit. Polym. J.* **1990,** *23*, 63–66.

Price, G. J.; Norris, D. J.; Peter, J. W. *Macromolecules* **1992,** *25*, 6447–6454.

Qiu, G.; Wang, Q.; Nie, M. *Macromol. Mater. Eng.* **2006,** *291*, 68–74.

Ranjbar, M.; Yousefi, M. *Int. J. Nanosci. Nanotechnol.* **2016,** *12*(2), 109–118.

Rogers, M. E.; Long, T. E. *Synthetic Methods in Step-Growth Polymers*; Wiley: New Jersy, 2003.

Roy, K.; Mondal, P.; Bayen, S. P.; Chowdhury, P. J. *Macromol. Sci., Part A: Pure Appl. Chem.* **2012,** *49*, 931–935.

Safarifard, V.; Morsal, A. *Coord. Chem. Rev.* **2015,** *292*, 1–14.

Singh, B.; Sharma, N. *Polym. Degrad. Stab.* **2008,** *93*, 561–584.

Sivakumar, M.; Gedanken, A. *Synth. Met.* **2005,** *148*, 301–306.

Sonawane, S. H.; Teo, B. M.; Broachi, A.; Grieser, F.; Ashokkumar, M. *Ind. Eng. Chem. Res.* **2010,** *49*, 2200–2205.

Suslick, K. S. *Ultrasound: Its Chemical, Physical, and Biological Effects*; VCH Publishers Inc.: New York, 1988.

Tabata, M.; Miyawaza, T.; Sohma, J.; Kobayashi, O. *Chem. Phys. Lett.* **1980,** *73*, 178–180.

Taghizadeh, M. T.; Mehrdad, A. *Ultrason. Sonochem.* **2003,** *10*(6), 309–313.

Tang, Q.; Lin, L.; Zhao, X.; Huang, K.; Wu, J. *Langmuir* **2012,** *28*, 3972–3978.

Tarek, S.; Salama M.; El-Sayed, A. A. *Indian J. Sci. Technol.* **2016,** *9*(17). DOI: 10.17485/ijst/2016/v9i17/87215.

Teo, B. M.; Prescott, S. W.; Ashokkumar, M.; Grieser, F. *Ultrason. Sonochem.* **2008,** 15, 89–94.

Trovati, G.; Sanches, E. A.; Claro, N. S.; Mascarenhas, Y. P.; Chierice, G. O. *J. Appl. Polym. Sci.* **2010,** *115*(1), 263–268.

Urban, M. W.; Salazar-Rojas, E. M. *Macromolecules* **1988,** *21*, 372–378.

Van der Hoff, B. M. E.; Gall, C. E. *J. Macromol. Sci. A: Macromol. Chem.* **1977,** *11*, 1739–1758.

Van der Hoff, B. M. E.; Glynn, P. A. R. *J. Macromol. Sci. A: Macromol. Chem.* **1974,** *8*, 429–499.

Vargas-Salazar, C. Y.; Ovando-Medina, V. M.; Ledezma-Rodríguez, R.; Peralta, R. D.; Martínez-Gutiérrez, H. *Iran. Polym. J.* **2015,** *24*, 41–50.

Vijayalakshmi, S. P.; Madras, G. *Polym. Degrad. Stab.* **2004,** *84*, 341–344.

Wang, Y.; Jing, X. *Mater. Sci. Eng. B* **2007,** *138*, 95–100.

Wang, Y.; Gai, L.; Ma, W.; Jiang, H.; Peng, X.; Zhao, L. *Ind. Eng. Chem. Res.* **2015,** *54*, 2279–2289.

Xia, H.; Wang, Q. *Chem. Mater.* **2002,** *14*, 2158–2165.

Xia, H.; Wang, Q. *J. appl. polym. Sci.* **2003,** *87*(11), 1811–1817.

Xia, H. S.; Wang, Q.; Liao, Y. Q.; Xu, X.; Baxter, S. M.; Slone, R. V.; Wu, S. G.; Swift, G.; Westmoreland, D. G. *Ultrason. Sonochem.* **2002,** *9*, 151–158.

Xia, H. S.; Wang, Q.; Qiu, G. H. *Chem. Mater.* **2003,** *15*, 3879–3886.

Xing, Y. C.; Li, L.; Chusuei, C. C.; Hull, R. V. *Langmuir* **2005,** *21*, 4185–4190.

Xiong, M.; Xu, Y.; Ren, Q.; Xia, Y. *J. Am. Chem. Soc.* **2008,** *130*, 7522–7523.

Xiuyuan, N.; Yuefang, H.; Bailin, L.; Xi, X. *Eur. Polym. J.* **2001,** *37*, 201–206.

Xu, H.; Brad, W. Z.; Kenneth, S. S. *Chem. Soc. Rev.* **2013,** *42*, 2555–2567.

Yawer, M.; Sanotra, S.; Kariem, M.; Sheikh, H. N. *J. Nanostruct.* **2014,** *4*, 249–257.

Zhao, X.; Bansode, S. R.; Ribeiro, A.; Abreu, A. S.; Oliveira, C.; Parpotf, P.; et al. *Ultrason. Sonochem.* **2016,** *31*, 506–511.

CHAPTER 8

WASTEWATER TREATMENT

ARPITA PANDEY[1], ARPITA PALIWAL[2], RAKSHIT AMETA[3]

[1]*Department of Chemistry, Sangam University, Bhilwara, India*
E-mail: pandeyarpita88@gmail.com

[2]*Department of Chemistry, M. L. Sukhadia University, Udaipur, India*
E-mail: paliwalarpita7@gmail.com

[3]*Department of Chemistry, J.R.N. Rajasthan Vidyapeeth (Deemed to be University), Udaipur, India E-mail: rakshit_ameta@yahoo.in*

CONTENTS

8.1 INTRODUCTION

The development of industrial sector has played a major role in the economic growth of several countries on one land, but various industrial processes generating wastewater have also caused so many environmental problems on the other. Among all the other pollutants, the pollution caused by dyestuff is considered as the main source of water contamination because synthetic dyestuffs are extensively used by several industries and these are not biodegradable.

Many dyeing, printing and textile industries dispose their wastewater directly to the stream without any proper treatment or with only some preliminary treatments. This negligence on the part of industries has caused harmful effects for the ecosystem. Such water pollution causes aesthetic problems and disrupts the incident sun rays to reach under water and as a consequence the process of photosynthesis is disturbed. Water pollution also causes lack of dissolved oxygen in water stream and, as a consequence, degradation of water quality is observed ultimately resulting in the death of various species in stream. Many of the pollutants are higher toxic and sometimes even carcinogenic in nature. These are harmful and because of bioaccumulation, they can also influence the whole food chain.

8.2 METHODS OF WASTEWATER TREATMENT

Various methodologies have been adopted for treating wastewater and these include biological treatment, oxidation, membrane filtration, sorption, coagulation/flocculation, and so on, which are discussed in detail in the following sections.

8.2.1 BIOLOGICAL TREATMENT

Biological treatment is the secondary treatment of water, where a wide variety of microorganisms, primarily bacteria, are used. These microorganisms convert biodegradable organic matter contained in wastewater into simple substances and additional biomass. Successful biological treatment depends on the development and maintenance of an appropriate, active, and mixed microbial population in the system. Biological treatments are those, which use organisms to break down organic substances in wastewater, and these are widely used worldwide. Biological treatments rely on bacteria, nematodes, or other small organisms to break down organic wastes using normal cellular processes.

The effect of salinity in the decolourization of azo dye by *Shewanella* sp. was investigated by Khalid et al. (2008). They reported that bacterium decolourized 100.0 $mg \cdot L^{-1}$ dye under static and low-oxygen conditions in presence of NaCl (60.0 $g \cdot L^{-1}$). The addition of glucose or ammonium nitrate inhibited the decolourization process, whereas the presence of yeast extract and $Ca(H_2PO_4)_2 \cdot H_2O$ were found to enhance the decolourization rate. Dawkar et al. (2008) reported 100% degradation of dye brown 3REL by *Bacillus* sp. VUS within 8 h at static anoxic condition. This bacterium was also able to degrade 80% of dye in textile effluent.

The simultaneous removal of Cr(VI) and azo dye acid orange 7 has been observed by Ng et al. (2010) using *Brevibacterium casei* under nutrient-limiting conditions, whereas Ghasemi et al. (2010) studied the decolourization of different azo dyes by *Phanerochaete chrysosporium* RP 78 under optimized conditions. They showed that the decolourization of azo dyes was due to an enzymatic degradation and almost 100% decolourization was acquired after 24 h. The potential of *Lemna minor* for the treatment of reactive dye polluted wastewater was also investigated by Kilic et al. (2010). Nearly 59.6% brilliant blue removal was found in samples containing 2.5 $mg \cdot L^{-1}$ dye and 1.0 $mg \cdot L^{-1}$ of triacontanol hormone concentration.

8.2.2 OXIDATION

Chemical oxidation oxidizes organic pollutants to less toxic or harmless substances. Complete oxidation or mineralization of organic substances will give CO_2 and H_2O. Oxidation can be used to remove some inorganic components such as cyanide. It can also be used in combination with biological purification and this process is known as partial oxidation. Chemical oxidation is used as a pretreatment technique, that is, either to break down recalcitrant components or to make them suitable for biological degradation and also for limiting sludge production by partly oxidizing the sludge.

Such an oxidation requires addition or generation of oxidants in wastewater. Most commonly used oxidants are oxygen, hydrogen peroxide, ozone, bleaching liquor or sodium hypochlorite, chlorine dioxide, chlorine gas, peroxy acetic acid, and so on. However, different combinations of oxidants are also possible.

The heterogeneous catalytic ozonation of methylene blue (MB) using volcanic sand showed a higher removal efficiency of MB as compared to homogenous system. One et al. (2000) studied the use of V_2O_5 catalyst in the oxidative dehydrogenation of propane. Hofmann et al. (2005) reported the degradation of chlorobenzene (CB), 4-chlorophenol, 4-chloroaniline and 1-nitro-4-CB using catalytic oxidation with hydrogen peroxide. The conversion of a wide spectrum of components including aliphatic and aromatic hydrocarbons as well as halogenated organic compounds into nontoxic or minor-toxic substances and finally into carbon dioxide and water can be obtained. It was demonstrated that the type of catalyst and oxidation agent as well as the reaction parameters influence the degradation rate.

Enhanced dye wastewater degradation efficiency was studied by Xu et al. (2008) using a flowing aqueous film photoelectrocatalytic reactor. It was observed that higher treatment efficiency can be achieved for the degradation of reactive brilliant blue X-BR by this process. An indirect electrochemical oxidation method for the decolourization of textile wastewater was explored by Maljaei et al. (2009). Here, graphite and sodium chloride were used as an electrode and electrolyte, respectively. Various variables affecting the efficiency of the decolourization and degradation process of reactive yellow 3 were studied.

Catalytic oxidation of simulated wastewater containing acid chrome blue K was investigated by Yu and Shi (2010) using chlorine dioxide as an oxidant. The activated carbon–MnO_2 catalyst was used for this catalytic oxidation and the removal efficiency of acid chrome blue K was observed to be 87.8%. Photocatalytic degradation of rhodamine B using thiocyanate complex of iron and H_2O_2 was reported by Lodha et al. (2011).

Wang et al. (2012) reported the use of amorphous $Fe_{78}Si_9B_{13}$ alloy as a heterogeneous Fenton catalyst in the degradation of phenol. The $Fe_{78}Si_9B_{13}$ showed better catalytic activity than iron powder and Fe^{2+}. Addition of *n*-butanol (hydroxyl radical scavenger) decreased the degradation rate of phenol, which demonstrates that hydroxyl radicals were mainly responsible for the removal of phenol. They observed that phenol may be degraded by hydroxyl radicals on the surface of $Fe_{78}Si_9B_{13}$. This amorphous alloy exhibited high stability in recycling experiments and showed excellent reuse performance even after continuous operations of 8 cycles.

Thermal catalytic oxidation of octachloronaphthalene (CN-75) using anatase TiO_2 nanomaterials was investigated at 300°C. Oxidation intermediates is classified into the following three types:

- naphthalene ring
- Single-benzene ring
- Completely ring-opened products

Reactive oxygen species on anatase TiO_2 surface, such as $O_2^{-\bullet}$ and O^{2-}, contributed to oxidative degradation. The naphthalene ring oxidative products with chloronaphthols and hydroxyl-pentachloronaphthalene-dione could be formed via attacking the carbon of naphthalene ring at one or more positions by nucleophilic O^{2-}. Lateral cleavage of the naphthalene ring at different carbon–carbon bonds by electrophilic $O_2^{-\bullet}$ may also occur. This will lead to the formation of tetrachlorophenol, tetrachlorobenzoic acid, tetrachlorophthalaldehyde, and tetrachloroacrolein benzoic acid, partially with further transformation into tetrachlorobenzene-dihydrodiol and tetrachloro-salicylic acid. Complete cleavage of naphthalene ring could produce the ring-opened products such as formic and acetic acids (Su et al., 2015).

8.2.3 MEMBRANE FILTRATION

Membranes are thin and porous sheets, which are able to separate contaminants from water on the application of a driving force. It was considered a viable technology for a long time for desalination, but these membrane processes are increasingly employed in both drinking water and wastewater treatment. It is useful in the removal of bacteria and other microorganisms, particulate material, micropollutants, and natural organic materials, which is responsible for imparting colour, taste and odour to water and can also react with disinfectants to form disinfection by-products (DBPs). As advancements are being made in membrane production, module design, capital and operating costs are continuing to decline. Membranes are becoming increasingly more popular for the production of potable drinking water from ground, surface and seawater sources, as well as for the advanced treatment of wastewater and desalination.

Many types of membranes are used for the treatment of water in the processes such as reverse osmosis (RO), nanofiltration (NF), dialysis and electrodialysis, ultrafiltration (UF) and microfiltration (MF), and so on. Reverse osmosis is primarily used to remove salts from salty water and it is capable for very high rejection of synthetic organic compounds. Nanofiltration is used to soften fresh waters and remove DBPs whereas UF and MF are used to remove turbidity, pathogens, and particles from fresh water.

Ultrafiltration, which typically operates pressures of between 5 and 15 bars, is widely used in chemical separation processes, because it can separate species according to molecular size. However, MF separates according to particle size. It is a process that can take place at low temperatures. This is mainly important because it enables the treatment of heat sensitive matter. It is a process with low-energy cost. Most of the energy that is required is used to pump liquids through the membrane.

The total amount of energy that is used is minor, compared to alternative techniques, such as evaporation. The membrane separation process is based on the presence of semipermeable membranes. The principle is quite simple. The membrane acts as a very specific filter that will let water flow through, but it catches suspended solids and other substances. The advantage of membrane technology is the fact that it works without the addition of chemicals, with a relatively low-energy use and easy- and well-arranged process conductions.

Novel asymmetric aromatic polyamide NF membrane has been used by Ren et al. (2010) for the treatment of dye aqueous solutions. This NF membrane was prepared using a phase inversion method. The copoly phthalazinone biphenyl ether sulphone UF membrane with low-molecular weight cut-off possesses excellent thermal resistance, and it is suitable for being used in dye wastewater treatment at high temperatures (Han et al., 2010). It showed 100% rejection for Congo red, sulphur black B, and gentian violet.

A new plate MF membrane was fabricated using low-cost cement and quartz to catalyse dissolved ozone. The membrane-catalysed ozonation process successfully increased the degradation efficiency of *p*-chloronitrobenzene (*p*-CNB) compared with the ozone-alone process in aqueous conditions under continuous flow. The ozone-membrane process decomposed 1.5 $mg \cdot L^{-1}$ of dissolved ozone and increased the *p*-CNB removal by 50% with little adsorption. It was confirmed that *p*-CNB degradation followed the mechanism of hydroxyl radical oxidation during the ozone-membrane process. The hybrid process was observed to maintain the ability of removal of *p*-CNB efficiently in different water sources, and it also had the stability of long-time operation. The cementitious membrane was an efficacious catalyser for *p*-CNB degradation by ozonation (Wang et al., 2015a).

Naidu et al. (2016) reported the removal of phenol from wastewaters using a novel nanoparticle adsorption and NF technique named as nanoparticle-assisted nanofiltration. Tri and di particle studies showed more phenol removal than that of their individual particles, particularly for using small particles on large membrane pore size and large particles at low concentrations.

8.2.4 SORPTION PROCESS

Sorption (adsorption and absorption) is a physical and chemical process by which one substance becomes attached to another substance. Absorption is the incorporation of a substance in one state into another of a different state (e.g. liquids being absorbed by a solid or gases being absorbed by a liquid). Adsorption is the physical adherence or bonding of ions and molecules onto the surface of another phase (e.g. reagents adsorbed to a solid catalyst surface).

Active manganese oxide was used for the treatment of wastewater containing dye Congo red (Chakrabarti et al., 2009). It was reported that the removal of Congo red by active manganese oxide is a reversible adsorption process. Asuha et al. (2010) carried out adsorption study using mesoporous TiO_2 (prepared by hydrothermal method) and showed that the maximum adsorption capacity for methyl orange (MO) and Cr(VI) were 454.5 and 33.9 $mg \cdot g^{-1}$, respectively. Here, cetyltrimethylammonium bromide (CTAB) was used as a structure-directing agent.

A study on the porous magnetic microspheres prepared with sulphonated macroporous polydivinylbenzene as a template and its ability to remove cationic dyes was conducted by Liu et al. (2010). It was interesting to note that basic fuchsin and methyl violet can be removed with high efficiency using these porous magnetic microspheres. These can be easily regenerated and used for wastewater treatment. Hollow Fe_3O_4 and magnetic nanospheres were used as an adsorbent for the removal of neutral red dye from aqueous solutions by Iram et al. (2010). Nanoscale particle size, magnetic property and hollow porous nature of Fe_3O_4 were influencing factors for better adsorption behaviour of neutral red.

Bottom ash and deoiled soya can also be used as adsorbents for the removal of Congo red (Mittal et al., 2009). The effects of pH, dye concentration and amount of adsorbents were studied for the degradation of Congo red on these adsorbents. Appreciably adsorptive tendencies for Congo red with 96.95% and 97.14% saturation were reported on bottom ash and deoiled soya, respectively. The possibility of producing an adsorbent for dye removal by calcinating waste drilling fluids was also explored by Song et al. (2009). Other low-cost adsorbents such as coconut and coffee husks were identified as potential biosorbents for hazardous dye quinoline yellow (Mittal et al., 2010).

8.2.5 COAGULATION/FLOCCULATION

Coagulation is the destabilization of colloidal particles brought about by the addition of a chemical reagent called as coagulant. Flocculation is the agglomeration of destabilized particles into microfloc and later on into bulky floccules, which can be settled as floc. The addition of another reagent called flocculant and flocculant aid may promote the formation of the floc. The factors which can promote the coagulation–flocculation are the velocity gradient, time, and pH. The time and the velocity gradient are important to increase the probability of particles to come together.

Decolourization of effluent-containing dye produced electrocoagulation, when mineral coagulant was used by Zidane et al. (2008). The inorganic coagulants used in this study were labelled as C1 (from the electrolysis of NaOH, 7.5×10^{-3} M), C2 (from the electrolysis of NaCl, 1.0×10^{-2} M), and C3 (from the electrolysis of NaOH, 7.5×10^{-3} M + NaCl, 10^{-2} M). Results showed that the best performance of reactive red 141 removals was obtained with coagulant C2.

Red mud is a by-product of bauxite processing during Bayer process and its role in gas cleaning and wastewater treatment has been explored by Wang et al. (2008). They revealed some promising results of using this red mud as a coagulant, an adsorbent and a catalyst for some industrial processes. It was indicated by leaching and eco-toxicological tests that red mud is not toxic in nature to the environment before and after reuse.

8.3 SONOCHEMICAL WASTEWATER TREATMENT

Several technologies have already been used for wastewater treatment. These are physical, chemical, biological or a combination of these techniques. Advanced oxidation processes have been found to be suitable methods for the degradation of various organic compounds. AOPs are characterized by production of $^{\bullet}OH$ (hydroxyl) radical as the primary oxidant. The $^{\bullet}OH$ radicals are extremely reactive species, nonselective and powerful oxidizing agent.

Sonochemistry is one of AOPs that utilizes the potential of ultrasound (US) in degrading dissolved organics in aqueous solution. Sonication is a relatively novel AOP and it is based on the application of low to medium frequency (20–1000 kHz) and high-energy US to catalyse the destruction of organic pollutants in aqueous solutions (Mahvi et al., 2009). Sonochemical treatment typically operates at ambient conditions and does not require the addition of extra chemicals or catalysts.

In the present era, US has been considered as a model process, which has the potential for being used in the treatment of water, wastewater and sludge. Ultrasonic sound irradiation can effectively degrade different contaminants present in water because of the presence of localized high concentrations of oxidizing species. Ultrasound enhances the degradation of various chemical pollutants such as herbicides, pesticides, dyes, chloroaromatics, nitro compounds, phenols, and so on from the industrial effluents. The US technique is a feasible means of disintegrating sewage sludge and degrading or transforming wastewater pollutants. This technique has also been used for anaerobic sludge degradation of aerobic or anaerobic wastewater.

The advantage of US is the use of ambient temperatures to preserve thermally sensitive substrates and to enhance the selectivity. Sonochemical process affects the reaction rate, activate metals and solids by US.

Ultrasound can enhance or promote chemical reactions and mass transfer and offer the potential for shorter reaction cycles, cheaper reagents, and less extreme physical conditions (Goel et al., 2004). Ultrasound has been applied in studies of cleaning, organic synthesis, catalysis, extraction, emulsification, material processing, food processing, and wastewater treatment (Saleh and Gupta, 2012).

The use of US has some advantages over other conventional methods. These are:

- Ultrasound technique has a good potential to enhance the rate of many reactions.
- It has also the ability to be used with not very pure solvent and reagents.
- Sonochemical reactions can be conducted at relatively milder conditions.
- It helps in improving conversion and yield of reactions and sometimes, it may also change reaction pathway.
- It can be applied for the preparation of emulsions, leaching, grinding, and so on and
- It has proved to be effective in the treatment of effluent containing highly toxic organic pollutants.

8.3.1 DYES

Several industries used dyestuff because these are low-cost, have long-term stability, and easy to use. But wastewater generated by textile dyeing and printing industries is colourful and hard to degrade. Azo dyes (–N=N–) are the major contributors and accountable for 60–70% of all dyestuffs. This is the most common chromophore in reactive textile dyes. Reactive dyes have high solubility in water and, therefore, their loss to wastewater is also in significant amount. An estimate indicates about 15–20% of the synthetic textile dyes used are lost in wastewater streams during manufacturing or processing operations.

Moholkar et al. (2003) studied the effect of application of US on the intensification of mass transfer in wet textile processes. It was noted that the formation of standing waves assists ultrasonic wet textile processing by raising the power consumption of the system. Industrial textile pretreatment and finishing process require a relatively long residence time, large amounts of water and chemicals, and these are also energy-consuming. EMPA 101-test fabric was selected as a model for the cleaning process. It focuses on two aspects: first is the mechanism of the US-assisted cleaning process, and second is effect of the presence of the cloth on the US wave field generated in a bath. It was found that the dissolved gas content in the system plays an important role in the cleaning process. The cleaning effects thus observed are explained by two different mechanisms:

- Small-amplitude acoustic bubble oscillations
- Microjets resulting from the collapse of acoustic bubbles in the boundary layer between the fabric and the bulk fluid giving rise to convective mass transfer in the intra-yarn pores.

Ince and Tezcanli (2001) studied the degradation of reactive dye using US. They observed that the rate of degradation increases by a combined process involving sonolysis (520 kHz) and ozonation. Rehorek et al. (2004) reported the use of US for the degradation of various industrial azo dyes such as acid orange 5 and 52, direct blue 71, reactive black 5 and reactive orange 16 and 107. Ultrasound was able to mineralize these azo dyes to same nontoxic and final products. All these dyes were decolourized and degraded within 3–15 h at 90 W and within 1–4 h at 120 W, respectively. It was shown that hydroxyl radicals degraded these azo dyes by the simultaneous process of azo bond scission, oxidation of nitrogen atoms and hydroxylation of aromatic ring structures.

Vajnhandl and Marechal (2007) reported the decolourization and mineralization of reactive dye, reactive black 5 at three different frequencies of ultrasonic irradiation, that is, 20, 279 and 817 kHz. It was observed that the decolourization increased with increasing frequency, acoustic power and irradiation time; however, an increase in dye concentrations resulted in a decrease of decolourization rates. Bejarano and Suarez (2007) studied the degradation of malachite green by sonolytic, photocatalytic and sonophotocatalytic methods.

Basic violet 16 (BV 16) is a cationic dye which is highly water-soluble and nonvolatile. It is widely used in textile, leather industries, in preparing carbon paper, bull pen, stamp-pad inks and paints. This dye is also a well-known water tracer fluorescent. BV 16 directly harms skin, eyes and gastrointestinal and respiratory tract. It provokes phototoxic and photo allergic reactions also. Its carcinogenicity, reproductive and developmental toxicity, neurotoxicity and chronic toxicity towards humans and animals have already been established. Rahmani et al. (2012) reported the degradation of BV 16 by ultraviolet radiation, ultrasonic irradiation (US), UV/H_2O_2, and US/H_2O_2 processes in a laboratory-scale batch photoreactor. This reactor was equipped with a 55 W immersed-type low-pressure mercury vapour lamp and a sonoreactor with high-frequency (130 kHz) plate-type transducer at 100 W of acoustic power. Basic violet 16 was almost degraded 99%, with UV/H_2O_2 within 8 min, whereas relatively low decolourization efficiency was observed using UV (23%), US (<6%) and US/H_2O_2 (<15%) processes.

Mishra and Gogate (2011) studied the degradation of rhodamine B using sonochemical reactors in combination with UV irradiations at operating capacity of 7 l. The degradation of naphthol blue black (NBB), a diazo dye widely used in the textile and soap industries, was carried out using US at 278 kHz. Sonochemical decolourization was enhanced by bicarbonate ions at low concentration of NBB, due to the formation of carbonate radicals, which are likely to migrate in the bulk of liquid and attack the substrate (Dalhatou et al., 2015).

Guzman-Duque et al. (2011) reported the ultrasonic degradation (800 kHz) of crystal violet (CV) under the effects of saturating gas (argon, carbon dioxide and air), concentration of CV 2.45–1225 $\mu mol \cdot L^{-1}$, pH (3–9) and power (20–80 W). The best performance was achieved at 80 W with argon as a saturating gas. Analyses of intermediates by gas chromatography–mass spectrometry (GC–MS) indicated several sonochemical by-products: *N,N*-dimethylaminobenzene, 4-(*N,N*-dimethylamino)-4′-(*N′,N′*-dimethylamino) benzophenone, and *N,N,N′,N′*-tetramethyl-4,4′-diaminodiphenylmethane. The presence of these aromatic structures showed that the main ultrasonic CV degradation pathway was due to oxidation by $^{\bullet}OH$ radicals. At the completion of the treatment, these intermediate products were finally converted into biodegradable organic by-products, which could be easily treated in a subsequent biological treatment.

The degradation of eosin Y and rhodamine B dyes was carried out using iron oxide nanoparticles under simultaneous irradiation with light and US (Gobouri, 2016). He et al. (2011) also carried out the degradation of MO using light and US. Similarly, Sandhyaa et al. (2013) reported the use of titania, titania-polyaniline and clay-titania-polyaniline composites for the degradation of acid blue 25, acid orange 7 and MB under visible light and US.

Wang et al. (2015b) carried out the degradation of MO with Fe_3O_4/polyaniline (Fe_3O_4/PANI) microspheres in near neutral solution (pH ~6) using US. The percentage of sonocatalytic degradation of MO with Fe_3O_4/PANI is about 4.8, 8.8 and 5.7 times that with Fe_3O_4, dedoped Fe_3O_4/PANI and ultrasonication (US) alone, respectively. Cui et al. (2011) also reported the degradation of MO using ultrasonic-enhanced ceramic membrane MF and TiO_2. Ceramic membrane MF could efficiently recover TiO_2 photocatalyst with a mean granular size of 0.33 μm from slurry reactor, thus, achieving 99.9% recovery rate.

Alahiane et al. (2014) reported the photocatalytic degradation of the synthetic textile dye reactive yellow 145 (RY 145) in aqueous solution, using TiO_2 coated nonwoven fibres as photocatalyst, under UV-lamp irradiation and US. The removal of reactive yellow 145 on TiO_2-coated nonwoven fibres followed Langmuir–Hinshelwood model. Degradation of MO was also carried out using CNT/TiO_2 composites and US (Wang et al., 2009). The polyaniline/zinc oxide (PANI/ZnO) nanocomposites were used for the degradation of MB dye. The PANI/ZnO nanocomposites showed three times higher photocatalytic activity to MB dye degradation as compared to pristine PANI, which may be attributed to high photogenerated electron–hole pair separation (Ameen et al., 2011).

Gemeay et al. (2012) also reported the degradation of acid blue 25 dye (AB-25) by hydrogen peroxide using polyaniline/manganese dioxide

($PANI/MnO_2$) composites and US. The decomposition of unhydrolysed and hydrolysed forms of reactive blue 19 dye (RB 19) was carried out using the US-assisted electrochemical process (Siddique et al., 2011).

A new hybrid material composed of $BiVO_4$ and HKUST-1 MOF (a metal organic framework made up of copper nodes with 1,3,5-benzenetricarboxylic acid struts between them, that is, HKUST-1-MOF–$BiVO_4$) was synthesized (Mosleh et al., 2016). It was found active under blue light irradiation. This sonophotocatalyst was used for the degradation of disulphine blue and rose bengal (RB) in the presence of light and US. FeS/ZnO nanoarrays also have high sonophotocatalytic activity for degrading MB in water under ultrasonic and solar irradiation (Guo et al., 2016).

Panneerselvam et al. (2014) reported the degradation of acid blue 113 in the presence of visible light-driven rare-earth nanoclusters loaded TiO_2 nanophotocatalysts. Gd^{3+}, Nd^{3+} and Y^{3+} were used for leading 1 wt. % using a low-frequency sonicator of rare earth nanoclusters (42 kHz). Khan et al. (2015) carried out the degradation of reactive blue 19 (RB 19) dye successfully using sulphur-doped TiO_2 (S-TiO_2) nanoparticles under visible light and US.

ZnO nanoparticles decorated on multiwalled carbon nanotubes (ZnO/MWCNTs composites) were used for the degradation of rhodamine B using US (Ahmad et al., 2014). The sonocatalytic degradation of some textile dyes such as MB, texbrite BAC-L, texbrite BBU-L, and texbrite NFW-L was carried out under ultrasonic irradiation, and compared with ZnO/graphene (ZG), graphene, ZnO and TiO_2, respectively. The results indicated that ZnO/graphene/TiO_2 (ZGT) displayed higher US-activated sonocatalytic activity than the other catalysts (Nuengmatcha et al., 2016). Tang et al. (2012) reported the use of TiO_2-coated activated carbon (TiO_2/AC) for the sonocatalytic degradation of rhodamine B dye.

8.3.2 HYDROCARBONS

Hydrocarbons are a class of compounds, which are primarily composed of carbon and hydrogen only. These are major components of oil, natural gas and pesticides. These hydrocarbons contribute not only to the greenhouse effect and global warming, and depleting the ozone but also increase in the occurrence of cancer and respiratory disorders along with reducing the photosynthetic ability of plants.

Aromatic hydrocarbons are produced from the combustion of coal, oil, tar and plant materials. Benzene is used as a solvent and in fuels. It depletes red blood cells, causes cancer in mammals and damages bone marrows. Marine oil pollution is becoming a major problem today. Oil and oil products spill, cause severe damages to the ecosystems in marine system and the amount of this pollution is expected to increase further in the future due to:

- An increase in number of ships
- The use of shelf zones for oil drilling
- Poor control to prevent oil pollution in the seas

Visscher et al. (1996) reported the degradation of benzene, ethylbenzene, styrene and *o*-chlorotoluene in aqueous solution using ultrasonic waves. The basic assumptions of the model are first-order pyrolysis in the cavitations yielding both reactive/volatile and inert/nonvolatile products, and lowering of the maximum cavitation temperature due to the presence of the organic compounds in the bubble phase.

Wu et al. (2004) reported the sonolytic degradation of cyclic hydrocarbons such as benzene, 1,3-cyclohexadiene, 1,4-cyclohexadiene and cyclohexene. Sonolysis was found to follow first-order kinetics and depends on the solubility of the substrates in water. The rate of degradation was inversely proportional to the concentration of substrate. Benzene showed more resistance against degradation as compared to cyclohexene.

Hydrocarbons in the middle distillate range (C8–C26) have been treated with US at 20 kHz, a frequency sufficient to drive acoustic cavitation. The high temperatures experienced as a result of the implosion of fuel vapour bubbles are sufficient to produce pyrolytic degradation and dehydrogenation. During this process, formation of small particles takes place that have similarities with the primary soot particles produced during diesel combustion. The particles are carbonaceous in character as it is a mixture of amorphous and graphitic-like carbon. The mass of material produced increases with the C/H atomic ratio of the hydrocarbon undergoing cavitation (Price et al., 2015).

Psillakis (2004) reported the degradation of naphthalene, acenaphthylene and phenanthrene in water using US. A horn-type sonicator was used to deliver the US energy, whereas immersion sampling solid-phase microextraction coupled with GC–MS was employed to follow concentration–time profiles of the selected polyaromatic hydrocarbons (PAHs).

Littl et al. (2002) also carried out the degradation of naphthalene, acenaphthylene and phenanthrene in water using US. The effects of increasing sonication time (60–150 min), calcium chloride (1–10 $g \cdot L^{-1}$), ferrous oxide (2·10 $mg \cdot L^{-1}$), aluminium oxide (2–10 $mg \cdot L^{-1}$) and sodium carbonate (2–12 $mg \cdot L^{-1}$), on the destruction of six polycyclic aromatic hydrocarbons and acute toxicity in a petrochemical industry wastewater, were investigated at a sonication intensity and a frequency of 251.48 $W \cdot cm^{-2}$ and 35 kHz, respectively. The contribution of 4–6 $mg \cdot L^{-1}$ $CaCl_2$ to the yields in more hydrophobic PAHs (benz[a]pyrene (BaP) and indeno[1,2,3-cd]pyrene (IcdP) was low (6–8%), whereas it was high (21–26%) in less hydrophobic PAHs such as acenaphthene, fluoranthene, benz[a]anthracene, benz[k]fluoranthene at 30°C after 150 min sonication (Sponza and Oztekin, 2010).

Okuno et al. (2000) reported the degradation of benzene and biphenyl in aqueous solutions using 200 kHz US at an intensity of 6 $W \cdot cm^{-2}$ under an argon atmosphere. Nearly 80–90% of initial amount of these compounds was degraded in 30–60 min of sonication, when the initial concentrations were 10–100 $\mu mol \cdot L^{-1}$. The degradation rate of these compounds increased with increase in their vapour pressures.

8.3.3 CHLORO COMPOUNDS

Chlorinated compounds are one of the most widespread pollutants in our environment (Vogel et al., 1987) and, because of this, a large number of techniques, including traditional techniques (Hitchman et al., 1995), sequential combinations of techniques (Entezari et al., 2005), hybrid (Adewuyi, 2005) and new technologies (Shannon et al., 2008) are continuously being developed to provide an efficient solution to this problem. Among these methods, sonochemical treatment has received a lot of attention due to its special features. Ultrasound is a unique means of interacting energy and matter, and it differs from traditional energy sources such as light, heat or ionizing radiation in duration, pressure and energy per molecule (Suslick, 1990).

Drijvers et al. (1996) reported that the degradation of trichloroethylene (TCE) was the fastest in basic solutions saturated with argon. TCE was not degraded in the bulk solution. Some volatile and nonvolatile organic degradation products were formed during sonication. The most important volatile compounds were identified as C_2HCl, C_2Cl_2, C_4Cl_2, C_2Cl_4, C_4HCl_3, C_4Cl_4, C_4HCl_5 and C_4Cl_6. Those products are typical in the pyrolysis of TCE. These intermediates are formed in single cavitation events and disappeared from the aqueous solution as well.

Degradation of CB in aqueous solution was carried out using US (Drijvers et al., 1998). The addition of the radical scavenger, benzoate, demonstrated that no significant degradation took place in the bulk solution. Different degradation products were identified as methane, acetylene, butenyne, butadiyne, benzene, chlorophenols, phenylacetylene and other chlorinated and nonchlorinated monocyclic and dicyclic hydrocarbons in air-saturated solutions, whereas same products were found, except for the chlorophenols in case of argon-saturated solutions. The presence of the chlorophenols in the case of air-saturated solutions only supported the interaction between the radicals formed and oxygen, and not direct degradation by $^{\bullet}OH$ radicals. The kinetics of several organic degradation products and chloride were also studied for the sonolysis of air- and argon-saturated solutions.

Griffing (2004) reported that the rate of the hydrolysis of CCl_4 is strongly dependent on the dissolved gas. Weissler et al. (1950) confirmed the effect of dissolved gases (O_2, N_2, He, CO_2) on the yields of iodine during the ultrasonic irradiation of KI solutions in the absence of CCl_4 and with a large excess of CCl_4. Ultrasonic degradation of carbon tetrachloride leads to chlorine formation. Carbon tetrachloride was used as an additive in this sonochemical degradation. It was shown that carbon tetrachloride was found to improve the degradation efficiency of the sonochemical process. Bhatnagar and Cheung (1994) reported that the degradation of TCE and carbon tetrachloride remained constant between −7 and 20 and 20–60°C, respectively.

Destaillats et al. (2000) indicated that the sonochemical degradation of CB and TCE, increased with increasing temperature. Degradation of dilute aqueous solutions of 2-, 3- and 4-chlorophenols and pentachlorophenol was carried out under air or argon atmosphere using US. This degradation follows first-order kinetics in the initial state with rates in the range of 4.5–6.6 $mM\cdot min^{-1}$ under air and 6.0–7.2 $\mu M\cdot min^{-1}$ under argon at a concentration of 100 μM of chlorophenols. The rate of $^{\bullet}OH$ radical formation from water is 19.8 $\mu M\cdot min^{-1}$ under argon and 14.7 $\mu M\cdot min^{-1}$ under air in the same sonolysis conditions (Nagata et al., 2000).

Pandit et al. (2001) carried out degradation of 2,4,6-trichlorophenol at 20 kHz using US in the presence of TiO_2. Cresol and 2-chloro-5-methylphenol (2C5MP) are used in the manufacture of resins, herbicides, pharmaceuticals and surfactants. Degradation of an aqueous solution of 2C5MP in the presence of TiO_2 and H_2O_2 in combination with US (33 kHz) was reported by Laxmi et al. (2009). The order of the effect of inorganic ions on degradation rate of 2C5MP was found as:

$$Cl^- > SO_4^{2-} > HPO_4^{2-} > HCO_3^-$$

Perchloroethylene is used as a solvent and degreaser. The degradation of perchloroethylene (PCE) in aqueous solution was carried out using ultrasonic irradiation and photochemical oxidation by ultraviolet rays and hydrogen peroxide (Kargar et al., 2012). Saez et al. (2010) also carried out the degradation of PCE using US. The combination of US and electrical fields provides a favourable reaction environment for PCE aqueous solution treatment, which significantly improves on the sonochemical and electrochemical treatments.

Barik and Gogate (2016) reported the degradation of 4-chloro-2-aminophenol (4C2AP) using ultrasonic reactors and combination of US with photolysis and ozonation. Sivasankar and Moholkar (2010) reported about the degradation of 2,4-dichlorophenol (2,4-DCP) using US. The degradation of 4-chlorophenol (4-CP) in aqueous media was carried out by 516-kHz ultrasonic irradiation. The degradation of concentrated 4-CP solution by means of US, UV irradiation, and their combination was also studied (Hamdaoui and Naffrechoux, 2008).

The use of US was made in the degradation of polychlorinated biphenyl (PCB)-contaminated sediments by Lu and Weavers (2002). Li and Zhu (2015) reported the degradation of 1,1,1-trichloroethane (TCA) using US. The degradation of 2-chloro-5-methyl phenol (2C5MP) was carried out in aqueous solution by US in the presence of TiO_2 and H_2O_2. Maximum degradation rate of 2.66×10^{-2} was accomplished by US/TiO_2/H_2O_2 process, when compared to US/TiO_2 (1.01×10^{-2}) and US/H_2O_2 (5.5×10^{-3}) systems. The impact of synthetic additives such as CCl_4 as hydrogen radical scavenger and CH_3OH as hydroxyl radical scavenger was studied, which confirmed the involvement of hydroxyl radicals in the degradation of 2C5MP (Laxmi et al., 2010).

Ultrasonic degradation of 2,4-dichlorophenol (2,4-DCP) was carried out under oxygen, air, argon, and nitrogen atmosphere in aqueous solutions for a liquid temperature of 20°C at 489 kHz. The degradation rate was found to increase in the presence of O_2 and argon, whereas it remains the lowest under nitrogen (Uddin and Hayashi, 2009). Experiments have also been performed at three different pH, 11.0, 6.3 and 2.0. A significant degradation was achieved at pH 6.3 under O_2 (0.86×10^{-3} s^{-1}), which is 1.9 and 4 times higher than the acidic (pH 2.0) and basic (pH 11.0) conditions, respectively. The degradation rates have decreased in the order:

$$O_2 > Ar > air > N_2 \text{ irrespective of pH.}$$

Chlorobenzene, 1,2-, 1,3-, 1,4-dichlorobenzene, and PCBs such as 2-,4-chlorobiphenyl and 2,2′-dichlorobiphenyl in aqueous solutions were subjected to sonolysis with 200 kHz US at an intensity of 6 $W \cdot cm^{-2}$ under an argon atmosphere. Nearly 80–90 % of initial amount of these compounds were degraded by 30–60 min of sonication when the initial concentrations were 10–100 $\mu mol \cdot l^{-1}$. The degradation rate of these compounds increased with increase in their vapour pressures. In all the cases of sonolysis of chlorinated organic compounds, an appreciable amount of liberated chloride ion was observed (Okuno et al., 2000).

Jiang et al. (2002) reported the ultrasonic degradation of chlorobenzene (ClBz), 1,4-dichlorobenzene and 1-chloronaphthalene at 500 kHz. The disappearance of ClBz on sonication is almost simultaneously accompanied by the release of chloride ions as a result of the rapid cleavage of carbon–chlorine bonds with a release of CO, C_2H_2, CH_4 and CO_2. The intermediates resulting from the attack of $^{\bullet}OH$ radicals were detected but in a quite low yield (<2 μM). The generation of H_2O_2 on sonolysis was not significantly affected by the presence of aqueous ClBz, whereas the generation of NO_2^- and NO_3^- is inhibited initially. The degradation of *p*-chlorophenol in water using high-frequency US (1.7 MHz) was carried out. The *p*-chlorophenol was degraded successfully in the high-frequency ultrasonic device with low consumption of energy. No products or intermediate products were detected in the reaction mixture by MS and ^{1}H-NMR after ultrasonic irradiation (Hao et al., 2004).

Neppolian et al. (2004) studied the degradation of *p*-chlorobenzoic acid (a model compound for free radical-mediated reactions) using US and Fenton-like oxidation (FeOOH–H_2O_2). The oxidation rate of *p*-CBA was measured at various concentrations of H_2O_2 and FeOOH particles and pH conditions at pH values <3.98, which is pK_a of *p*-CBA, showed significantly better degradation of *p*-CBA than at higher pH values from 5 to 9. The rates of degradation of *p*-CBA by FeOOH–H_2O_2 were enhanced in the presence of US. This rate enhancement for the degradation of *p*-CBA may be due to the continuous cleaning and chemical activation of the FeOOH surfaces by acoustic cavitation and the accelerated mass transport rates of reactants and products between the solution phase and the FeOOH surface.

Lin and Ma (2000) studied the use of US (20 kHz) for the degradation of 2-chlorophenol in the presence of Fe^{2+} and H_2O_2. This degradation reaction was monitored by measuring its redox potential that increases slightly with increasing amounts of Fe^{2+}. The main intermediate formed during the decomposition was identified as 2-chloro-p-benzoquinone, which readily decomposed during further treatment.

Goskonda et al. (2002) have reported the degradation of chlorophenols with US (920 kHz). It was found that the rate of mineralization was dependent on various factors such as the structure and concentration of substrate, temperature, pH and the presence of other solutes (detergents and humic acids).

Aqueous solutions of CB and PCBs have been degraded under 200 kHz sonication under an atmosphere of argon. More than 80% of the initial amounts of substrates degraded after 30 min sonication, and it was found that degradation rates increased with their increasing vapour pressure. The effect of US on CB in water containing iron and palladium sulphates was observed by Stavarache et al. (2002). The degradation of CB begins with pyrolysis inside the cavitation bubble giving, initially, benzene radicals, which migrate outside the bubble and are converted into benzene.

Pentachlorophenol (PCP) has been used for a long period as an important component of biocides, pesticides, insecticides and wood preservatives, but now it has become a source of priority pollutant. Weavers et al. (2000) reported the degradation of aqueous solutions of PCP using 20 and 500 kHz US as well as in the presence of ozone. Here, pyrolysis and oxidation by hydroxyl radical both processes occur. It was reported that the addition of ozone did not affect the degradation constant of PCP.

Peters (2001) studied remediation of ground water contaminated with some simpler, but more volatile chlorinated alkanes and alkenes. They concluded that chlorinated hydrocarbon (1, 2-dichloroethane) was completely decomposed in 60–120 min in pure and natural water using US at frequencies of 20, 206, 361, 620, and 1086 kHz. The destruction rate in deionized water and in ground water was found to depend only on the dose rate and not on the frequency of US. However, the energy efficiency decreases with higher frequency.

8.3.4 SULPHUR-CONTAINING COMPOUNDS

Crude oil contains numerous aromatic sulphur-containing compounds such as sulphides, thiophenes, benzothiophene, dibenzothiophene (DBT), and their substituted derivatives. During distillation and refining of crude oil, these compounds end up in the gasoline and diesel fractions. Combustion of these compounds in vehicle engines results in the generation of SO_2 and particulate matter emission. Growing concerns over air pollution created by these emissions have led to stringent restrictions on the sulphur content of the liquid fuels (Bolla et al., 2014).

Sulphur compounds in transportation fuels are converted to SO_x by combustion, leading to acid rain and air pollution. In many countries, the average sulphur content in diesel has been restricted to 10 $\mu g \cdot ml^{-1}$, and it may be reduced even to zero level in the near future. Hydrodesulphurization process is a very efficient technique for removing simple sulphur compounds, but it is not effective for DBT and its derivatives. This method is not suitable for deep desulphurization of diesel, because it requires severe operation conditions, high-energy and hydrogen consumption (Faghihian and Sadeghinia, 2014).

Chemical warfare agent (CWA) is a chemical substance, which has toxic properties to kill, injure, or incapacitate human beings. Decontamination of CWA involves the conversion of toxic chemicals into harmless products by degradation. Solvent extraction process is considered one of the best methods for decontamination of sensitive CWAs. This is a type of physical decontamination, which is relatively fast and nondestructive. The efficiency increases on combining its ultrasonics with higher temperature.

Dimethyl methylphosphonate (DMMP) is a simulant for nerve CWAs. The role of 20 kHz US in photocatalytic oxidation of DMMP was studied in a batch reactor by Chen et al. (2003). It was observed that DMMP did not undergo mineralization under low-frequency (20 kHz) ultrasonic irradiation. An increase in the rate of DMMP photocatalytic mineralization in the presence of US was achieved. It was not due to deagglomeration of titania, but it was associated with an enhanced mass transport of reagents.

Similar intermediate products were detected in both photocatalytic and sonophotocatalytic degradations. Apparent rate constants of DMMP mineralization increase under sonication in all stages. A reaction route for DMMP mineralization under US was found to be without intermediate products. Such a behaviour was attributed to enabling mass transport of DMMP into micropores and to the surface of TiO_2.

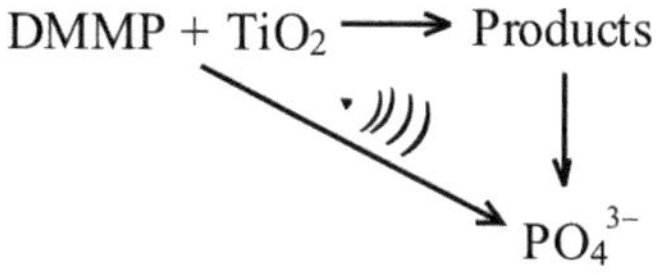

An effort was made to develop some novel photocatalysts for the destruction of some CWAs such as HD, G-type (GA, GB, GD) and VX to harmless products in gas and liquid phases (Smirniotis, 2002). Some visible light/solar active photocatalysts were synthesized, characterized and finally used for the degradation of CWA. Ultrasound was found to enhance the photochemical transformation and sidewise, it also avoids the formation of any toxic intermediates. Photodegradation of diethylsulphide (DES) in gas phase as a simulant of mustard gas (HD) over various titanias was carried out utilizing UV light. It was revealed that main routes of DES degradation are C–S cleavage, S oxidation and α-C and β-C oxidation.

Vorontsov et al. (2002) carried out photocatalytic and sonophotocatalytic treatment of some CWAs such as DMMP, diethyl phosphoramidate (DEPA), pinacolyl methylphosphonate (PMP), butylaminoethanethiol (BAET) in aqueous suspensions of TiO_2. It was observed that complete conversion of these imitants to inorganic products was obtained within 600 min for DMMP, DEPA, PMP, but it required a longer time to degrade BAET. Ultrasound was found to enhance photodegradation of DMMP.

Decontamination of CWAs is usually carried out either by oxidation or by basic hydrolysis (Koskela et al., 2007). Most decontaminants such as hypochlorite bleach and super tropical bleach solutions have limited ability to solve organic CWAs and remove them from contaminated surfaces, as well as they are mostly corrosive in nature (Kim et al., 2011). Decontamination solution 2 (DS2) is an organic medium, but it has certain unwanted effects on other organic parts of contaminated objects such as its paints, and so on (Fallis et al., 2009).

Wagner et al. (2008) reported hydrolysed reactions of *o*-ethyl S-[2-(diisopropylamino)ethyl] methylphosphonothioate) (VX), pinacolyl methylphosphonofluoridate (GD), and bis(2-chloroethyl) sulphide (HD) on nanotubular titania and anatase TiO_2 as a reactive sorbent. The result is significant so that VX is typically the least reactive, most persistent, and most toxic CWA. The use of semiconductors photocatalysis for the degradation of CWAs and corresponding simulants was reported by Vorontsov et al. (2003). Prasad et al. (2011) reported the decontamination of HD on ZnO nanoparticles by US whereas decontamination of sulphur mustard was carried out using titania nanoparticles (Ramacharyulu et al., 2012). Decomposition of 2-chloroethyl ethyl sulphide was carried out using TiO_2 and US by Thompson et al. (2004).

Ultrasound-assisted oxidative desulphurization process (UAOD) was used to reduce sulphur compounds of gas oil containing various types of sulphur contents. UAOD is a promising technology with lower operating cost and higher safety and environmental protection. The typical phase-transfer agent (tetraoctyl-ammonium-bromide) was replaced for the first time with isobutanol as it is much more economical than TOAB, and imposing no contamination (Shayegan et al., 2013).

Margeta et al. (2016) reported the degradation of hydrocarbon fuels, diesel fuels and produced wastewater using US. The US-assisted oxidative desulphurization (UAOD) process of diesel fuel has gained an increasing attention due to the regulation of sulphur content in the fuel. The conversions of DBT ODS and UAOD tests were 36 and 87%, respectively in 30 min. Subsequent extraction with acetonitrile resulted in the sulphur removal of 96.5%.

The impact of US application for oxidative desulphurization (ODS) of hydrocarbon fuels was investigated by Margeta et al. (2016). The model diesel fuel with initial sulphur concentration of 1220–3976 mgL^{-1} was used for ODS and UAOD tests, and hydrogen peroxide/acetic acid was applied as the oxidant/catalyst system, respectively.

8.3.5 NITROGEN-CONTAINING COMPOUNDS

Pyridine and its derivatives are of major concern as environmental pollutants due to their recalcitrant, toxic, and teratogenic nature (Stapleton et al., 2007). Pyridine is produced from coal tar and as a by-product of the coal gasification. Increased demand for pyridine resulted in the development of more economical methods of synthesis from acetaldehyde and ammonia, and more than 20,000 tons per year are manufactured worldwide. Thus, researchers have long sought to develop effective, economically feasible techniques for cleaning the atmosphere of wastes such as pyridine (Mittal et al., 2008). Ultrasonication not only promotes oxidative degradation of pyridine by hydroxyl radicals but also provides a possible route for its thermal decomposition in the gas phase (Naffrechoux et al., 2000).

Kotronarou et al. (1991) investigated the influence of the changes in the initial pH of the *p*-nitrophenol (PNP) solutions on the decay of PNP under US. PNP decayed exponentially with time at all pH values. The pseudo-first-order rate constant decreased with the increase of pH (pH 5 to 8), and remained constant up to pH 10.

A simple and an efficient sonochemical method was developed for the degradation of organic matter and ammonia nitrogen in azodicarbonamide (ADC) wastewater. The removal of organic matter and ammonia nitrogen in ADC wastewater by a combination of power US radiation and hydrogen peroxide was investigated by Wenjun et al. (2012). The effects of reaction time, initial pH, ultrasonic format and peripheral water level on the chemical oxygen demand (COD) reduction and ammonia removal efficiency were studied. Higher initial pH was found favourable to the sonolysis of hydrazine, but this effect on the degradation of urea is relatively small.

Song et al. (2007) reported the degradation of aniline using US in the presence of ozone. It was found that only a little decomposition was observed with US only, whereas using a combination of US and ozone gives complete removal after 25 min. By-products such as nitrobenzene (NB), *p*-benzoquinone, *p*-aminophenol and low-molecular weight carboxylic acids were observed during this degradation.

Mishra and Gogate (2012) carried out the degradation of PNP using ultrasonic reactors at pilot-scale operation. The effect of air sparging, addition of CCl_4, solid particles such as TiO_2 and CuO, and ozone on the extent of degradation of PNP was studied. The sonochemical reactor used was an ultrasonic longitudinal horn-type reactor operating at 25 kHz frequency, power output of 1 kW, and capacity of 7 l. It was observed that the extent of degradation increased for all the additives used due to enhanced cavitational activity and generation of additional free radicals or oxidizing species.

Morteza and Tabatabae (2011) reported the degradation of 2-butyl-4,6-dinitrophenol in aqueous solution using titanium dioxide nanoparticles and US. Oxidative degradation of dinitrotoluene and 2,4,6-trinitrotoluene in wastewater was carried out using electrochemical and electro-Fenton processes, respectively, combined with ultrasonic irradiation. The influence of various operating variables on the sonoelectrolytic behaviour, such as electrode potential, sonoelectrolytic temperature, acidity of wastewater, oxygen dosage, and dosage of ferrous ions, was also observed. The nitrotoluene contaminants could be completely decomposed by sonoelectro-Fenton method, where hydrogen peroxide was in situ generated from cathodic reduction of oxygen and was also supplied partially by the anodic oxidation of water (Chen and Huang, 2014).

Verma et al. (2013) reported the degradation of 4-chloro-2-nitrophenol (4C2NP) using heterogeneous TiO_2 and US. Experiments were performed in slurry mode with artificial UV-125 W medium-pressure mercury lamp coupled with US (100 W, 33+3 kHz) for sonication of the slurry. The degradation follows the trend:

Sonophotocatalysis > Photocatalysis > Sonocatalytic > Sonolysis.

The sonocatalytic degradation of NB in aqueous solution was carried out using H_2O_2 oxidant and nanostructured metal oxide catalysts. Nanostructured metal oxide catalysts was prepared in the absence of CTAB (series I), (Fe_2O_3-I, CuO-I, NiO-I, and Co_3O_4-I) and nanostructured metal oxide catalysts prepared in the presence of CTAB (series II), (Fe_2O_3-II, CuO-II, NiO-II, and Co_3O_4-II). The potential degradation feasibility for NB in US/nano-metal oxide (series I and/or series II)/H_2O_2 systems was measured. The effect of various operational parameters such as ultrasonic irradiation time, solution pH value, H_2O_2 concentration, different weights of the nanosized catalysts, NB concentration, leaching of the nanosized catalysts, and reaction kinetics was examined. H_2O_2 oxidant can effectively assist the sonocatalytic degradation of NB in the presence of nanosized oxides (series I) and with more efficiency in the presence of nanosized oxides (series II) (Yehia et al., 2015a). The effect of efficiency in degradation was in the order of:

$$Fe_2O_3 > CuO > NiO > Co_3O_4$$

The influence of surface modification of some metal oxides (Fe_2O_3, CuO, NiO, and Co_3O_4) with amino acids was investigated in the sonocatalytic degradation of NB (Yehia et al., 2015b). Glycine, arginine and glutamic acid were used as surface modifiers. This modification resulted in a considerable enhancement of sonochemical degradation of NB at pH>7, with respect to the pristine catalysts before modification. Catalysts modified with arginine exhibited the highest degradation of NB. The enhanced degradation efficiency was attributed to the promoted coupling via electrostatic attraction between the negatively charged NB molecules and the positively charged arginine layer on the surface of Fe_2O_3, CuO, NiO and Co_3O_4. The attraction capability at all modified metal oxides increased via secondary forces such as hydrogen bonding, n–π and π–π interactions.

Nakui et al. (2007) studied the sonolysis of 4-nitrophenol in argon-saturated water using frequency of 321 kHz. 4-Nitrophenol was fully deprotonated at pH 10 and therefore, it was concluded that the degradation proceeded by hydroxyl radical action.

8.3.6 PHENOLS

Phenol is released to the environment from effluents discharge by industries such as petroleum refining, coal tar, steel, dyestuff, synthetic resins, coal gasification and liquefaction, surface runoff from coal mines, by-products of agricultural chemicals, paper and pulp mills, tanning, fibreboard production, paint stripping operations, pesticides, medications, pharmaceuticals, and even from food-processing industries (Lathasreea et al., 2004).

Phenol has to be handled with utmost care because it causes immediate white blistering on the skin. The major hazard of phenol is its ability to penetrate through the skin rapidly, particularly when in liquid state, causing severe injury, which can also be fatal sometimes. Phenol also has a strong corrosive effect on body tissue causing severe chemical burns. Phenols have attracted public attention because their exposure can result in acute and chronic effects to human health. Due to its toxicity, hazardous character and increasing social concern on the environment, the discharge of such compounds in the environment has been restricted. Therefore, there is more demand for some improved methods of treatment containing phenol (Beltran et al., 2005).

The rates of ultrasonic degradation of phenol can be increased by using simple sodium chloride addition to its aqueous phenol (Seymore and Gupta, 1997). They also reported that pyrolytic destruction of phenol in the gas phase is negligible, and the degradation occurs mainly in the bulk solution, which is accelerated with the addition of Fenton's reagent. The most effective ultrasonic frequency for destroying phenol in solution was found to be 200 kHz, which has been explained by a larger potential for the ejection of radical species to the bulk solution at this frequency (Petrier and Francony, 1997).

The rate of degradation by the 487 kHz was 10 times more due to the much faster hydroxyl radical transfer from the gas phase to the aqueous solution at this frequency as compared to 20 kHz (Petrier et al., 1994). They reported that the larger efficiency observed at 487 kHz is due to the fact that radicals are concentrated at the surface of the solution at these higher-frequency conditions, whereas at 20 kHz frequency, they are located intensely around the tip of the transducer.

Wu et al. (2001) have studied the effect of UV irradiation on sonochemical degradation of phenol. They reported that total phenol destruction was possible, the degree of mineralization was not more than 20.6%; however, this could be effectively improved by the addition of Fenton's reagent to the solution.

Singh et al. (2016) carried out the degradation of phenol using ultrasonically dispersed nano-metallic particles (NMPs). The probable mechanism of phenol degradation by ultrasonically dispersed NMPs was the oxidation of phenol caused by the hydroxyl radicals. These radicals are produced by the reaction between H_2O_2 and the NMPs during the US process.

Some phenolic compounds, such as model pollutants in aqueous solution using ZnO-immobilized horseradish peroxidase (HRP), were studied (Zhang et al., 2016). The ZnO-immobilized HRP exhibited higher removal efficiencies for these phenolic compounds. The irradiation of microwaves and US can further promote the oxidation of primary intermediates in the conversion experiment, whereas the immobilization showed an obvious protective effect on HRP against the inactivation/inhibition effects.

Yehia et al. (2016) reported the degradation of phenol using nanosized zero-valent copper or nanosized zero-valent nickel in combination with H_2O_2 and 20 kHz US irradiation. Catalytic wet air oxidation of phenol over copper catalysts supported by CeO_2-ZrO_2 under atmospheric pressure of oxygen at 160°C in a stirred batch reactor using US was carried out (Parvas et al., 2014a). They also reported the degradation of phenol using nonnoble metal Ni with different loadings coated on precipitated CeO_2–ZrO_2 using US and catalytic wet air oxidation Parvas et al. (2014b).

The combination of peracetic acid and heterogeneous catalyst (MnO_2) was used for the degradation of phenol in an aqueous solution in the presence of US irradiation (Rokhina et al., 2013). Ren et al. (2014) carried out the degradation of triclosan in aqueous solutions with high-frequency US (850 kHz) and various electrodes. Diamond-coated niobium electrode showed the best results as an electrode, leading to effective degradation and positive synergistic effect.

The combination of 50 kHz US and ozone was used for the degradation of phenol (Yang et al., 2010). The effect and different conditions of degradation of phenol in wastewater samples was studied using US and H_2O_2 (Wu et al., 2015). Anju et al. (2012) reported the sonocatalytic activity of ZnO, TiO_2 and ZnO-TiO_2 for the degradation of phenol pollutant in water. The efficiency of the catalysts for the degradation was found to be in the following order:

$$ZnO\text{-}TiO_2 > ZnO > TiO_2$$

At lower concentrations of ZnO, the percentage degradation in the presence of coupled ZnO-TiO_2 was more than the sum of the degradation achieved in the presence of these individual oxides under identical conditions, implying that there is a synergistic effect. Mahvi and Maleki (2010) reported the degradation of phenol in water using light and US. Identification of the first intermediates of the reaction (hydroquinone, catechol, benzoquinone, and resorcinol) indicated that hydroxyl radicals are involved in the photosonochemical degradation mechanisms.

Entezari et al. (2003) reported the degradation of phenol using hydrogen peroxide, US and copper sulphate (Fenton-like conditions). The degradation of phenol was higher at 500 kHz than at 35 or 20 kHz, when water alone was used. When hydrogen peroxide and copper sulphate (Fenton-like conditions) were used, the most efficient system was at 35 kHz frequency. Thus, additives like hydrogen peroxide and copper sulphate can be used to enhance the sonodegradation of phenol. The degradation of 2-butyl-4,6-dinitrophenol in aqueous solution was carried out using titanium dioxide nanoparticles and US (Morteza and Tabatabaei, 2011).

Cravotto et al. (2008) have studied an integrated system by using a combination of sonochemical and microbial treatment for the removal of nonylphenols from water. Polluted water containing 1000 ppm of nonylphenol was subjected to the sono-Fenton process and then it was followed by biosorption by the fungus *Paecilomyces lilacinus*. A decrease in nonphenol content of 90% was achieved by treating the sample with 300 kHz US in the presence of Fe_3O_4/H_2O_2 for 1 h and then inoculating the fungus and culturing it for 7 days.

8.3.7 PESTICIDES

Fresh water pollution with insecticides, pesticides, herbicides, weedicides, and so on is an environmental problem due to the high toxicity and nonbiodegradability of these compounds.

Diazinon and malathion are highly effective pesticides, which are extensively applied in agriculture (Liu et al., 2008). However, their residues have negative effects on the environment, even on human health because of their toxicity and stability (Fadaei et al., 2012). If the organophosphorous pesticides are released in the suitable environmental conditions, it is possible for them to persist in many environmental compartments for a long period of time (Evgenidou et al., 2005). These pollutants are usually poorly biodegradable or persistent to environmental conditions.

Different types of synthetic pesticides are mostly used worldwide for the agricultural activities and increasing the yield. These pesticides are categorized into various groups such as insecticides, micro-biocides, fungicides, herbicides and so on. Nearly 98% of sprayed insecticides and 95% of herbicides reach a destination other than their target species including nontarget species, air, water, bottom sediments, and food (Miller, 2004). The most shocking and dangerous effects of exposure to pesticide, that is, mutated DNA, have been found in blood samples of farmers in some areas.

Many organophosphate insecticides (esters of phosphoric acid) such as monocrotophos (MCP), parathion, malathion and methylparathion are well known for their bioaccumulation and neurotoxic properties, because they act on the enzyme acetyl cholinesterase. Organophosphate pesticides inactivate acetylcholinesterase irreversibly, which is essential for nerve function in insects, humans, and many other animals. Therefore, they have full potential for neurotoxicity (Randhawa and Kullar, 2011).

MCP is one of the highly hazardous organophosphate insecticides that is widely used and easily available. It is acutely toxic to birds and humans, and kills bird like Swainson's Hawks (Bai and Reddy, 1977). It can be absorbed through injection, inhalation and skin contact. When inhaled, it also affects the respiratory system. Many organophosphate pesticides are not completely degraded during conventional treatment technology due to their low biodegradability. It has been shown that due to incomplete mineralization, some amount of pesticides are released in aquatic environment and mix up into ground water and become a part of drinking water.

Degradation of pesticides was also carried out using US (Chowdhury and Viraraghavan, 2009). Degradation of MCP was carried out using US, TiO_2 and Fe^{3+} ions (Madhavan et al., 2010A) whereas degradation of MCP pesticides was carried out in the presence of TiO_2 and US (Agarwal and Das, 2014). Degradation of diazinon pesticide was carried out using high-frequency US wave. It was observed that as the concentration of diazinon increases, the degradation also increases, but when the volume of solution was increased, the ability to degrade pesticides decreases (Matouq et al., 2008).

Alachlor is an acetanilide herbicide which is widely used to control most annual grasses and many broad leaf weeds. It has been classified as carcinogen of B2 group due to its strong carcinogenic effects on the animals. The degradation of alachlor was carried out using sonolysis (US), photocatalysis (UV) and sonophotocatalysis (US/UV) using three different photocatalysts, namely TiO_2 (mixture of anatase and rutile), TiO_2 (anatase) and ZnO. TiO_2 (a mixture of anatase and rutile) showed highest photocatalytic activity as compared to TiO_2 (anatase) and ZnO due to the enhanced generation of hydroxyl radicals and slow electron hole recombination (Bagal et al., 2013).

Azinphos-methyl and chlorpyrifos are the organophosphorous pesticides, which pose serious threats to the environment including their detrimental effect on humans and, therefore, their removal from the environment is a must. Ultrasonic irradiation is a promising process for the removal of azinphos-methyl and chlorpyrifos from aqueous solutions (Agarwal et al., 2016).

Shayeghi et al. (2010) studied the degradation of diazinon using US and acoustical processor reactor. Acoustical processor reactor with 130 kHz was used to study the degradation of pesticide solution. Samples were analysed using high-performance liquid chromatography (HPLC) at different time intervals. Khan et al. (2015) reported the use of zinc oxide nanoparticles for the degradation of pesticide chlorpyrifos by sonochemical method. ZnO exhibits good photocatalytic efficiency under UV light irradiation but when combined with membrane filtration, degradation rate increases several times as compared to ZnO-mediated photocatalysis.

Dehghani and Fadaei (2013) studied the degradation of organophosphorus pesticides, diazinon, and malathion in aqueous solution by ultrasonic irradiation. The degradation of diazinon and malathion by US followed first- and second-order model. Formetanate hydrochloride (FMT) is an insecticide–acaricide, quite widely used in agriculture. The degradation of FMT in the presence of homogeneous (Fe^{3+}) and heterogeneous photocatalysts (TiO_2) was carried out using US (Madhavan et al., 2010b).

Carbaryl (1-naphthyl-*N*-methyl carbamate) is a chemical of the carbamate family used as an insecticide. It is a cholinesterase inhibitor and is also toxic to humans. It is classified as a likely human carcinogen. The degradation of the carbaryl pesticide was carried out using US. A combination of high-frequency US wavelength and UV irradiation was considerably more effective than US or ultraviolet irradiation alone for degradation of carbaryl in water (Khoobdel et al., 2010).

Shayeghi et al. (2011) reported the sonochemical degradation of malathion using acoustic wave technology (AWT). AWT can be used to reduce the concentration of dissolved insecticide using high frequency. MCP, present as a severe contaminant in surface water, was studied in slurry batch photoreactor by using aeroxide P-25 and laboratory reagent (LR)-grade TiO_2 as catalysts. Aeroxide P25 showed better degradation efficiency than LR grade TiO_2 in both UV and sunlight conditions under optimized parameters. The degradation of MCP was carried out using US (Sraw et al., 2014).

The insecticide dichlorodiphenyltrichloroethane (DDT) is widely used in indoor residual spraying for malaria control owing to its longer residual efficacy in the field compared to other World Health Organization (WHO) alternatives. The degradation of DDT was carried out using US (Sibanda et al., 2011). Norflurazon (4-chloro-5-(methylamino)-2-[3-(trifluoromethyl) phenyl]pyridazin-3(2H)-one) is an excellent weed-controlling agent used in the agricultural lands. The excessive addition or the undissolved norflurazon enters into the aquatic environment and causes the adverse effects associated with its high concentration. To avoid such problem, visible light-assisted photocatalysis setup coupled with the 42 kHz US producing bath type sonicator was used to degrade the norflurazon (Panneerselvam et al., 2016).

The herbicide (4-chloro-2-methylphenoxy) acetic acid (MCPA) is normally used in paddy fields and, therefore, wastewater may be contaminated. Ultrasonic degradation of aqueous solutions of MCPA was investigated with 500 kHz under argon atmosphere by monitoring liberation of chloride ion, total organic carbon (TOC) and by-product production. Kojima et al. (2005) reported the degradation of MCPA under sonication in the presence of nitrogen, air, oxygen, argon, and argon/air (60/40 %v/v). The 4-chloro-2-methylphenoxy acetic acid was fully decomposed after 180 min sonication with a power of 21 W, but 60 % of the TOC remained in solution even after 6 h. The rate of decomposition of MCPA was found to be greater in oxygen-enriched atmospheres as radical production was more there, but dechlorination and TOC removal rates were relatively higher with argon.

Diuron, (*N*-(3,4-dichlorophenyl)-*N*,*N*-dimethyl-urea), is a water-soluble herbicide belonging to the phenylamide and the subclass of phenylurea. It is used to control a wide variety of annual and perennial broadleaf and grassy weeds, as well as mosses. Diuron can contaminate the surface and ground water by agricultural runoffs and also by wastewaters generated from diuron manufacturing plants. The degradation of (*N*-(3,4-dichlorophenyl)-*N*,*N*-dimethyl-urea) by electrochemical oxidation at a boron-doped diamond anode was enhanced by low-frequency (20 kHz) US (Bringas et al., 2011).

8.3.8 PHARMACOLOGICAL DRUGS

Pharmaceutical manufacturers are under strict regulations by the food and drug agencies in different countries so as to maintain acceptable water quality standards for use, discharge, or reuse in the plant. These pharmaceuticals reach the environment through usage and inappropriate disposal from the manufacturing units as such plants generate a large amount of waste during manufacturing, purification, cleaning, washing, and maintenance (Gadipelly et al., 2014).

Madhavan et al. (2010c) have studied US-assisted photocatalytic degradation of diclofenac (DCF) in an aqueous environment and reported that the degradation of dichlofenac using a sonophotocatalytic process showed a slight synergy, when TiO_2 was present as a photocatalyst under UV light irradiation and in the presence of Fe-ZnO also, degradation showed an additive effect.

Safari et al. (2015) reported the sonochemical degradation of tetracycline (TC) in aqueous solution in the presence of $S_2O_8^{2-}$. The activation energy value of this reaction was 31.71 kJ·mol^{-1}, which means that this degradation process is thermodynamically feasible. Ultraviolet-visible spectra before and after US irradiation in the presence of $S_2O_8^{2-}$ indicated that the proposed degradation pathway for TC involves loses of *N*-methyl, hydroxyl, and amino groups.

The degradation of 17β-estradiol (E2) and 17α-ethinylestradiol (EE2) in water and wastewater was carried out at ultrasonic frequency of 850 kHz. The degradation process followed the pseudo-second-order kinetic model and the rate constant was 1.719×10^{-2} min^{-1} at 25°C. The value for activation energy was found to be 15.21 kJ·mol^{-1}, indicating that the ultrasonic degradation of steroid hormones is thermodynamically feasible, and does not progress only on radical reactions but other intermediate reaction processes. The higher dissolved organic carbon reduces the effectiveness of degradation of the E2 and EE2 in wastewater and, therefore, US treatment may be more effective as a tertiary treatment option in wastewater applications (Ifelebuegu et al., 2014).

Atenolol is a β-blocker drug and it is an identified emerging pollutant. The degradation of atenolol was carried out using US with frequencies ranging from 200 kHz to 1 MHz. The degradation was monitored by TOC and COD reduction. It was observed that the degradation depends on the frequency and power of the US. Studies were carried out in the presence of various additives generally present in contaminated water (Nejumal et al., 2014).

The degradation of seven estrogen hormones (17α-estradiol, 17β-estradiol, estriol, 17α-ethinylestradiol, 17α-dihydroequilin, estrone and equilin) was carried out using US (Andaluri et al., 2012). They used artificial neural networks (ANNs) as a tool to identify the correlations between process parameters. ANN enabled the establishment of relationship between sonication parameters such as power density, power intensity, US amplitude, as well as the reactor design parameters. The antiepileptic drug "carbamazepine (CBZ)" is one of the most abundant pharmaceuticals in the aquatic environment. The degradation of CBZ was also carried out using US (Braeutigam et al., 2012).

The degradation of aqueous solutions of antipyrine was carried out using homogeneous sonophotocatalytic oxidation process (H_2O_2/UV/Fe/Ultrasound) with an artificial UV lamp. Photodegradation of antipyirine proceeds mainly through a radical mechanism. Probably, it begins with cleavage of the *N–N* bond of penta heterocycle leading to the formation of aromatic acids (anthranilic and 1,4-benzenedicarboxylic acids), which is then followed by the opening of phenyl ring to form small molecular organic acids (mainly 2-butenedioic, 4-oxo-pentanoic and butanedioic acids). These acids may be decomposed further into CO_2 (Duran et al., 2013).

Guo et al. (2010) carried out the degradation of cephalexin in aqueous solution under US irradiation. Biodegradability of the solution was evaluated by the BOD_5/COD ratio, which was increased from 0 to 0.36 after US treatment. It indicates that the US irradiation process is a successful pretreatment step to improve the biodegradability of cephalexin solution. The degradation of cephalexin follows a pseudo-first-order kinetics.

Guyer and Ince (2011) also reported the degradation of a widely used anti-inflammatory drug, DCF, using high-frequency US and iron species. Occurrence and fate of pharmaceutically active compounds (PhACs) in the aquatic environment has been recognized as one of the emerging issues in environmental chemistry and has become a matter of public concern. Residues of PhACs have been found as contaminants in sewage, surface, ground and drinking water samples (Heberer et al., 2002).

Hirsch et al. (1999) reported the degradation of various water samples for 18 antibiotic substances, from the classes of macrolid antibiotics, sulphonamides, penicillins and TCs. Samples were preconcentrated via lyophilization and quantified using HPLC-electrospray-tandem-mass spectrometry. The investigated sewage treatment plant effluents and surface water samples showed frequent appearance of an erythromycin degradation product, roxithromycin and sulphamethoxazole with concentrations up to 6 $mg \cdot L^{-1}$.

Acetaminophen (AAP) and naproxen (NPX) are used as analgesics. The degradation of a pharmaceutical (PhAC) mixture of acetaminophen and naproxen was carried out using US and single-walled carbon nanotube (SWNT)-catalysed ultrasonic (US/SWNT) in water. In the absence of SWNTs, maximum degradation of AAP and NPX was found to occur at a high-frequency (1000 kHz), acidic conditions (pH=3) and different solution temperatures (25°C at 28 kHz and 35°C at 1000 kHz) during US reactions (Im et al., 2013).

Paracetamol is a widely used nonsteroidal anti-inflammatory recalcitrant drug found in water and levodopa, the most frequently used drug for the treatment of Parkinson disease. The degradation of paracetamol (PARA) and levodopa was carried out using US (Isariebel et al., 2009).

Jelic et al. (2013) carried out the degradation of the antiepileptic CBZ by sonolysis, TiO_2-based heterogeneous photocatalysis under UV-A and simulated solar irradiation, and also by the combined use of UV-A and US irradiation (sonophotocatalysis) in demineralized water, ground water and effluent wastewater.

Ibuprofen is ordinarily and widely used as a pain reliever, but it is a persistent organic pollutant it is not biodegradable and passes through normal sewage treatment process. Mendez-Arriya et al. (2008) made use of 300 kHz US to increase the extent of degradation from 30% to 98% in 30 min by keeping initial concentration low. The rate of degradation of ibuprofen was found to increase under air or oxygen and it was highest in acidic pH. Although, complete removal of IBP was accomplished in this case, but the presence of intractable degradation products was observed, which means that there was still residual TOC.

Langenhoff et al. (2013) reported the degradation of ibuprofen and DCF using US. Ibuprofen was completely removed, whereas DCF yielded a residual concentration, showing a necessity of post-treatment to remove DCF using activated carbon.

The combination of US and heat was used for the oxidation of some ethyl 3,4-dihydropyrimidin-2(1H)-one-5-carboxylates to their corresponding ethyl pyrimidin-2(1H)-one-5-carboxylates by using potassium peroxydisulphate in aqueous acetonitrile. The degradation of dihydropyrimidinone was carried out by Memarian and Farhadi (2008). Rahmani et al. (2014) reported the degradation of aqueous solutions of the tinidazole (TNZ) using the combination of US irradiation and H_2O_2. It was observed that the degradation of TNZ was enhanced with decreasing both TNZ the initial concentrations and pH. Furthermore, TNZ removal efficiency in the case of actual wastewater was less than of synthetic wastewater (75 and 68% of synthetic and actual wastewaters, respectively).

Fluoxetine (FLX) is one of the most widely used antidepressants in the world. It is an emergent pollutant found in natural waters that causes disrupting effects on the endocrine systems of some aquatic species. The total elimination of FLX was carried out using sonochemical treatment coupled with a biological system. The biological process alone was shown to be unable to remove the pollutant, even under favourable conditions of pH and temperature. Sonochemical treatment (600 kHz) was able to remove this pharmaceutical (Serna-Galvis et al., 2015).

Suri et al. (2007) reported the use of US to destroy many estrogen compounds in water including 17α-estradiol, 17β-estradiol, estrone, estriol, equilin, 17α-dihydroequilin, 17α-ethinyl estradiol and norgestrel. Cetirizine dihydrochloride is a pharmaceutical drug of the class antihistamines and it is frequently detected in wastewater samples. The degradation of cetirizine dihydrochloride was carried out using a novel technique of laccase enzyme as a catalyst under the influence of US irradiation (Sutar and Rathod, 2015).

A hybrid advanced oxidation process combining sonochemistry (US) was developed and electrochemistry (EC) for the batch-scale degradation of ibuprofen (Thokchom et al., 2015). The performance of this hybrid reactor system was evaluated by quantifying the degradation of ibuprofen under the variation in electrolytes, frequency, applied voltage, ultrasonic power density and temperature in aqueous solutions with a platinum electrode.

The degradation of CBZ in synthetic solution and in municipal effluent was carried out using a sonoelectrochemical reactor. Sonoelectrolysis was conducted using two concentrical electrodes installed in a cylindrical reactor containing a ceramic transducer (Tran et al., 2015). The N-doped TiO_2 was used for the photocatalytic degradation (PCD) of pharmaceutical micropollutants, namely ciprofloxacin HCl (CFX), NPX and PARA and it was found that the rate of degradation of CFX and NPX is higher than that of PARA (Shetty et al., 2016).

Bis(indol-3-yl) methane (BIM) derivative is a potential drug in curing cancerous cells and disorders in the nervous system, and the degradation of BIM derivative by US was carried out by Li et al. (2010).

8.3.9 SURFACTANTS

Surfactant is a group of chemicals widely used as detergents and other cleaning products. The residual surfactants are discharged into sewage systems or directly into surface waters after use. As a result, most of them end up via dispersion creating different kinds of environmental pollution such as water, soil, sediment, and so on. The toxic effects of these surfactants on various aquatic organisms are well established. Although most of surfactants are readily biodegradable in nature and their amount is significantly reduced with secondary treatment in wastewater treatment plants, but more concern arises about the release of untreated wastewater or wastewater that has undergone only primary treatment. Such a discharge of wastewater polluted with large quantities of surfactants could have deleterious effects on our ecosystem. Therefore, it is a demand of the day to replace them with some more environmental-friendly surfactants or to search for some alternate technologies to degrade them, where sonochemistry can play an important role.

The degradation of sodium dodecylbenzene sulphonate from aqueous solutions was carried out by Manousaki et al. (2004) using ultrasonic irradiation. Ultrasound was used for the degradation of nonvolatile surfactants, sodium 4-octylbenzene sulphonate (OBS) and dodecylbenzenesulphonate (DBS) and a nonvolatile nonsurfactant 4-ethylbenzene sulphonic acid (EBS), as single components and binary mixtures by (Yang et al., 2006). OBS exhibited a faster degradation rate than DBS at shorter pulse intervals as it has faster rate of transfer to the cavitation bubble interfaces, but at longer pulse intervals, DBS (which is more surface active) degrades faster than OBS due to the increased amounts of DBS accumulation on the bubble surfaces.

The application of pulsed US for the degradation of the nonvolatile surfactants, OBS and DBS was investigated at a frequency of 354 kHz (Yang et al., 2005). The enhanced degradation of surfactants by pulsed US was due to the accumulation of surfactants on cavitation bubble surfaces. As compared to shorter pulse intervals, longer pulse intervals enhanced DBS degradation, indicating that DBS, a more surface-active compound, accumulated and equilibrated with the bubble interface more slowly.

Perfluorosulphonates and carboxylates, C4 to C8 surfactants, were sonolytically degraded primarily at the vapour–water interface of cavitation bubbles (Campbell and Hoffmann, 2015). Pseudo-first-order rate constants for the elimination of the perfluorosurfactants indicated that the rates should be proportional to the degree of interfacial bubble–water partitioning. Sonochemical degradation rates of the more soluble and less hydrophobic perfluorobutane sulphonate, PFBS, and perfluorobutanoate, PFBA, were found to be slower than longer chain analogs as they have greater water solubility as compared to their C8 counterparts. It was observed that the degradation rate constants were found to increase with increasing power density. Enhanced degradation rates were observed for the C8 surfactants, PFOS (12 %), and perfluorooctanoic acid (PFOA; 23 %) under simultaneous irradiation at 20 and 202 kHz at a power-combined power density of 250 $W\,L^{-1}$.

Singla et al. (2010) reported sonochemical degradation of the cationic surfactant, laurylpyridinium chloride (LPC), in water at concentrations of 0.1–0.6 mM, much below its critical micelle concentration (15 mM). It has been found that the initial step in the degradation of LPC occurs primarily by a pyrolysis pathway. Propionamide and acetamide were identified as two of the degradation intermediates. These are formed due to opening of the pyridinium ring following addition of $^{\bullet}OH$ radicals. Most of the LPC is eventually converted into carboxylic acids. The complete mineralization of these carboxylic acids by sonolysis is comparatively a slow process due to the hydrophilic nature of these low molecular weight products.

PFOA is an aqueous anionic surfactant, which is a persistent organic pollutant. It can be easily adsorbed onto the bubble–water interface and both mineralized and degraded by US cavitation at room temperature. The effect of US on the degradation of PFOA in solution can be enhanced by the addition of surfactant like cationic (hexadecyl trimethyl ammonium bromide, CTAB), a nonionic (octyl phenol ethoxylate, Triton-X-100), and an anionic (sodium dodecyl sulphate, SDS) surfactant. It was found that the addition of CTAB increased the degradation rate followed by Triton X-100 whereas SDS inhibited the degradation rate. The removal efficiency of PFOA increased with increasing CTAB concentration, with the efficiency reaching 79% after 120 min at 25°C with 0.12 mM CTAB dose (Lin et al., 2015).

8.3.10 MISCELLANEOUS

Beckett and Hua (2003) investigated the degradation of 1,4-dioxane using Fe(II) and US. Sodium bicarbonate is a hydroxyl radical scavenger and its addition reduced the rate of decomposition; however, it could not stop the degradation completely. Son et al. (2006) studied the effect of adding Fe(0), Fe (II) and $S_2O_8^{2-}$ during the sonication of aqueous 1,4-dioxane. Nakajima et al. (2007) reported the sonochemical oxidation of 1,4-dioxane by a combination of TiO_2 and SiO_2. TiO_2 proved to be more effective as compared to SiO_2 but the photocatalytic destruction through sonoluminescence was not the main mechanism. It seems that the process of sonication produced thermally excited holes in the TiO_2 and oxygen vacancies in the TiO_2 surface help in the decomposition by enhancing cavitation.

Wen et al. (2011) reported the use of noble metal nanoparticles such as Au/TiO_2, Au/ZrO_2, Ag/AgCl as efficient sonocatalysts for the degradation of environmental pollutants including aldehydes, alcohols, and acids using US. Nowadays, ultrasonic irradiation has received a considerable interest as an advanced oxidation process.

Liu et al. (2005) studied the application of US for enhancing organic pollutant biodegradation in biological-activated carbon membrane reactor. Ultrasonic treatment at 10 W for 24 h could increase organic load of the bioreactor and the removal efficiency of organic substances. The ultrasonic effect on biodegradation was studied through the organic load of the bioreactor, 2,3,5-triphenyl tetrazoliumchloride-dehydrogenase activity (TTC-DHA) of biological-activated carbon and the apparent molecular size distribution of the organic compounds in influent and effluent water. The ultrasonic field in the reactor was analysed, and the mechanism of biological activity enhancement by ultrasonics was investigated.

The decomposition of complex organic compounds to much simpler compounds during cavitation process was reported by Mahvi (2009). Doosti et al. (2012) reported the degradation of halomethanes, total suspended solid, algae and [(1,1,1,-trichloro-2,2-bis(*p*-chlorophenyl)ethane] DDT using US. The results show that this technique could improve the water treatment process environmentally. Various parameters could affect the efficiency of US technique such as power density, frequency and irradiation time. So, it becomes necessary to obtain the optimum power density, frequency and irradiation time to be cost-effective.

Yaqub et al. (2012) observed that the concentration of lead (Pb) decreased from 11.5 to 0.6 ppm at 80 kHz from battery industries, 95% chromium (Cr) was removed from tannery industry using lead cathode, and high COD removal was observed using steel anode at 80 kHz. Thus, use of US is quite effective in the removal of heavy metals and organic pollutants from industrial wastewater. Karaboga et al. (2007) observed the effect of application of US in enzymatic pretreatment processes of cotton fabrics. It was reported that the use of US during enzymatic treatment of cotton fabrics significantly improved the efficiency of enzymes without affecting the strength of the fabric as well as it increases mass transfer toward the textile material.

Water is the basic need for all plants, animals and human beings, but its quality and quantity is degrading day by day. There is an urgent need to improve the quality of water as it is required for survival of life in enormous amount for human beings, animals and plant kingdoms. Various methods have been employed for wastewater remediation. Ultrasound is one such advanced oxidation technique, which can be used in wastewater treatment and that too without many complications. Sonochemistry finds great potential in degradation of several pollutants like dyes, pesticides, phenols, chloro compounds, surfactants, and so on under ambient conditions and can be carried out without adding any chemical. It can be thus considered as a green chemical pathway for wastewater treatment.

REFERENCES

Adewuyi, Y. G. *Environ. Sci. Technol.* **2005,** *39*, 3409–3420.

Agarwal, V.; Das, S. *J. Environ. Waste Manage.* **2014,** *1*, 2–10.

Agarwal, S.; Tyagi, I.; Gupta, V. K.; Dehghani, M. H.; Bagheri, A.; Yetilmezsoy, K.; Amrane, A.; Heibati, B.; Rodriguez-Couto, S. *J. Mol. Liq.* **2016,** *221*, 1237–1242.

Ahmad, M.; Ahmed, E.; Hong, Z. L.; Ahmed, W.; Elhissi, A.; Khalid, N. R. *Ultrason. Sonochem.* **2014,** *21*, 761–773.

Alahiane, S., Qourzal, S., Mahmoud, E. O., Abaamrane, A., Assabbane A. *Am. J. Anal. Chem.* **2014,** *5*, 445–454.

Ameen, S. M.; Akhtar, S.; Kim, Y. S.; Yang, O. B.; Shin, H. S. *Colloid Polym. Sci.* **2011,** *289*, 415–421.

Andaluri, G.; Rokhina, E. V.; Suri, R. P. S. *Ultrason. Sonochem.* **2012,** *19*, 953–958.

Anju, S. G.; Jyothi, K. P.; Joseph, S.; Suguna, Y.; Yesodharan, E. P. *Res. J. Recent Sci.* **2012,** *1*, 191–201.

Asuha, S.; Zhou, X. G.; Zhao, S. *J. Hazard. Mater.* **2010,** *181*, 204–210.

Bagal, M. V.; Gogate, P. R. *Adv. Environ. Res.* **2013,** *2*, 261–277.

Bai, A. R. K.; Reddy, C. C. *J. Apic. Res.* **1977,** *16*, 161–162.

Barik, A. J.; Gogate, P. R. *Ultrason. Sonochem.* **2016,** *28*, 90–99.

Beckett, M. A.; Hua, I. *Water Res.* **2003,** *37*, 2372–2376.

Bejarano-Perez, N. J.; Suarez-Herrera, M. F. *Ultrason. Sonochem.* **2007,** *14*, 589–595.

Beltran, F. J.; Rivas, F. J.; Gimeno, O. *J. Chem. Technol. Biotechnol.* **2005,** *80*(9), 973–984.

Bhatnagar A.; Cheung H. M. *Environ. Sci. Technol.* **1994,** *28*, 1481–1486.

Bolla, M. K.; Bolla, V. S.; Rekkabu, L. *Int. J. Innovative Res. Adv. Eng.* **2014,** *1*, 27–36.

Braeutigam, P.; Franke, M.; Schneider, R.J.; Lehmann, A.; Stolle, A.; Ondruschka, B. *Water Res.* **2012,** *46*, 2469–2477.

Bringas, E.; Saiz, J.; Ortiz, I. *Chem. Eng. J.* **2011,** *172*, 1016–1022.

Campbell, T.; Hoffmann, M. R. *Sep. Purif. Technol.* **2015,** *156*, 1019–1027.

Chakrabarti, S.; Dutta, B. K.; Apak, R. *Water Sci. Technol.* **2009,** *60*, 3017–3024.

Chen, W. S.; Huang, C. P. *Ultrason. Sonochem.* **2014,** *21*, 840–845.

Chen, Y. C.; Vorontsov, A. V.; Smirniotis, P. G. *Photochem. Photobiol. Sci.* **2003,** *2*(6), 694–698.

Chowdhury, P.; Viraraghavan, T. *Sci. Total Environ.* **2009,** *407*, 2474–2492.

Cravotto, G.; Carlo, D. S.; Binello, A.; Mantegna, S.; Girlanda, M.; Lazzari, A. *Water Air Soil Pollut.* **2008,** *187*, 353–359.

Cui, P.; Chen, Y.; Chen, G. *Ind. Eng. Chem. Res.* **2011,** *50*, 3947–3954.

Dalhatou, S.; Pe´trier, C.; Laminsi, S.; Baup S. *Int. J. Environ. Sci. Technol.* **2015,** *12*, 35–44.

Dawkar, V. V.; Jadhav, U. U.; Jadhav, S. U.; Govindwar, S. P. *J. Appl. Microbiol.* **2008,** *105*, 14–24.

Dehghani, M. H.; Fadaei, A. *Ind. J. Sci. Technol.* **2013,** *6*, 3876–3881.

Destaillats H.; Colussi A.; Joseph J. M.; Hoffmann M. R. *J. Phys. Chem A* **2000,** *104*, 8930–8935.

Doosti, M. R.; Kargar, R.; Sayadi, M. H. *Int. Acad. Ecol. Environ. Sci.* **2012,** *2*, 96–110.

Drijvers, D.; De Baets R.; De Visscher A.; Van L. H. *Ultrason. Sonochem.* **1996,** *3*, S83–S90.

Drijvers, D.; Van L. H.; Vervaet K. *Ultrason. Sonochem.* **1998,** *5*, 13–19.

Duran, A.; Monteagudo, J. M.; Sanmartin, I.; Garcia-Diaz, A. *Appl. Catal. B Environ.* **2013,** *138*, 318–325.

Entezari, M. H.; Petrier, C.; Devidal, P. *Ultrason. Sonochem.* **2003,** *10*, 103–108.

Entezari, M.H.; Mostafai, M.; Sarafraz-yazdi, A. *Ultrason. Sonochem.* **2005,** *12*, 137–141.

Evgenidou, E.; Fytianos, K.; Poulios, I. *Photochem. Photobiol.* **2005,** *175*, 29–38.

Fadaei, A. M.; Dehghani, M. H.; Nasseri, S.; Mahvi, A. H.; Rastkari, N.; Shayeghi, M. *Contam. Toxicol.* **2012,** *88*, 867–869.

Faghihian, H.; Sadeghinia, R. *Adv. Chem. Eng. Res.* **2014,** *3*, 18–26.

Fallis, I. A.; Griffiths, P. C.; Cosgrove, T.; Dreiss, C. A.; Govan, N.; Heenan, R. K. et al. *J. Am. Chem. Soc.* **2009,** *131*, 9746–9755.

Gadipelly, C.; Perez-Gonzalez, A.; Yadav, G. D.; Ortiz, I.; Ibanez, R.; Rathod, V. K.; Marathe, K. V. *Ind. Eng. Chem. Res.* **2014,** *53*, 11571–11592.

Gemeay, A. H.; El-Sharkawy, R. G.; Mansour, I. A.; Zaki, A. B. *Bull. Mater. Sci.* **2012,** *35*, 585–593.

Ghasemi, F.; Tabandeh, F.; Bambai, B.; Sambasiva Rao, K. R. S. *Int. J. Environ. Sci. Technol.* **2010,** *7*, 457–464.

Gobouri, A. A. *Res. Chem. Intermed.* **2016,** *42*, 5099–5113.

Goel, M.; Hongqiang, H.; Mujumdar, A. S.; Ray, M. B. *Water Res.* **2004,** *38*, 4247–4261.

Goskonda, S.; Catallo, W. J.; Junk, T. *Waste Manage.* **2002,** *22*, 351–356.

Griffing V. *J. Chem. Phys.* **2004,** *20*, 939–942.

Guo, W. L.; Wang, H. Z.; Shi, Y. H.; Zhang, G. Y. *Water SA* **2010,** *36*, 651–654.

Guo, X.; Fu, Y.; Hong, D.; Yu, B.; He, H.; Wang, Q.; Xing, L.; Xue, X. *Nanotechnology* **2016,** 27. DOI: 10.1088/0957-4484/27/37/375704.

Guyer, G. T.; Ince, N. H. *Ultrason. Sonochem.* **2011,** *18*, 114–119.

Guzman-Duque, F.; Pe´trier, C.; Pulgarin, C.; Pen˜uela, G.; Torres-Palma, R. A. *Ultrason. Sonochem.* **2011,** *18*, 440–446.

Hamdaoui, O.; Naffrechoux, E. *Ultrason. Sonochem.* **2008,** *15*, 981–987.

Han, R.; Zhang, S.; Xing, D.; Jian, X. *J. Membr. Sci.* **2010,** *358*, 1–6.

Hao, H.; Chen, Y.; Wu, M.; Wang, H.; Yin, Y.; Lu, Z. *Ultrason. Sonochem.* **2004,** *11*, 43–46.

He, Y.; Grieser, F.; Muthupandian, A. K. *Ultrason. Sonochem.* **2011,** *18*, 974–980.

Heberer, T.; Reddersen, K.; Mechlinski, A. *Water Sci. Technol.* **2002,** *46*, 81–88.

Hirsch, R.; Ternes, T.; Haberer, K.; Kratz, K. L. *Sci. Total Environ.* **1999,** *225*, 109–118.

Hitchman, M. L.; Spackman, R. A.; Ross, N. C. *Chem. Soc. Rev.* **1995,** *24*, 423–430.

Hofmann, J.; Freier, U.; Wecks, M.; Demund, A. *Top. Catal.* **2005,** *33*, 243–247.

Ifelebuegu, A. O.; Onubogu, J.; Joyce, E.; Mason, V. *Int. J. Environ. Sci. Technol.* **2014,** *11*, 1–8.
Im, J. K.; Heo, J.; Boateng, L. K.; Her, N.; Flora, J. R. V.; Yoon, J.; Zoh, K. D.; Yoon, Y. *J. Hazard. Mater.* **2013,** *254*, 284–292.
Ince, N. H.; Tezcanlı, G. *Dyes Pigm.* **2001,** *49*, 145–153.
Iram, M.; Guo, C.; Guan, Y.; Ishfaq, A.; Liu, H. *J. Hazard. Mater.* **2010,** *181*, 1039–1050.
Isariebel, Q. P.; Carine, J. L.; Ulises-Javier, J. H.; Anne-Marie, W.; Henri, D. *Ultrason. Sonochem.* **2009,** *16*, 610–616.
Jelic, A.; Michael, I.; Achilleos, A.; Hapeshi, E.; Lambropoulou, D.; Perez, S.; Petrovic, M.; Fatta-Kassinos, D.; Barcelo, D. *J. Hazard. Mater.* **2013,** *263*, 177–186.
Jiang, Y.; Petrier, C.; Waite, T. D. *Ultrason. Sonochem.* **2002,** *9*, 317–323.
Karaboğa, C.; Körlü, A. E.; Duran, K.; Bahtiyari, M. İ. *Fibres Text.* **2007,** *15*(4), 97–100.
Kargar, M.; Nabizadeh, R.; Naddafi, K.; Nasseri, S.; Mesdaghinia, A.; Mahvi, A. H.; Alimohammadi, M.; Nazmara, S.; Pahlevanzadeh, B. *Iran. J. Environ. Health Sci. Eng.* **2012,** *9*, 32.
Khalid, A.; Arshad, M.; Crowley, D. E. *Appl. Microbiol. Biotechnol.* **2008,** *79*, 1053–1059.
Khan, M. A.; Siddique, M.; Wahid, F.; Khan, R. *Ultrason. Sonochem.* **2015,** *26*, 370–377.
Khan, S. H.; Rajendran, S.; Pathak, B.; Fulekar, M. H. *Front. Nanosci. Nanotechnol.* **2015,** *1*, 25–29.
Khoobdel, M.; Shayeghi, M.; Golsorkhi, S.; Abtahi, M.; Vatandoost, H.; Zeraatii, H.; Bazrafkan, S. *Iran. J. Arthropod Borne Dis.* **2010,** *4*, 47–53.
Kilic, N. K.; Duygu, E.; Donmez, G. *J. Hazard. Mater.* **2010,** *182*, 525–530.
Kim, K.; Tsay, O. G.; Atwood, D. A.; Churchill, D. G. *Chem. Rev.* **2011,** *111*(5), 345–5403.
Kojima, Y.; Fujita, T.; Ona, E. P.; Matsuda, H.; Koda, S.; Tanahashi, N.; Asakura, Y. *Ultrason. Sonochem.* **2005,** *12*, 359–365.
Koskela, H.; Rapinoja M. L.; Kuitunen, M. L.; Vanninen, P. *Anal. Chem.* **2007,** *79*, 9098–9106.
Kotronarou, A.; Mills, G. ;Hoffmann, M. R. *J. Phys. Chem.* **1991,** *95*, 3630–3638.
Langenhoff, A.; Inderfurth, N.; Veuskens, T.; Schraa, G.; Blokland, M.; Kujawa-Roeleveld, K.; Rijnaarts, H. *BioMed Res. Int.* **2013,** *2013*, 1–9. Article ID 325806.
Lathasreea, S.; Nageswara, R. A.; SivaSankarb, B.; Sadasivamb, V.; Rengarajb, K. J. *J. Mol. Catal. A: Chem.* **2004,** *223*, 101–108.
Laxmi, P. N. V.; Saritha, P.; Ranbabu, N.; Himabindu, V.; Anjaneyulu, Y. (2009). *J. Hazard. Mater.* **2009,** *10*, 245–252.
Laxmi, P. N. V.; Saritha, P.; Ranbabu, N, Himabindu, V.; Anjaneyulu, Y. *J. Hazard. Mater.* **2010,** *174*, 1–3.
Li, B.; Zhu, J. *Chem. Eng. J.* **2015,** *284*, 750–763.
Li, J. T. Zhang, X. H.; Song, Y. L. *Int. J. Chem. Techol. Res.* **2010,** *2*, 341–345.
Lin, J. C.; Hu, C. Y.; Lo, S. L. *Ultrason. Sonochem.* **2015,** *28*,130–135.
Lin, J. G.; Ma, Y. S. *J. Environ. Eng.* **2000,** *126*, 130–137.
Littl, C.; Hepher, M. J.; Sharif, E. M. *Ultrasonics.* **2002,** *40*, 667–674.
Liu, H.; He, Y. H.; Quan, X. C.; Yan, Y. X.; Kong, X. H. Lia, A. J. *Process Biochem.* **2005,** *40*, 3002–3007.
Liu, Q.; Wang, L.; Xiao, A.; Gao, J.; Ding, W.; Yu, H.; Huo, J.; Ericson, M. *J. Hazard. Mater.* **2010,** *181*, 586–592.
Liu, Y. N.; Jin, D.; Lu, X. P.; Han, P. F. *Ultrason. Sonochem.* **2008,** *15*, 775–760.
Lodha, S.; Jain, A.; Punjabi, P. B. *Arabian J. Chem.* **2011,** *4*, 383–387.

Lu, Y.; Weavers, L. K. *Environ. mental Sci. Technol.* **2002,** *36,* 232–237.
Madhavan, J.; Grieser, F.; Ashokkumar, M. *Sep. Purif. Technol.* **2010b,** *73,* 409–414.
Madhavan, J.; Kumar, P. S.; Anandan, S.; Zhou, M.; Grieser, F.; Ashokkumar, M. Chemosphere **2010c,** *80*(7), 747–752.
Madhvan J.; Selvam, P.; Kumar, S. *J. Hazard. Mater.* **2010a,** *177,* 944–949.
Mahvi, A. H. *Iranian J. Public Health* **2009,** *38,* 1–17.
Mahvi, A. H.; Maleki, A. *Desalin. Water Treat.* **2010,** *20,* 197–202.
Mahvi, A. H.; Maleki, A.; Razaee, R.; Safari, M. *Iranian J. Environ. Health Sci. Eng.* **2009,** *6,* 233–240.
Maljaei, A.; Arami, M.; Mahmoodi, N. M. *Desalin.* **2009,** *249,* 1074 –1078.
Manousaki, E.; Psillakis, E.; Kalogerakis, N.; Mantzavinos, D. *Water Res.* **2004,** *38,* 3751–3759.
Margeta, D.; Grčić, I.; Papić, S.; Bionda, K. S.; Foglar, L. *Environ. Technol.* **2016,** *37,* 293–299.
Matouq, M. A.; Al-Anber, Z. A.; Tagawa, T.; Aljbour, S.; Al-Shannag, M. *Ultrason. Sonochem.* **2008,** *15,* 869–874.
Memarian, H. R.; Farhadi, A. *Ultrason. Sonochem.* **2008,** *15,* 1015–1018.
Mendez-Arriaga, F.; Torres-Palma, R. A.; Petrier, C.; Esplugas, S.; Gimenez, J.; Pulgarin, C. *Water Res.* **2008,** *42,* 4243–4248.
Miller, G. T. Chapter 9.In *Sustaining the Earth: An Integrated Approach,* 6th Ed. Thompson Learning, Inc.: Pacific Grove, California, 2004; pp. 211–216.
Mishra, K. P.; Gogate, P. R. *J. Environ. Manage.* **2011,** *92,* 1972–1977.
Mishra, K. P.; Gogate, P. R. *Ind. Eng. Chem. Res.* **2012,** *51,* 1166–1172.
Mital, A.; Jain, R.; Mittal, J.; Shrivastava, M. *Fresenius Environ. Bull.* **2010,** *19,* 1171–1179.
Mittal, A.; Gupta, V.; Malviya, A.; Mittal, J. *J. Hazard. Mater.* **2008,** *151,* 821–832.
Mittal, A.; Mittal, J.; Malviya, A.; Gupta, V. K. *J. Colloid Interface Sci.* **2009,** *340,* 16–26.
Moholkar, V. S.; Nierstrasz, V. A.; Warmoeskerken, M. M. C. G. *AUTEX Res. J.* **2003,** *3,* 129–138.
Morteza, A.; Tabatabaei S. M. *The Appl. Chem. Environ. Spring* **2011,** *2,* 1–10.
Mosleh, S.; Rahimi, M. R.; Ghaedi, M.; Dashtian, K. *RSC Advances* **2016,** *6,* 61516–61527.
Naffrechoux E.; Chanoux, S.; Petrier, C.; Suptil, J. *Ultrason. Sonochem.* **2000,** *7,* 255–259.
Nagata, Y.; Nakagawa, M.; Okuno, H.; Mizukoshi, Y.; Yim, B.; Maeda, Y. *Ultrason. Sonochem.* **2000,** *7,* 115–120.
Naidu, L. D.; Saravanan, S.; Goel, M.; Periasamy, S.; Stroeve, P. *J. Environ. Health Sci. Eng.* **2016,** *14*(9) DOI: 10.1186/s40201-016-0249-8.
Nakajima, A.; Sasaki, H.; Kameshima, Y.; Okada, K.; Harada, H. *Ultrason. Sonochem.* **2007,** *14,* 197–200.
Nakui, H.; Okitsu, K.; Maeda, Y.; Nishimura, R. *Ultrason. Sonochem.* **2007,** *14,* 191–196.
Nejumal, K. K.; Manoj, P. R.; Aravind, U. K.; Aravindakumar, C. T. *Environ. Sci. Pollut. Res. Int.* **2014,** *21,* 4297–4308.
Neppolian, B.; Park, J. S.; Choi, H. *Ultrason. Sonochem.* **2004,** *11,* 273–279.
Ng, T. W.; Cai, Q.; Wong, C. K.; Chow, A. T.; Wong, P. K. *J. Hazard. Mater.* **2010,** *182,* 792–800.
Nuengmatcha, P.; Chanthai, S.; Mahachai, R.; Oh, W. H. *Dyes Pigm.* **2016,** *134,* 487–497.
Okuno, H.; Yim, B.; Mizukoshi, Y.; Nagata, Y.; Maeda, Y. *Ultrason. Sonochem.* **2000,** *7,* 261–264.

One, T.; Tanakas, Y.; Takeuchi, T.; Yanamoto, K. *J. Mol. Catal.* **2000,** *159A*, 293–300.
Pandit, A. B.; Gogate, P. R.; Mujumdar, S. *Ultrason. Sonochem.* **2001**, *8*, 227–231.
Panneerselvam, S. K.; Mangalaraja, R. V.; Rozas, O.; Mansilla, H. D.; Gracia-Pinilla, M. A.; Anandan, S. *Ultrason. Sonochem.* **2014,** *17*, 1675–1681.
Panneerselvam, S.; Mangalaraja, R. V.; Rozas, O.; Sambandam, A. *Chemosphere* **2016,** *146*, 216–225.
Parvas, M.; Haghighi, M.; Allahyari, S. *Arabian J. Chem.* **2014a.** DOI: org/10.1016/j.arabjc.2014.10.043.
Parvas, M.; Haghighi, M.; Allahyari, S. *Environ. Technol.* **2014b,** *35*, 1140–1149.
Peters, D. *Ultrason. Sonochem.* **2001,** *8*, 221–226.
Petrier, C.; Francony, A. *Ultrason. Sonochem.* **1997,** *4*, 295–299.
Petrier, C.; Lamy, M.; Francony, A.; Benahcene, A.; David, B.; Renaudin, V.; Gondrexon, N. *J. Phys. Chem.* **1994,** *98*(41), 10514–10520.
Prasad, G. K.; Ramacharyulu, P. V. R. K.; Singh, B.; Batra, K.; Srivastava, A. R.; Ganesan, K.; Vijayaraghavan, R. *J. Mol. Catal. A: Chem.* **2011,** *349*, 55–62.
Price, R. J.; Blazina, D.; Smith, G. C.; Davies, T. J. *Fuel* **2015,** *156*, 30–39.
Psillakis, E. *J. Hazard. Mater.* **2004,** *108*(1–2), 95–102.
Rahmani, H.; Gholami, M.; Mahvi, A. H.; Alimohammadi, M.; Azarian, G.; Esrafili, A.; Rahmani, K.; Farzadkia, M. *Bull. Environ. Contam. Toxicol.* **2014,** *92*, 341–346.
Rahmani, Z.; Kermani, M.; Gholami, M.; Jafari, A. J.; Mahmoodi, N. J. *Iran. J. Environ. Health Sci. Eng.* **2012,** *9*. DOI: 10.1186/1735-2746-9-14.
Ramacharyulu, P. V. R. K.; Prasad G. K.; Ganesan K.; Singh B. *J. Mol. Catal. A: Chem.* **2012,** *353*, 132–137.
Randhawa, G. K.; Kullar, J. S. *ISRN Pharmacol.* **2011,** DOI: 10.5402/2011/362459.
Rehorek, A.; Tauber, M.; Gubitz, G. *Ultrason. Sonochem.* **2004,** *11*, 177–182.
Ren, X.; Wang, T.; Zhou, C.; Du, S.; Luan, Z.; Wang, J.; Hou, D. *Fresenius Environ. Bull.* **2010,** *19*, 1441–1446.
Ren, Y. Z.; Franke, M.; Anschuetz, F.; Ondruschka, B.; Ignaszak, A.; Braeutigam, P. *Ultrason. Sonochem.* **2014,** *21*, 2020–2025.
Rokhina, E. V.; Makarova, K.; Lathinen, M.; Golovina, E. A.; As, H. V.; Virkutyte, J. *Chem. Eng. J.* **2013,** *221*, 476–486.
Sáez, V.; Esclapez, M. D.; Tudela, I.; Bonete, P.; Louisnard, O.; Gonzá-lez-García, J. Proceedings of 20th International Congress on Acoustics, ICA, Sydney, 2010.
Safari, G. H.; Nasseri, S.; Mahvi, A. H.; Yaghmaeian, K.; Nabizadeh, R.; Alimohammadi, M. *J. Environ. Health Sci. Eng.* **2015,** *13*, 76-1-15.
Saleh, T. A.; Gupta, V. K. *Environ. Sci. Pollut. Res.* **2012,** *19*, 1224–1228.
Sandhya, Haridas, S.; Sugunan, S. *Bull.Chem. React. Eng. Catal.* **2013,** *8*, 145–153.
Serna-Galvis, E. A.; Silva-Agredo, J.; Giraldo-Aguirre, A. L.; Torres-Palma, R. A. *Sci. Total Environ.* **2015,** 524–525, 354–360.
Seymore, J. D.; Gupta, R. B. *Industrial Eng. Chem. Res.* **1997,** *36*(9), 3954–3959.
Shannon, M. A.; Bohn, P. W.; Elimelech, M.; Georgiadis, J. G.; Mariñas, B. J.; Mayes, A. M. *Nature* **2008,** *452*, 301–310.
Shayegan, Z.; Razzaghi, M.; Niaei, A.; Akbari, A. N. *Korean J. Chem. Eng.* **2013,** *30*, 1751–1759.

Shayeghi, M.; Dehghani, M. H.; Mahvi, A. H.; Azam, K. *Iran. J. Arthropod-Borne Dis.* **2010,** *4*, 11–18.
Shayeghi, M.; Dehghani, M. H.; Fadaei, A. M. *Iran. J. Public Health* **2011,** *40*, 122–128.
Shetty, R.; Chavan, V. B.; Kulkarni, P. S.; Kulkarni, B. D.; Kamble, S. P. *Indian Chem. Eng.* **2016,** 1–23.
Sibanda, M. M.; Focke, W. W.; Labuschagne, F. J. W. J.; Moyo, L.; Nhlapo, N. S.; Maity, A. et al. *Malar. J.* **2011,** *10*, DOI: 10.1186/1475-2875-10-307.
Siddique, M.; Farooq, R.; Khan, Z. M.; Khan, Z.; Shaukat, S. F. *Ultrason. Sonochem.* **2011,** *18*, 190–196.
Singh, J.; Yang, J. K.; Chang, Y. Y. *J. Environ. Manage.* **2016,** *175*, 60–66.
Singla, R.; Grieser, F.; Muthupandian, A. K. *Ultrason. Sonochem.* **2010,** *18*, 484–488.
Sivasankar, T.; Moholkar, V. S. *Environ. Technol.* **2010,** *3*, 1483–1494.
Smirniotis, P. G. Defense Technical Information Center, Fort Belvoir, Virginia, 2002.
Son, H. S.; Choi, S. B.; Khan, E.; Zoh, K. D. *Water Res.* **2006,** *40*, 692–698.
Song, L.; Chen, J.; Gao, J.; He, B.; Wei, M. Wang, Y. *Drill. Fluid Completion Fluid* **2009,** *26*, 86–92.
Song, S.; He, Z.; Chen, J. *Ultrason. Sonochem.* **2007,** *14*, 84–88.
Sponza, D. T.; Oztekin, R. *J. Chem. Technol. Biotechnol.* **2010,** *85*(7), 913–925.
Sraw, A.; Wanchoo, R. K.; Toor, A. P. *Am. Inst. Chem. Eng. Environ Prog.* **2014,** *33*, 1201–1208.
Stapleton D. R.; Mantzavinos, D.; Papadaki, M. *J. Hazard. Mater.* **2007,** *146*, 640–645.
Stavarache, C.; Yim, B.; Vinatoru, M.; Maeda, Y. *Ultrason. Sonochem.***2002,** *9*, 291–296.
Su, G.; Li, Q.; Lu, H.; Zhang, L.; Huang, L.; Yan, L.; Zheng, M. *Sci. Rep.* **2015,** *5*, 17800.
Suri, R. P. S.; Nayak, M.; Devaiah, U.; Helmig, E. *J. Hazard. Mater.* **2007,** *146*, 472–478.
Suslick, K. S. *Science* **1990,** *247*, 1439–1445.
Sutar, R. S.; Rathod, V. K. *Ultrason. Sonochem.* **2015,** *24*, 80–86.
Tang, S. K.; Teng, T. T.; Abbas, F. M.; Alkarkhi, Li, Z. APCBEE Procedia **2012,** *1*, 110–115.
Thokchom, B.; Kim, K.; Park, J.; Khim, J. *Ultrason. Sonochem.* **2015,** *22*, 429–436.
Thompson, T. L.; Panayotov, D. A.; Yates, J. T.; Martyanov, I.; Klabunde, K. *J. Phy. Chem.: B* **2004,** *108*, 17857–17865.
Tran, N.; Drogui, P.; Brar, S. K. *Environ. Chem. Lett.* **2015,** *13*, 251–268.
Uddin, H.; Hayashi, S. *J. Hazard. Mater.* **2009,** *170*, 1273–1276.
Vajnhandl, S.; Le Marechal, A. M. *J. Hazard. Mater.* **2007,** *141*, 329–335.
Verma, A.; Kaur, H.; Dixit, D. *Arch. Environ. Prot.* **2013,** *39*, 17–28.
Visscher, D. A.; Eenoo, V. P.; Drijvers, D.; Langenhove, V. H. *J. Phys. Chem.* **1996,** *100*, 11636–11642.
Vogel, T. M.; Criddle, C. S. McCarty, P. L. *Environ. Sci. Technol.* **1987,** *21*, 722–737.
Vorontsov, A. V.; Lion C.; Savinov, E. N.; Smirniotis, P. G. *J. Catal.* **2003,** *220*, 414–423.
Vorontsov, A. V., Davydov, L., Reddy, E. P., Lion, C., Savinov, E. N., Smirniotis, P. G. *New J. Chem.* **2002,** *26*, 732–744.
Wagner, G. W.; Chen, Q.; Wu, Y. *J. Phys. Chem.: C* **2008,** *112*, 11901–11906.
Wang, P.; Bian, X.-F.; Li, Y. X. *Chin. Sci. Bull.* **2012,** *57*(1), 33–40.
Wang, S.; Ang, H. M. Tade, M. O. *Chemosphere.* **2008,** *72*, 1621–1635.
Wang, S.; Gong, Q.; Liang, J. *Ultrason. Sonochem.* **2009,** *16*(2), 205–208.
Wang, X.; Wu, G.; Lu, C.; Wang, Y.; Fan, Gao, Y. H.; Ma, J. *Colloids Surf. B: Biointerfaces* **2011,** *86*(1), 237–241.

Wang, Y.; Gai, L.; Ma, W.; Jiang, H.; Peng, X.; Zhao, L. *Ind. Eng. Chem. Res.* **2015b,** *54,* 2279–2289.

Wang, Z.; Chen, Z.; Chang, J.; Shen, J.; Kang, J.; Chen, Q. *Chem. Eng. J.* **2015a,** *262,* 904–912.

Weavers, L. K.; Malmstadt, N.; Hoffmann, M. R. *Environ. Sci. Technol.* **2000,** *34,* 1280–1285.

Weissler, A.; Cooper H. W.; Snyder S. *J. Am. Chem. Soc.* **1950,** *72,* 1769–1775.

Wen, B.; Ma, J. H.; Chen, C. C.; Ma,W. H.; Zhu, H. Y.; Zhao, J. C. *Sci. China Chem.* **2011,** *54,* 887–897.

Wenjun, L.; Di, W.; Xin, W.; Lixion, W.; Lei, S. *Chin. J. Chem. Eng.* **2012,** *20,* 754–759.

Wu, C.; Liu, X.; Wei, D.; Fan, J.; Wang, L. *Water Res.* **2001,** *35*(16), 3927–3933.

Wu, J.; Jia, R.; Liu, C.; Wang, H. *Int. Forum Energy, Environ. Sci. Mater.* **2015,** 584–587.

Wu, Z.; Lifka, J.; Ondruscha, B. *Ultrason. Sonochem.* **2004,** *11,* 187–190.

Xu, Y. L.; Zhong, D. J.; Jia, J. P.; Chen, S.; K. Li. *J. Environ. Sci. Health A* **2008,** 43, 1215–1222.

Yang, L. P.; Hu, W. Y.; Huang, H. M.; Yan, B. *Desalin. Water Treat.* **2010,** *21,* 87–95.

Yang, L.; Rathman, J. F.; Weavers, L. K. *J. Phys. Chem. B* **2006,** *110*(37), 18385–18391.

Yang, L.; Rathman, J. F.; Weavers, L. K. *J. Phys. Chem. B* **2005,** *109*(33), 16203–16209.

Yaqub, A.; Ajab, H.; Isa, M. H.; Jusoh, H.; Junaid, M.; Farooq, R. *J. New Mater. Electrochem. Syst.* **2012,** *15,* 289–292.

Yehia, F. Z.; Badawi, A. M.; Eshaq, Gh.; Dimitry, O. I. H. *Egypt. J. Pet.* **2015a,** *24,* 265–276.

Yehia, F. Z.; Eshaq, Gh.; Rabie, A. M.; Mady, A. H.; ElMetwally, A. E. *Desalin. Water Treat.* **2016,** *57,* 2104–2112.

Yehia, F. Z.; Gh. Eshaq; El Metwally, A. E. *Desalin. Water Treat.* **2015b,** *56,* 2160–2167.

Yu, F.; Shi, L. *Water Sci. Technol.* **2010,** *61,* 1931–1940.

Zhang, F.; Zhang, W.; Zhao, L.; Liu, H. *Desalin. Water Treat.* **2016,** *57,* 24406–24416.

Zidane, F.; Drogui, P.; Lekhlif, B.; Bensaid, J.; Blais, J. F.; Belcadi, S.; Kacemi, K. E. *J. Hazard. Mater.* **2008,** *155,* 153–163.

CHAPTER 9

FOOD TECHNOLOGY

SANYOGITA SHARMA[1], NEETU SHORGAR[2]

[1]*Department of Chemistry, PAHER University, Udaipur, India*
E-mail: sanyogitasharma22@gmail.com

[2]*Department of Chemistry, PAHER University, Udaipur, India*
E-mail: nshorgar@gmail.com

CONTENTS

9.1 INTRODUCTION

Nowadays food scientists are focused on the development of products that have long storage life and are not only microbiologically safe but at the same time, these products must also have fresh-like characteristics and high quality in taste, flavour and texture. This is the requirement of the consumers, and it has become one of the major areas of research in the so-called emerging technologies for food science. Traditionally, thermal treatments have been used for a long time to produce safe food products. Pasteurization of juice, milk, beer and wine is a common process, where the final product can gain a long storage life of some weeks under refrigeration. Although it is known that vitamins, taste, colour, and other sensorial characteristics are decreased with such treatment. High temperature is responsible for causing these effects. It is observed that in the loss of nutritional components and changes in flavour, taste and texture, it often creates the need for additives to improve the quality of a product. Thus, one of the main challenges in food science today is to develop new technologies that can ensure high-quality properties and long storage life of food simultaneously.

In recent years, the most popular emerging technologies being tested in food science laboratories have been high pressure, electric-pulsed fields, ultraviolet light, irradiation, light pulses and ultrasound, and some of these technologies are already being used in the food industry. The use of ultrasonics has a wide scope in the food industry for the analysis and amendment of food products. The possibility of using low-intensity ultrasound to characterize foods was first realized almost six decades ago; however, the full potential of this technique has been realized by Povey and McClements (1988). The food industry is becoming increasingly more aware of the importance of developing new analytical techniques such as ultrasonic techniques to study complex food materials and to monitor properties of foods during processing. McClements (1995) suggested that sonication is a nondestructive and noninvasive method which can easily be adapted for online applications and is used to analyse systems particularly that are optically opaque. The use of ultrasound in processing develops some novel and interesting methodologies which are often balancing to conventional techniques.

Various areas have been identified with immense potential for future progress such as freezing, extraction, drying, crystallization, homogenization, filtration, meat tenderization, degassing, sterilization and so forth. There is a wide-growing area of research into the use of ultrasound in food processing both from an industrial and academic perspective. Ultrasound technology has not only been applied in the food industry mostly as a processing aid but also for the cleaning or disinfecting of factory surfaces. Kentish and Feng (2014) discussed the physical and chemical effects of power ultrasound treatments, which are based on the actions of aural cavitation and also the biological effects of ultrasonication biological systems present in food such as microorganisms and food enzymes. Carcel et al. (2012) also reported the ultrasonic applications in various media such as liquid, gas and supercritical fluid. These are dependent on the travelling medium of ultrasound and also on the material to be used.

In various areas of science, high-frequency ultrasound has more applications as compared to power ultrasound. Low-intensity ultrasound has been used more in the food industry than power ultrasound due to it is nondestructive, rapid and traditional technique. Ultrasound is a novel technology adapted in many countries. Chemat and Hoarau (2004) proposed that ultrasound operations in food engineering require the establishment of a hazard analysis and critical control point. McClements (1995) has reviewed the role of low-intensity ultrasound technique in food processing for providing information about physicochemical properties such as composition, structure, physical state and flow rate whereas high-intensity ultrasound is used to modify physically or chemically, the properties of foods for the generation of emulsions, cell disruption, chemical reaction, enzyme inhibition, tenderize meat and modify crystallization processes. Rastogi (2011) found thermosonication, manosonication and manothermosonication more relevant energy-efficient process for alternative in the food industry and concluded role of ultrasound in the inactivation of microorganisms and enzymes, crystallization, drying, degassing, extraction, filtration, homogenization, meat tenderization, oxidation, sterilization and so forth. Simal et al. (1998) observed the use of ultrasound to increase mass transport rates during osmotic dehydration. Osmotic dehydration of 1-cm apple cubes in sucrose solution carried out using ultrasound treatment at various temperatures mass transfer was described by Fick's unsteady-state diffusion equation, and measured water diffusivity coefficients (D_w) ranged from 2.6 to 6.8×10^{-10} $m^2 \cdot s^{-1}$ at 40–70 °C and found a significant decrease in water and solute transport rate.

9.2 MATERIALS

Quality assurance is an area that is important for food technologists because it deals with the consumer's desire for quality foods. Longer treatment times are required, and destructive techniques are used in some tests. Here, ultrasound appears to be quite helpful in some of these tests. Low-intensity ultrasound has been used to assess the quality of some fruits such as mangoes, avocados and melons according to their ripeness by various ultrasonic parameters such as velocity and reduction in relation to the physical characteristics of the medium. Benedito et al. (2002) observed the quality of beef, chicken, cod, pork meat, milk, wine, sugar solutions and oils with ultrasonic parameters while textures of products such as cheese and cooked vegetables and the ripeness of fruits have also been determined by Coupland (2004) using ultrasonic waves.

9.2.1 BEVERAGES

Although power ultrasound has been capable to achieve pasteurization standards for some beverages such as fruit juices, milk and ciders, low-intensity ultrasound can also be useful for quality evaluation purposes. Ultrasound has been used for removing oxygen from orange juice. The final vitamin A content in product was higher during storage in ultrasonicated juice than in conventionally heat-treated juice. Lemonin content, brown pigments and colour were evaluated in orange juice after sonication at 500 kHz, 240 W; low temperature, 5 and 14 °C, where only minor changes were formed in these characteristics in the product (Valero et al., 2007). Ultrasound has been successfully used to process strawberry and blackberry juice and apple cider as well. It was observed that enzymes such as pectin methylesterase and polyphenoloxidase were inactivated along with pathogenic bacteria. Minor changes in ascorbic acid and anthocyanins were also detected in sonicated juice by Valdramidis et al. (2010). Beer yield was also improved by using ultrasound at the beginning of the mashing process (Knorr et al., 2004).

This improvement in yield, occur with beer as well as cheese production. The yield increases after the milk is pasteurized with power ultrasound. It shows economic benefits for the dairy and beer industries. Ultrasound can develop various characteristics of products such as colour, texture and storage life, as well as maintain some of the same nutritional values in fresh products.

Oxidation process is widely used in the alcoholic beverage industry. This process is assisted by ultrasound as it enhances oxidation in fermented products, which leads to characteristic flavour and early maturation. Mason (1996) stated that ultrasound of 1 MHz alters the alcohol/ester balance producing an apparent aging in the product which has been used for wines, whiskeys and spirits.

9.2.2 BREAD

There are always some innovation and improvement in bread quality by food scientists. The texture of the bread is an important parameter. Sonication offers the benefit of being a nondestructive method that enables to know the texture of breads through acoustic parameters. Physical structure of foods is quite tough to study due to the nonhomogeneity of foods, as in case of breads, but sound waves can give intensity information about the structure of bread crumbs. Velocity and reduction of ultrasonic waves have been used to explain certain quality factors of bread such as changes in the microstructure of freeze-dried bread crumbs due to changes in density (Elmehdi et al., 2003). They showed that the size and shape of the gas cells in bread are susceptible to ultrasound measurements, whereas the signal amplitude increases linearly with density. This nondestructive system is a very helpful tool in the cereal industry.

9.2.3 CHEESE AND TOFU

Cheese industry is one of the most important food industries in most of the countries as evident from their sales reports of cheese products and high consumer demand. Production of cheese can take several hours or several years and it depends on the place of manufacture and the type of cheese. The desirable characteristics in cheese depend on its variety or type. Cheese manufacturing consists of various stages such as coagulation, drainage, salting and ripening. The low-intensity ultrasound has been used to monitor these different stages of the process, particularly in the evaluation of internal cracks due to bad fermentation and the determination of the optimum rennet cut time. This nondestructive technique also helped in improving packaging and storage of check. Ultrasound has been directly involved in the production of soft cheese by promoting the coagulation of proteins and oils (Mason, 1996). High-frequency ultrasound has also been used to assess the quality of tofu during its manufacture. It is a water-based gel composed mainly of soy protein and its quality is based a lot on the final texture of product. Ting et al. (2009) observed the ultrasound can be used to monitor the development of texture during the gelation process of tofu. The propagation of ultrasonic waves through the structure of the tofu gel provides information regarding the whole process in relating the ultrasound parameters to the enzyme activity in milk during the coagulation process. Ultrasonic characteristics of Torta del Casar-type cheese have been investigated by Jiménez et al. (2010). This method was used to assess the quality of many foods in a nondestructive manner. This noninvasive ultrasonic method also controls the change in physical properties of organic cheese made from the sheep milk.

Vilkhu et al. (2008) reviewed the use of ultrasound in the extraction process in food industries. They concluded worthwhile gain can be accomplished in extraction efficiency and extraction rate. In addition, it also provides an opportunity for enhanced extraction of heat-sensitive bioactive and food component at lower processing temperatures. There is also a potential for achieving simultaneous extraction and encapsulation of extracted components to provide protection using ultrasound.

9.2.4 MEAT

The use of ultrasound for envisaging fat and muscle content in live cattle has been investigated by Wild (1950).Wilson (1992) reported that the ultrasound technology can be used in the beef industry for evaluating seed stock, palatability and cutability in carcasses, identifying dates to slaughter cattle, and also predicting quality. One of the most important quality aspects of beef considered by the consumer is its tenderness. Koohmaraie (1994) evaluated that inconsistency in beef softness has been one of the major problems faced by the meat industry. Jayasooriya et al. (2004) reported that the tenderness is influenced by composition, structural organization and the integrity of skeletal muscle.

Jayasooriya et al. (2007) determined the tenderness of meat by two major components of the skeletal muscle that are contractile tissue, which is largely the myofibrillar fraction, and connective tissue fraction. Traditional ageing relies on endogenous proteases (Koohmaraie 1994), though it is timeconsuming and its effectiveness differs between animals. Meat tenderness can be controlled by manipulating pre- and post-slaughter situations through the use of physical methods, such as electrical stimulation (Hwang and Thompson, 2001) and tender stretch of the pre-rigor carcass. Post-rigor meat tenderness can also be improved by mechanical techniques such as blade tenderization (Hayward et al., 1980) and high-pressure technology (Cheftel and Culioli, 1997). The use of ultrasound can cause physical disruption of materials through cavitation-related mechanisms such as high shear, pressure and temperature, and formation of free radicals. Ultrasound was also tested for its ability to induce membrane cell disruption that could increase meat tenderness either directly, through the physical weakening of muscle structure, or indirectly by the activation of proteolysis either by the discharge of cathepsins from lysosomes or of Ca^{2+} ions from intracellular stores so that it may activate the calpains.

Jayasooriya et al. (2004) reported that the sonication treatment of meat has produced incompatible effects on meat tenderness. The use of high-power ultrasound to disrupt muscle structure may prove effective for reducing both myofibrillar and collagenous toughness. The experiment was carried out with *longissimus lumborum etthoracis* and *semitendinosus* muscles from 3–4-year-old steers. Alterations in muscle structure, particularly the loss of the typical myofibrillar structure, were observed for sonicated horse semimembranosus muscle pumped with brine. Dolatowski (1988) also observed that the ultrasound treatment caused fragmentation of myofibrils and disintegration of other cellular components. Dolatowski and Stasiak (1995) reported that the ultrasound-assisted process of meat tumbling caused the major improvement in the yield, tenderness and juiciness of the end product. Sonication resulted in decreased drip loss and sheer force of pale, soft and exudative (PSE) meat.

Got et al. (1999) evaluated the effect of high-frequency ultrasound on meat texture, pre- and post-rigor meat on treatment with high frequency. Ultrasound-treated meat was having a slightly softer raw meat texture after 3–6 days of ageing. The difference between control and ultrasound treated sample texture had disappeared after 14 days of ageing.

Alarcon-Rojo et al. (2015) studied the application of ultrasound in meat processing. The effects on quality and technological properties such as texture, water retention, colour, curing, marinating, cooking yield, freezing, thawing and microbial inhibition were observed. They concluded that ultrasound is a useful tool for the meat industry as it helps in tenderization, accelerates maturation and mass transfer, reduces cooking energy and increases shelf life of meat without affecting other quality properties. It also improves functional properties of emulsified products, eases mould cleaning and improves the sterilization of equipment surfaces.

9.3 PROCESSES

9.3.1 DRYING

Garcia-Perez et al. (2006) reported that the application of high-power ultrasound for dehydration of porous materials may be very effective in processes, when heat-sensitive materials such as foodstuffs have to be treated. Ultrasound helps in reducing temperature as well as treatment time. The effect of air flow rate, ultrasonic power and mass loading on hot air drying assisted by a new power ultrasonic system was studied. An aluminium-vibrating cylinder was used as the drying chamber, which is able to create a high-intensity ultrasonic field in the gas medium. Drying of carrot cubes and lemon peel cylinders was carried out at 40 °C for different air velocities, with and without ultrasound. It was observed that the effect of ultrasound on drying rate is affected by air flow rate, ultrasonic power and mass loading. The acoustic field inside the chamber is disturbed and at high air velocities the effect of ultrasound on drying kinetics diminishes.

The effect of osmotic dehydration and ultrasound pretreatment on melon tissue structure at atmospheric pressure for different time was evaluated by Fernandes et al. (2008a). It was observed that the ultrasound induced the formation of microscopic channels in the fruit structure but it did not induce breakdown of the tissue. The changes observed on the structure of the fruit show the effects of these two pretreatments on the water diffusivity of the subsequent air-drying step. Ultrasound treatment enhances water diffusivity due to the formation of microscopic channels, which also offers lower resistance to water diffusion.

Fernandes et al. (2008b) investigated the effect of ultrasonic pretreatment and ultrasound-supported osmotic dehydration on the dehydration of pineapple (*Ananascomosus*) before air drying. They concluded that the water diffusivity increased after the application of ultrasound and that the overall drying time was reduced by 8% over 1 h of air-drying time. They showed that the water loss increased with increasing soluble solids content of the osmotic solution and that the ultrasound-assisted osmotic dehydration incorporated more sugar than the conventional osmotic dehydration. The water-effective diffusivity of the pineapples during the air-drying process was influenced by the pretreatment, increasing the water-effective diffusivity application of ultrasound.

Azoubel et al. (2010) studied the drying kinetics of fresh and ultrasonic-pretreated banana cv Pacovan using the diffusional model. These diffusivities increased with increasing temperature and with the application of ultrasound, but the process time reduced which represents an economy of energy, as air drying is cost intensive.

The potential nonthermal technique of power ultrasound is also applicable to inactivate the microorganism relevant to fruit juices. Adekunte et al. (2010) observed the modeling of yeast inactivation in tomato juice. Tomato juice samples were sonicated at amplitude level. The sonication is an effective process to achieve the desired level of yeast inactivation in tomato juice.

Oliveira et al. (2011) studied the dehydration of Malay apple (*Syzygiummalaccense* L.). Using ultrasound as pretreatment, they examined drying of a large variety of tropical and subtropical fruits (also called exotic fruits), diverse drying techniques, drying kinetics and key quality parameters of dried fruits.

9.3.2 FREEZING AND CRYSTALLIZATION

Every day newer and newer applications of ultrasound are being reported and tested in the food industry. High-power ultrasound is used in thawing processes for the thawing of beef, pork and fish, with frequencies and intensities around 500 kHz and 0.5 $W \cdot cm^{-2}$ (Miles et al., 1999). Zheng and Sun (2006) have reported that cavitation process can not only lead to the production of gas bubbles but also is responsible of microstreaming. The use of ultrasound for freeze preservation of fresh food stuffs results in the improvement of quality of product. The ultrasound is beneficial to control crystal size distribution in the frozen product and reducing crystal size, which prevents incrustation on freezing surface.

Power ultrasound is also useful in the formation of ice crystals during the freezing of water. The mechanism involved in this process is acoustic cavitation, where the acoustically generated bubbles act as nuclei for crystal growth (Sun and Li, 2003). Sonication is a helpful tool in the control of crystallization processes. It increases the nucleation rate and crystal growth rate, and, therefore, generates new and fresh nucleation sites. They observed that the freezing rate of potato assisted by ultrasound was very fast, with an output power of 15.85 W and a treatment time of 2 min, and a better microstructure was attained under these freezing conditions.

Ultrasound plays a very important role in some crystallization methods in the initial stage of freezing. As a result, the size of crystal is smaller than conventional freezing for products such as strawberries. Therefore, there is reduction in damage to the microstructure of product. Mason (1996) studied the use of ultrasound on ice lollipops. A better adherence of the lollipop to the wooden stick was observed, and the lollipop became harder than the common product. Sonication has also been used to monitor the cooling processes of gelatin, chicken, salmon, beef and yogurt by measuring the ultrasonic pulse time-of-travel to the cool surface. Sigfusson et al. (2004) measured the movement of the ice front by the time-of-flight of an ultrasonic pulse by recording echoes as a function of time, and it was correlated with the percentage of frozen food.

9.3.3 EXTRACTION

The use of ultrasound in extraction processes has certain benefits. There is greater diffusion of the solvent into cellular materials, upgrading in mass transfer, which improves discharge of contents due to the disruption of cell walls. Extraction of sugar from sugar beets, medicinal compounds such as helicid, berberine hydrochloride and protein from defatted soybeans and tea are some good examples of extraction using sonication. Extraction of the enzyme rennin by ultrasound for cheese-making achieved a higher yield of the enzyme as compared to normal extraction. The other advantages of ultrasound in extraction methods include the decrease of temperatures and the shortening of treatment times consequential in a pure extract.

One specific advantage of ultrasound is in the extraction of tea solids from water and that too with increased yield. Compounds from *Salvia officinalis* were also extracted. The extraction of antioxidant carnosic acid from the culinary herb *Rosmarinus officinalis* was found to be increased and extraction times were reduced. A reduction in maceration time was also achieved with ultrasound in the extraction of alkaloid reserpine from *Rauwolfia serpentina*. Albu et al. (2004) observed the extraction using ultrasound combined with temperature from 47 to 53 °C and at different time intervals. Only 15 min were sufficient to extract most of the materials. Karki et al. (2010) extracted sugar and protein from soy flakes and obtained excellent results after a few seconds of sonication only. The total sugar and protein release was up to 50 and 46%, respectively, as compared to untreated samples. Vilkhu et al. (2008) reported the extraction of almond oils, herbal extracts (marigold, fennel, hops mint), ginseng saponins, ginger, rutin, carnosic acid from rosemary, polyphenols, amino acid and caffeine from green tea and pyrethrins from flowers. In general, all these processes using ultrasound show higher extraction yields and rates, shorter processing times and, on the whole, very efficient processing. Sonochemical reactions could be an important tool for food extraction, if these reactions are successfully controlled. Muthupandian et al. (2008) reported some promising effects of ultrasound in the extraction of gingerol from ginger, homogenization of milk and generation of high-quality emulsions from food ingredients at laboratory-scale trials. They concluded that undesired reactions between radicals generated by ultrasonication and food ingredients could be neglected by using lower ultrasonic frequencies, while high-frequency ultrasound is used for the identification of the potential of sonochemical hydroxylation of phenolic compounds to enhance the antioxidant properties of certain food materials.

The quality and the efficiency of specific products in food and chemical industries can be improved by the use of ultrasonication. Easy formulation and lower energy requirement are other additional advantages. Rastango et al. (2003) compared the mix-stirring extraction with ultrasound extraction of daidzin, glycitin, genistin and malonylgenistin from freeze-dried ground soybeans by the variation of solvents and extraction temperatures. They found that ultrasound improved the extraction, but was dependent on the solvent used.

9.3.4 CLEANING

Ultrasound is also a proficient method for cleaning apart from other uses. It can also remove dirt and bacteria from surfaces and reach crevices that are difficult to find using conventional methods. Medical, surgical, dental and food-processing instruments and surfaces can be cleaned with ultrasound. Ultrasound has also been applied in combination with a bactericide to clean the surfaces of hatchery eggs. Mason (1996) observed that the activities of chemical biocides are also enhanced with ultrasound. A number of studies have revealed that ultrasound can be used to remove biofilm, which causes sometime fouling in some food industry equipment. Fouling of tubes used to pasteurize and process certain products is the main cause for the contamination of milk in dairy industry, because some microorganisms can be added to the walls of these tubes. Oulahal-Lagsir et al. (2000) showed that the ultrasound was twice as effective in removing such biofilm from some of the evaluated surfaces compared to the common process of swabbing to clean the equipment. Ultrasound is very useful for cleaning in dairy membrane processes during ultrafiltration activities, when applied at a constant low frequency (Muthukumaran et al., 2007). It is also being studied as an alternative to chlorination for disinfection of water. Some studies carried out in water have focused on the emergent pathogen *Cryptosporidium parvum* due to the generation of free radicals in aqueous solutions during the ultrasound treatment. The use of ultrasound as a disinfectant technique could avoid some of the troubles associated with carcinogenic products formed in chlorination processes. Tsukamoto et al. (2004a, 2004b) reported that *Saccharomyces cerevisiae* has a structure similar to the pathogenic microorganisms and ultrasound treatment at 27.5 MHz for longer treatment times with a sodium hypochlorite solution, yielded a better effect against the cells.

9.4 PRODUCT IDENTIFICATION

Food quality assurance refers to the detection of foreign bodies in the final product so as to ensure the safety of foods. All food industries have specific detectors for glass, bones, metals and other materials. Haeggström and Luukkala (2001) observed the presence of animal bones in meat products, fragments of glass in glass jars and metal in the products as manufacturing practices were poor. Knorr et al. (2004) showed that foreign bodies such as glass and plastic pieces and other raw materials can be detected in yogurt, fruit juices and tomato ketchup by ultrasonic signals in a time–frequency analysis. Strange bodies of materials such as stone, wood, plastic, glass and steel spheres were detected in cheese and marmalade using ultrasound. Ultrasound technique is nondestructive and highly sensitive method and it is viable for use in a homogeneous product with 20–75 mm of probing depth. Ultrasonic pulse compression is used to detect variations in the uniformity of several liquids, liquid leveling in polymer-based soft drink bottles and foreign objects present in containers.

The air-coupled ultrasonics is a new method to be used to estimate the level of water within a polymer drink bottle (Gan et al., 2002). The volume fractions of some components in foods such as fruit juices, syrups and alcoholic beverages were also determined using ultrasonic waves. Sonication technique was also used by Saggin and Coupland (2001) to detect the thickness of some foods, including cheddar cheese, luncheon meat and cranberry sauce. Bamberger and Greenwood (2004) also demonstrated that ultrasound affects product quality. The concentration of solids is an important parameter in slurries processing industry, where the parameters density and attenuation of ultrasonic waves have proved to be useful tools for quality assurance and process control of the thick slurry suspensions, because both these parameters are related to concentration of the medium. Schöck and Becker (2010) used ultrasound in fermentation as the concentration of sugars changes with the production of ethanol because of the yeast activity. The concentrations of ethanol and sugar were estimated based on the sound wave velocity travelling through the medium. Gestrelius et al. (1993) tested the use of ultrasound in microbial control such as packaged milk. They showed that the acoustic streaming induced by ultrasound in a liquid is affected by the microbial activity. Therefore, bacteria affecting physicochemical parameters can be easily detected using ultrasonic waves.

9.5 INACTIVATION

Some of the inactivation studies are discussed in the following sections.

9.5.1 MICROORGANISMS

Ultrasound was studied as a microbial inactivation technique about six decades back (Piyasena et al., 2003). This killing effect was first observed when ultrasonic waves were applied in a US army experiment to investigate their use in antisubmarine warfare, but fishes died in large numbers. The goal of such emerging technologies in food processing is to inactivate the initial population of microorganisms to a safe level so as to minimize damage to the product's quality features. However, microorganisms become more resistant over time to the action of a specific factor such as sound waves, pressure or electricity. Rodríguez et al. (2003) reported that the ultrasound treatments can have deleterious effects on microorganisms, when applied with sufficiently high frequencies above 18 kHz. Raso and Barbosa-Cánovas (2003) reported that the use of ultrasound, along with pressure, heat and low water activity isa good combination for reducing microbial populations in foods. Guerrero et al. (2001) reported no change in the sensitivity of *S. cerevisiae*, when ultrasound and heat were applied with different pH values, but in Sabouraud broth at pH 5.6, the addition of chitosan enhanced the inactivation of yeast under thermosonication (Guerrero et al., 2005).

Knorr et al. (2004) reported that inactivation of *Bacillus stearothermophilus* and *Escherichia coli* K12DH5α was improved with the use of direct steam injected into the ultrasound treatment, which leads to reduction in temperature and process time. On the other hand, a different type of microorganism *Lactobacillus acidophilus* was more resistant to the ultrasound-assisted thermal combination process. In ultrasound technology, a new area called manosonication is being tested, where the use of moderate and high pressures along with ultrasound reduces the initial level of microbial population effectively. Manosonication and heat action have also been reported to have an additive effect in achieving the inactivation of *Listeria monocytogenes*. It was observed that on subjecting the microorganisms to ultrasound energy at 20 kHz and 117 μm at ambient conditions, no reductions were found; however, when pressure was added as an obstacle, the microbial inactivation was observed to be quite high (Piyasena et al., 2003).

Raso and Barbosa-Cánovas (2003) developed a mathematical model to describe the resistance of various bacteria to manosonication treatments. They came to a conclusion that with the combination of heat and ultrasound, substantial reductions in *D* values are obtained for *Listeria innocua* and *Salmonella*. The studies on *Zygosaccharomyces bailii* have shown that thermosonication is independent of the treatment medium and good reductions in D values were observed with the combination of temperature and sonication. Pagán et al. (1999) showed the combination of ultrasound with pressure clearly improves the inactivation of microorganisms as for *L. innocua*. Raso et al. (1998) also showed that the use of pressure in combination with thermosonication enhanced the inactivation of *Bacillus subtilis* spores compared to thermal treatment. Various results of thermoultrasonic treatments on moulds such as *Aspergillus flavus* and *Penicillium digitatum* also showed the effect of adding antimicrobials on increasing the inactivation. An interesting study was made on the use of ultrasound to delay fruit decay and to maintain fruit quality. Ultrasound was used at different frequencies at maximum 59 kHz and temperature 20 °C to process strawberries immersed in water. Decay was delayed and the number of microorganisms also decreased. Cao et al. (2010) observed that the firmness of the fruit and the levels of total soluble solids, titratable acidity and vitamin C were also retained after processing.

The most common techniques currently used to inactivate microorganisms in food products are conventional thermal pasteurization and sterilization. Thermal processing does kill vegetative microorganisms and some spores; however, its effectiveness is dependent on the treatment temperature and time. However, the magnitude of treatment, time and process temperature is also proportional to the amount of nutrient loss, development of undesirable flavours and deterioration of functional properties of food products. High-power ultrasound is known to damage or disrupt biological cell walls, which will result in the destruction of living cells. Unfortunately, very high intensities are needed if ultrasound alone is to be used for permanent sterilization. However, the use of ultrasound coupled with other decontamination techniques, such as pressure, heat or extremes of pH looks quite promising technology. Thermosonic (heat plus sonication), manosonic (pressure plus sonication) and manothermosonic (heat plus pressure plus sonication) treatments are likely to be the best methods to inactivate microbes, as these are more energy-efficient and effective in killing microorganisms.

The advantages of ultrasound over heat pasteurization are minimization of flavour loss, greater homogeneity and significant energy savings (Mason et al., 1996; Piyasena et al., 2003). A considerable amount of data is available regarding the impact of ultrasound on the inactivation of microorganisms. The effectiveness of an ultrasound treatment is dependent on the type of bacteria being tested. Other factors include amplitude of the ultrasonic waves, exposure time, volume of food being processed, the composition of food and the treatment temperature. Bactericidal effects of ultrasound were observed while suspended in culture medium (Davies 1959). *Salmonellae* attached to broiler skin were reduced upon sonication in peptone at 20 kHz for 30 min (Lillard, 1993). Dolatowski and Stasiak (2002) proved that ultrasound processing do have a significant influence on microbiological contamination of meat. There are a large number of potential applications of high-intensity ultrasound in the food industry. Applications of both high- and low-frequency ultrasound in the food industry have considerable potential for either modifying or characterizing the properties of foods. At many places, techniques based on ultrasound have considerable advantages over the existing technologies.

9.5.2 ENZYMES

High-intensity ultrasound is used for the degassing of liquid foods, the extraction of enzymes and proteins, the induction of oxidation/reduction reactions, the inactivation of enzymes and the induction of crystallization processes. Ultrasound is also combined with other treatments to enhance the efficiency of these treatments. Enzymatic inactivation using different technologies is widely studied to attain low residual enzymatic activities in several products. Earlier studies in enzyme inactivation were carried out almost 60 years ago, when pure pepsin was inactivated by ultrasound possibly due to cavitation. Inhibition of sucrose inversion can also be achieved by cavitation. Vercet et al. (2002) observed the use of ultrasound in combination with heat and pressure and proved its effectiveness in the inactivation of enzymes such as soybean lipoxygenase, horseradish peroxidase, tomato pectic enzymes, orange pectin methyl esterase and water cressperoxidase. Some of these enzymes have been tested in a number of different food products and it was observed that the medium is a significant factor affecting the intensity of inactivation of enzymes. One of the big challenges of using ultrasound is the heterogeneity of the food matrix, because it showed assist in enzyme inactivation not only after processing but also during storage of food as well.

They observed that the manothermosonication is responsible for particle size decrease and molecular breakage. It induces the breakage of pectin molecules in a purified pectin solution. Ultrasound can also lead to denaturation of proteins, maybe due to changes in pressures producing stretching and compression effects in the cells and tissues. It is known that the free radical production is promoted by ultrasound and such free radicals as $^{\cdot}H^{+}$ and $^{\cdot}OH^{-}$ could recombine with amino acid residues of the enzymes. These residues are associated with structure stability, substrate binding, and catalytic functions. In the case of enzymes, the main mechanism is free radical formation, which allows changing some characteristics of the enzymes. Disorder of tissue is much important because it generates improved surface contact between the enzymes and free radicals. Mason (1996) observed that the oxidases are usually inactivated by sonication, while catalases are affected at low concentrations and reductases and amylases are highly resistant to sonication. Manothermosonication has inactivated several enzymes at lower temperatures and/or in a shorter time than thermal treatments. Sensitivity of the enzymes to manothermosonication treatment is independent of the medium of treatment. The substrates, small co-solutes and other proteins are unable to shield the enzymes during treatments (Vercet et al., 2001).

Food is one of the most basic needs of mankind. Till date, many methods have been used for preparing, sterilizing, dehydrating, freezing and tenderizing food products. Scientists have found ultrasound to be versatile in applications; it has also played significant role in the field of food technology. One of the most important advantages of ultrasound is its property of nondestructiveness even at high frequencies. Nowadays, ultrasound is substituting some of the traditional methods of food processing for monitoring the quality of food. Inactivation of microbes and enzymes using sound waves, integrated with parameters like temperature, pressure, and so on is proving to be of great significance. The other advantages of food products processed by ultrasound also include improvement in yield, texture and colour along with reduced process time. Thus, it can be concluded that ultrasound proves to be a boon for the development of food technology.

Nowacka et al. (2012) investigated the utilization of ultrasound as a mass transfer enhancing method prior to drying of apples tissue. Apple cubes were dried using traditional technology at 70 °C and at air velocity of 1.5 m s^{-1}. The ultrasound treatment caused reduction in the drying time by 31 % as compared to untreated tissue. The ultrasound-treated apples showed around 9–11 % higher shrinkage, and porosity of 9–14 % higher than untreated samples. Ultrasound application caused modification of rehydration properties in comparison to untreated sample.

Lida et al. (2008) developed a method for controlling the viscosity of starch (polysaccharide) solutions and it is one of the most promising processes. Power ultrasound can effectively decrease the viscosity of starch solutions after gelatinization. At high starch concentrations (20–30 %), starch gel can be liquidized by sonication. The viscosity of the starch solution of moderate concentration (5–10 %) can be reduced by two orders of magnitude to 100 mPa·s by the ultrasonic irradiation for 30 min. The treated solution can be efficiently powdered by a spray-dryer after the sonication. The ultrasonic process is applicable for many kinds of starches (corn, potato, tapioca and sweet potato) and polysaccharides.

There has been a growing interest in the development of processes using waste or residual biomass as source of cellulose nanocrystals (CNCs).Tang et al. (2015) extracted the CNCs from old corrugated container fibre using phosphoric acid and enzymatic hydrolysis using ultrasound treatment. The effect of enzymatic hydrolysis on the yield and microstructure of resulting CNC has been reported. The enzymatic hydrolysis was found effective in increasing CNC yield after phosphoric acid hydrolysis. The presence of enzymatic hydrolysis gave CNC with improved dispersion, increased crystallinity and thermal stability.

The combination of sonication and dispersive liquid–liquid microextraction based on the solidification of floating organic drop (SDLLME–SFO) has been developed by Pirsaheb et al. (2015) for the extraction and determination of imidacloprid and diazinon from apple and pear samples. Target pesticides were determined by high-performance liquid–liquid chromatography–ultraviolet detector. Extraction recoveries for different fruits are in the range of 58–67 % under optimum conditions. This method was applied with satisfactory results to the analysis of these pesticides in apple and pear samples. They showed that DLLME–SFO is a very simple, rapid, sensitive and efficient analytical method for the determination of trace amount of pesticides in fruit samples.

Saeeduddin et al. (2016) investigated the effect of sonication on physicochemical parameters and microorganisms of pear juice. Ultrasound processing of fresh pear juice was done at fixed amplitude and frequency for different time intervals at room temperature. pH, titratable acidity total soluble solids and Ca and Mn remained stable, whereas the total flavonoids, total antioxidant capacity, cloud value, ascorbic acid, total phenols, sugar contents and Na, K, Fe and Mg showed a considerable increase during this process. Decreases in microbial population and P and Cu were also observed. Ultrasound processing for 60 min exhibited optimum results in terms of physicochemical and microbial quality.

9.6 OTHER APPLICATIONS

Variation in ultrasonic parameters is often coupled with the changes in the different food products during processing. Hæggström and Luukkala

(2000) examined the transmission of ultrasound in crushed beef during simulated automated roasting, which gave good results. Here, temperatures ranged from 45 to 74 °C, with ultrasonic wave (300 kHz), and only small pieces of crushed beef were used with thicknesses between 7.15 and 15 mm. Sonication has also been used to study changes in flow actions and thermophysical properties of whey protein isolate (WPI) and whey protein concentrate (WPC). Water solubility of both WPI and WPC was significantly increased after sonication as compared to control samples. The apparent viscosity, flow behaviour index and consistency coefficient of WPI and WPC were significantly changed after sonication.

Kresic et al. (2008) reported also that thermophysical properties, such as initial freezing and initial thawing temperatures, were changed on sonication. The effects of ultrasound on the solubility and foaming properties of whey proteins were studied by Jambrak et al. (2008). Some important effects on these properties were observed, on exposing whey proteins at low frequency (20 kHz) compared to high frequency (40 kHz); however, higher frequencies (500 kHz) did not show any significant effect. An important effect with increase in temperature was observed in sonicated sample at low frequency. Mason (1996) stated that the stable emulsions were generated by the use of ultrasound requiring little if any surfactant, which include tomato ketchup and mayonnaise. Ultrasonic velocity and attenuation spectra were also measured as a function of frequency in emulsions treated with ultrasound at 1–5 MHz.

The emulsions were made of corn oil in water and the velocity and attenuation was found to increase with frequency in all emulsions having different particle sizes (Coupland and McClements, 2001). Ultrasound is also being tested to inactivate microorganisms in combination with heat and pressure. Wu et al. (2001) suggested that the ultrasound can also stimulate living cells and it is very useful in the production of yogurt, where up to 40 % of production time is reduced by using ultrasound. Milk is homogenized better using ultrasound than by traditional methods. Additional positive effects in yogurt due to ultrasound are improvements in uniformity and texture with added sweetening effects. Mason (1996) also observed faster germination of agricultural crops as well, when seeds are subjected to ultrasound treatment. Moreover, ultrasound can be used in some wastewater treatment plants within the food industry.

The use of sonic waves has assisted the bioprocess and an improved biological activity has been observed by Schläfer et al. (2002). The rate of drying was increased in drying processes, and the final moisture content was reduced by ultrasonic treatments carried out at lower temperatures. Jambrak et al. (2007) have shown that ultrasound pretreatment of vegetables for drying decreases the drying time significantly, because of the increased water loss. The rehydration properties of some vegetables such as mushrooms, sprouts, cauliflower and so on are better than others. Similar results were achieved during the drying of bananas. A sonication pretreatment to air drying had significant effect on the banana tissue. Fernandes and Rodrigues (2007) also reported that the drying time of banana was reduced by up to 11%, due to ultrasound, which allowed the elimination of larger amounts of water from the fruit. Ultrasound has been shown to be effective during osmotic dehydration processes in speeding up and enhancing the diffusivity of water and solids.

Cárcel et al. (2007) observed that sonication was able to increase the water diffusivity up to 117% and dry matter diffusivity up to 137% in apple–sucrose solution. The application of ultrasound was found effective in removing the water from the cell structure of the apple tissue, as well as in taking solids from the sucrose solution. Garcia-Perez et al.(2012) reported the feasibility of power ultrasound at low-temperature drying processes for kinetics of carrot, eggplant and apple cubes in the presence and absence of ultrasound. Similar effects have been observed for all products such as reduction in drying time (between 65% and 70%). Further ultrasonication results into the increase in mass transfer coefficient and effective moisture coefficient and effective moisture diffusivity increased by 96–170% and 407–428%, respectively.

Jambrak et al. (2007) used ultrasound treatment for drying of mushrooms, Brussels sprouts and cauliflower and found reduction in drying time. The effect of the ultrasound in mass transfer process was also observed. They investigated the effect of ultrasound and blanching pretreatments on weight and moisture loss/gain, upon drying and rehydration. The drying time after ultrasound treatment was shortened for all samples, as compared to untreated one. The rehydration properties (weight gain, %) were found to be the best for freeze-dried samples, which showed weight gains for mushrooms, Brussels sprouts and cauliflower. It was observed the rehydration properties for ultrasound-treated samples were higher than those for untreated samples.

Deng and Zhao (2008) studied the effect of pulsed vacuum and ultrasound pretreatments on glass transition, texture, rehydration, microstructure, and other selected properties of air- and freeze-dried apples were investigated. Apple cylinders (15 mm height × 15 mm diameter) were first osmo concentrated in a 60 g/100 g high-fructose corn syrup solution containing 7.5 g/100 g Gluconal Cal combined with agitation, under ultrasound for 3 h. It is followed by hot-air or freeze drying. Ultrasound led to a higher glass transition temperature, lower water activity, moisture content and rehydration rate.

Food is one of the most basic needs of mankind. Till date, many methods have been used for preparing, sterilizing, dehydrating, freezing and tenderizing food products. Ultrasound is a versatile energy source and it has played a significant role in the field of food technology. One of the most important advantages of ultrasound is its property of nondestructiveness even at high frequencies. Nowadays, ultrasound is substituting some of the traditional methods of food processing for monitoring the quality of food. Inactivation of microbes and enzymes using sound waves, integrated with parameters such as temperature, pressure and so on is proving to be of great importance. The other advantages of food products processed by ultrasound also include improvement in yield, texture and colour along with reduced processing time. Thus, it can be concluded that ultrasound proves to be a boon for the development of food technology.

REFERENCES

Adekunte, A.; Tiwari, B. K.; Scannell, A.; Cullen, P. J.; Donnell, C. O. *Int. J. Food Microbiol.* **2010,** *137,* 116–120.

Alarcon-Rojo, A. D.; Janacua, H.; Rodriguez, J. C.; Paniwanvk, L.; Mason, T. J. *Meat Sci.* **2015,** *107,* 86–93.

Albu, S.; Joyce, E.; Paniwnyk, L.; Lorimer, J. P.; Mason, T. J. *Ultrason. Sonochem.* **2004,** *11,* 261–265.

Azoubel, P. M.; Baima, M. D. A. M.; Amorim, M. D. R.; Oliveira, S. S. B. *J. Food Eng.* **2010,** *97*(2) 194–198.

Bamberger, J. A.; Greenwood, M. S. *Food. Res. Int.* **2004,** *37,* 621–625.

Benedito, J.; Carcel, J. A.; Gonzalez, R.; Mulet, A. *Ultrasonics* **2002,** *40,* 19–23.

Cao, S.; Hu, Z.; Pang, B.; Wang, H.; Xie, H.; Wu, F. *Food Control.* **2010,** *21*(4), 529–532.

Cárcel, J. A.; Benedito, J.; Rosselló, C.; Mulet, A. *J. Food Eng.* **2007,** *78,* 472–479.

Carcel, J. A.; Perez, J. V.; Benedito, J.; Mulet, A. *J. Food Eng.* **2012,** *110*(2), 200–207.

Cheftel, J. C.; Culioli, J. *Meat Sci.* **1997,** *46*(3), 211–236.

Chemat, F.; Hoarau, N. *Ultrason. Sonochem.* **2004**, *11*, 257–260.
Coupland, J. N. *Food. Res. Int.* **2004**, *37*, 537–543.
Coupland, J. N.; McClements, D. J. *J. Food Eng.* **2001**, *50*, 117–120.
Davies, R. *Biochem. Biophys. Acta.* **1959**, *33*, 481–493.
Deng, Y.; Zhao, Y. *LWT: Food Sci. Technol.* **2008**, *41*(9), 1575–1585.
Dolatowski, Z. J. *Fleischwirtschaft.* **1988,** *68*(10), 1301–1303.
Dolatowski, Z. J.; Stasiak, D. M. *The 9th Congress of Food Science and Technology*, Budapest, **1995**; p.153
Dolatowski, Z. J.; Stasiak, D. M. *Acta Sci. Pol., Technol. Aliment.* **2002,** *1*(1), 55–65.
Elmehdi, H. M.; Page, J. H.; Scanlon, M. G. *J. Cereal Sci.* **2003**, *38*, 33–42.
Fernandes, F. A. N.; Rodrigues, S. *J. Food Eng.* **2007**, *82*(2), 261–267.
Fernandes, F. A. N.; Gallao, M. L.; Rodrigues, S. *LWT: Food Sci. Technol.* **2008a**, *41*(4), 604–610.
Ferenandes, F. A. N.; Linhares Jr., F. E.; Rodrigues, S. *Ultrason. Sonochem.* **2008b**, *15*(6), 1049–1054.
Gan, T. H.; Hutchins, D. A.; Billson, D. R. *J. Food Eng.* **2002**, *53*, 315–323.
Garcia-Perez, J. V.; Carcel, J. A.; Fuente-Blanco, S. D. L.; Sarabia, E. R. F. D. *Ultrasonics* **2006**, *44*, 539–543.
Garcia-Perez, J. V.; Carcel, J. A.; Riera, E.; Rosello, C.; Mulet, A. *Drying Technol. Int. J.* **2012**, *30*(11–12), 1199–1208.
Gestrelius, H.; Hertz, T. G.; Nuamu, M.; Persson, H. W.; Lindström, K. *LWT* **1993**, *26*, 334–339.
Got, F.; Culioli, J.; Berge, P.; Vignon, X.; Astruc, T.; Quideau, J. M.; Lethiecq, M. *Meat Sci.* **1999**, *51*, 35–42.
Guerrero, S.; López-Malo, A.; Alzamora, S. M. *Innovative Food Sci. Emerg. Technol.* **2001**, *2*, 31–39.
Guerrero, S.; Tognon, M.; Alzamora, S. M. *Food Control.* **2005**, *16*, 131–139.
Haeggström, E.; Luukkala, M. *LWT–Food Sci. Tech.* **2000**, *33*(7), 465–470.
Haeggström, E.; Luukkala, M. *Food Control* **2001**, *12*, 37–45.
Hayward, L. H.; Hunt, M. C.; Kastner, L. C.; Kropf, D. J. *J. Food Sci.* **1980**, *45*(4), 925–935.
Hwang, I. H.; Thompson, J. M. *Meat Sci.* **2001**, *58*, 135–144.
Jambrak, A. R.; Mason, T. J.; Paniwnyk, L.; Lelas, V. *J. Food Eng.* **2007**, *81*(1), 88–97.
Jambrak, A. R.; Mason, T. J.; Lelas, V.; Herceg, Z.; Herceg, I. L. *J. Food Eng.* **2008**, *86*(2), 281–287.
Jayasooriya, S. D.; Bhandari, B. R.; Torley, P.; D'Arcy, B. R. *Int. J. Food Prop.* **2004**, *7*(2), 301–319.
Jayasooriya, S. D.; Torley, P. J.; D'Arcy, B. R.; Bhandari, B. R. *Meat Sci.* **2007**, *75*, 628–639.
Jiménez, A.; Crespo, A.; Piedehierro, J.; Rufo, M. M.; Guerrero, M. P.; Paniagua, J. M. et al. Preliminary Study to Assess Ultrasonic Characteristics of Torta del Casar-Type Cheese. *Proceedings of 20th International Congress on Acoustics*, ICA, Sydney, **2010**.
Karki, B.; Lamsal, B. P.; Jung, S.; van Leeuwen, J.; Pometto A. L., III; Grewell, D.; Khanal, S. K. *J. Food Eng.* **2010**, *96*, 270–278.
Kentish, S.; Feng, H. *Annu. Rev. Food Sci. Technol.* **2014**, *5*(1), 263–284.
Knorr, D.; Zenker, M.; Heinz, V.; Lee, D.-U. *Trends Food Sci. Technol.* **2004,** *15*, 261–266.
Koohmaraie, M. *Meat Sci.* **1994**, *36*(1–2), 93–104.

Kresic, G.; Lelas, V.; Jambrak, A. R.; Herceg, Z.; Brncic, S. R. *J. Food Eng.* **2008,** *87*(1), 64–73.
Lida, Y.; Tuziuti; Yasui, K.; Towata, A.; Kozuka, T. *Innovative Food Sci. Emerg. Technol.* **2008,** *9*(2) 140–146.
Lillard, H. S. *J. Food Prot.* **1993,** *56*, 716–717.
Mason, T. J. Power Ultrasound in Food Processing–The Way Forward. In *Ultrasound in Food Processing*; Povey, M. J. W, Mason, T. J., Eds.; Springer: New York, **1996**; pp. 105–126.
Mason, T. J.; Paniwnyk, L.; Lorimer, J. P. *Ultrason. Sonochem.* **1996**, *3*, S253–S260.
Mc Clements, D. J. *Trends Food Sci. Technol.* **1995**, *6*, 293–299.
Miles, C. A.; Morley, M. J.; Rendell, M. *J. Food Tech.* **1999**, *39*, 151–159.
Muthukumaran, S.; Kentish, S. E.; Stevens, G. W.; Ashokkumar, M.; Mawson, R. *J. Food Eng.* **2007**, *81*, 364–373.
Muthupandian, A.; Sunartio, D. D.; Kentish, S.; Mawson, R.; Simons, L.; Vilkhu, K. et al. *Innovative Food Sci. Emerg. Technol.* **2008**, *9*(2), 155–160.
Nowacka, M.; Wiktor, A.; Sledz, M.; Jurek, N.; Rajchert, D. W. *J. Food Eng.* **2012**, *113*(3), 427–433.
Oliveira, F. I. P.; Gallao, M. L.; Rodrigues, S.; Fernande, F. A. N. *Food Bioprocess Technol.* **2011**, *4*(4), 610–615.
Oulahal-Lagsir, N.; Martial-Gros, A.; Bonneau, M.; Blum, L. J. *J. Appl. Microbiol.* **2000**, *89*, 433–441.
Pagán, R.; Mañas, P.; Alvarez, I.; Condón, S. *Food Microbiol.* **1999**, *16*, 139–148.
Pirsaheb, M.; Fattahi, N.; Pourhaghighat, S.; Shamsipur, M.; Sharafi, K. *LWT-Food Sci. Technol.* **2015**, *60*(2), 825–831.
Piyasena, P.; Mohareb, E.; McKellar, R. C. *Int. J. Food Microbiol.* **2003**, *87*, 207–216.
Povey, M. J. W.; McClements, D. J. *J. Food Eng.* **1988**, *8*, 217–245.
Raso, J.; Barbosa-Cánovas, G. V.*Crit. Rev. Food Sci. Nutr.* **2003**, *43*(3), 265–285.
Raso, J.; Palop, A.; Pagán, J.; Condón, S. *J. Appl. Microbiol.* **1998**, *85*, 849–854.
Rastogi, N. K. *Crit. Rev. Food Sci. Nutr.* **2011**, *51*(8), 705–722.
Rodríguez, J. J.; Barbosa-Cánovas, G. V.; Gutiérrez-López, G. F.; Dorantes-Álvarez, L.; Yeom, H. W.; Zhang, Q. H. An Update on Some Key Alterative Food Processing Technologies: Microwave, Pulsed Electric field, High Hydrostatic Pressure, Irradiation and Ultrasound. In *Food Science and Food Biotechnology*; Gutiérrez-López, G. F., Barbosa-Cánovas, G. V., eds.; CRC Press, **2003**; pp 279–312.
Rostagno, M. A.; Palma, M.; Barroso, C. G. *J. Chromatogr. A.* **2003**, *1012*(2) 119–128.
Saeeduddin, A. M.; Jabbar, S.; Hu, B.; Hashim, M. M.; Khan, M. A.; Xie, M.; Wu,T.; Zeng, X. *Food Sci. Technol.* **2016**, *51*(7), 1552–1559.
Saggin, R.; Coupland, J. N. *Food Res. Int.* **2001**, *34*, 865–870.
Schläfer, O.; Onyeche, T.; Bormann, H.; Schröder, C.; Sievers, M. *Ultrasonics* **2002,** *40*, 25–29.
Schöck, T.; Becker, T. *Food Control.* **2010**, *21*(4), 362–369.
Sigfusson, H.; Ziegler, G. R.; Coupland, J. N. *J. Food Eng.* **2004**, *62*, 263–269.
Simal, S.; Benedito, J.; Sanchez, E. S.; Rosello, C. *J. Food Eng.* **1998,** *36*(3), 323–336.
Sun, D. W.; Li, B. *J. Food Eng.* **2003**, *57*, 337–345.
Tang, Y.; Shen, X.; Zhang, J.; Guo, D.; Kong, F. Zhang, N. *Carbohydr.Polym.* **2015**, *125*, 360–366.
Ting, C. H.; Kuo, F. J.; Lien, C. C.; Sheng, C. T. *J. Food Eng.* **2009**, *93*, 101–107.

Tsukamoto, I.; Constantinoiu, E.; Furuta, M.; Nishimura, R.; Maeda, Y. *Ultrason. Sonochem.* **2004a**, *11*, 61–65.
Tsukamoto, I.; Constantinoiu, E.; Furuta, M.; Nishimura, R.; Maeda, Y. *Ultrason. Sonochem.* **2004b,** *11*, 167–172.
Valdramidis, V. P.; Cullen, P. J.; Tiwari, B. K.; O'Donnell, C. P. *J. Food Eng.* **2010**, 96, 449–454.
Valero, M.; Recrosio, N.; Saura, D.; Muñoz, N.; Martí, N.; Lizama, V. *J. Food Eng.* **2007**, *80*, 509–516.
Vercet, A.; Burgos, J.; Crelier, S.; Lopez-Buesa, P. *Innovative Food Sci. Emerg. Technol.* **2001**, *2*, 139–150.
Vercet, A.; Sánchez, C.; Burgos, J.; Montañés, L.; Lopez Buesa, P. *J. Food Eng.* **2002**, *53*, 273–278.
Vilkhu, K.; Mawson, R.; Simons, L.; Bates, D. *Innovative Food Sci. Emerg. Technol.* **2008,** *9*, 161–169.
Wild, J. J. *Surgery* **1950**, *27*, 183–188.
Wilson, D. E. *J. Anim. Sci.* **1992**, *70*, 973–983.
Wu, H.; Hulbert, G. J.; Mount, J. R. *Innovative Food Sci. Emerg. Technol.* **2001**, *1*, 211–218.
Zheng, L.; Sun, D. W. *Trends Food Sci. Technol.* **2006**, *17*, 1, 16–23.

CHAPTER 10

ANAEROBIC DIGESTION

SANGEETA KALAL[1], SATISH KUMAR AMETA[2], ABHILASHA JAIN[3]

[1]*Department of Chemistry, M. L. Sukhadia University, Udaipur, India*
E-mail: sangeeta.vardar@yahoo.in

[2]*Department of Environmental Science, PAHER University, Udaipur, India, E-mail: skameta2@gmail.com*

[3]*Department of Chemistry, St. Xavier's College, Mumbai, India*
E-mail: jainabhilasha5@gmail.com

CONTENTS

10.1 INTRODUCTION

Wastewater treatment has become a necessity of the day as the world is in cancerous grip of water pollution and is facing a scarcity of potable water in many developing and undeveloped countries. Many techniques are used for the treatment of wastewater, and biological treatment of wastewater is one of these widely used techniques. Excess sludge production is one of the most serious challenges in biological wastewater treatment plants. Biodegradation of excess sludge is a commonly used method in biological treatment by anaerobic digestion. Anaerobic digestion is quite attractive waste treatment practice, where two objectives can be accomplished and these are pollution control and energy recovery. It is a suitable technology to treat not only solid waste but wastewater also.

Major progress has been made in all areas of waste management but the use of anaerobic digestion has emerged as one of the most successful and newer technology developments in the field of waste management in the past few decades (De Baere, 2000). Anaerobic digestion offers distinct advantages in the production of biogas compared to some other forms of bioenergy production. As carbon dioxide and some other emissions are placed under limitation by emission regulations and the carbon taxes, anaerobic digestion has become relatively a more attractive and competitive technology for the management of waste.

Formation of methane in anaerobic digestion involves basically four different steps and these are:

- Hydrolysis
- Acidogenesis
- Acetogenesis
- Methanogenesis

In the near future, bioenergy will be the most significant renewable energy source because it is economical and is an alternative to fossil fuels. Thus, anaerobic digestion has various advantages that are as follows:

- Less consumption of energy
- Low sludge production
- Requirements of smaller space
- Reduction in the volume of waste
- Production of biofertiliser and valuable soil conditioners

10.2 ANAEROBIC DIGESTION

Anaerobic digestion is a series of biological processes, where microorganisms break down biodegradable material in the absence of oxygen. One of

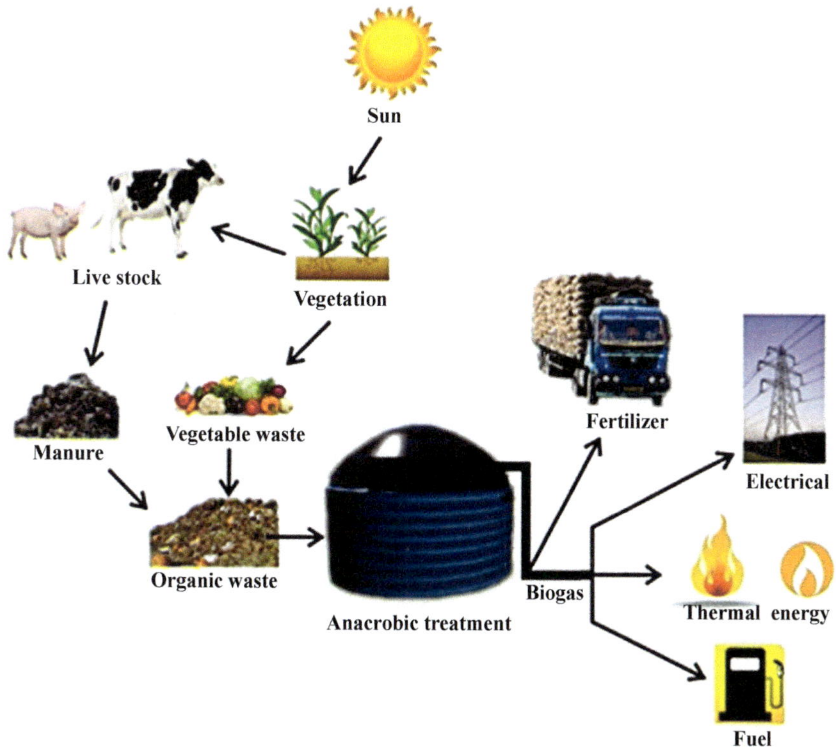

FIGURE 10.1 The process of anaerobic digestion.

the final products of this process is biogas which is composed of methane as a major component and carbon dioxide. This biogas may be utilized to generate electricity and heat through combustion or it can be processed into some other renewable natural gas and transportation fuels (Fig. 10.1).

Traditional anaerobic digesters have a residence time of about 20 days and they also require large digesters. The rate-limiting step can be defined as the step that causes process failure in an anaerobic digestion process under imposed kinetic stress. Sludge hydrolysis is the rate-limiting step of anaerobic digestion (Eastman and Ferguson, 1981; Malina and Pohland, 1992), and it is all due to the low biodegradability of the cell walls and the presence of extracellular biopolymers (Tiehm et al., 2001).

There are various already known pretreatments for waste treatment such as mechanical, thermal, chemical and biological interventions (Dewil et al., 2007; Climent et al., 2007) and all these pretreatments result in a certain degree of lysis or disintegration of sludge cells. These processes release and solubilize intracellular material into the water phase and transform refractory nonbiodegradable organic compounds into biodegradable material. Dewil et al. (2006a) have demonstrated that ultrasound (US) causes disruption of sludge flocs, thus changing the structure of its constituent and solubilizing the organics. Any improvement method in the biodegradability of a particular substrate is mainly based on the fact that the substrate is better accessible to microorganisms.

Therefore, interest of researchers, technology designers and manufacturers of relevant equipment was focused on using US for disintegration (Tiehm et al., 2001). Such ultrasonic disintegration of waste-activated sludge (WAS) was examined to improve the anaerobic stabilization process. It was observed that:

- Ultrasonic pretreatment enhances the subsequent anaerobic digestion resulting in a better degradation of volatile solids (VSs) and increased production of biogas.
- VSs concentration in the effluent of the digester fed with disintegrated sludge was 31% less than in the conventional process.
- Total biogas production was almost two times higher for disintegrated WAS.

10.3 ANAEROBIC DIGESTERS

Anaerobic digester is an air-tight, oxygen-free container, where anaerobic bacteria digest sewage producing methane and CO_2. Anaerobic digesters have been used around the world for decades for utilizing organic waste and its products, that is, biogas to increase self-sufficiency. It may be of the following types.

10.3.1 STANDARD-RATE (COLD) DIGESTER

This type of anaerobic digester is simple digester, but it requires a long digestion period of 30–60 days. This type of digester is schematically represented in Figure 10.2. Normally, the sludge content is neither heated

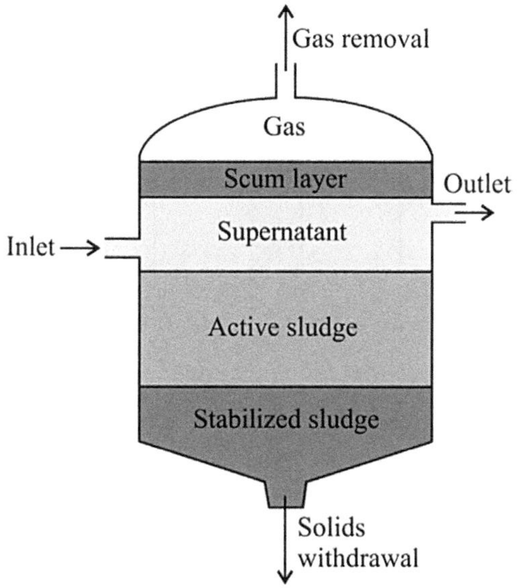

FIGURE 10.2 Standard-rate sludge digester.

nor mixed in this digester. However, the biogas generated provides some form of mixing. Stratification occurs in four zones and these are:

- Scum layer
- Liquid layer (or supernatant)
- Layer of digesting solids
- Layer of digested solids

10.3.2 HIGH-RATE DIGESTER

This digester was designed with major improvement in standard-rate digester (Fig. 10.3). In this case, the sludge is heated and completely mixed. The raw sludge is thickened and there is a uniform feeding. All these elements combined create a uniform environment and as a consequence, the tank volume is reduced and the process stability and efficiency are improved (Turovskiy and Mathai, 2006). The sludge is mixed by gas recirculation, pumping or draft-tube mixers and it is mostly heated by external heat exchangers because of their flexibility and ease of

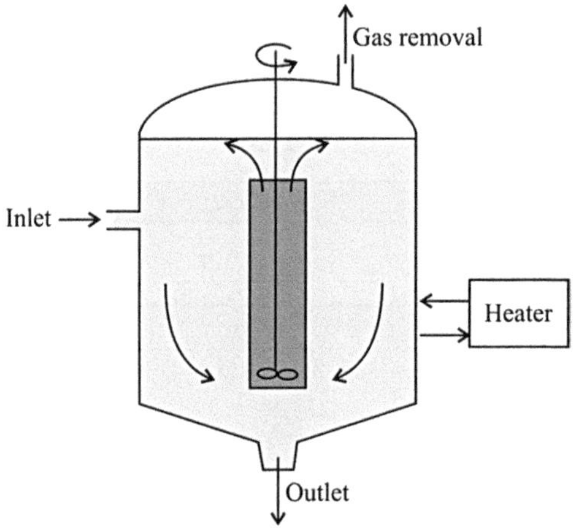

FIGURE 10.3 High-rate sludge digester.

maintenance. Other ways of heating include internal heat exchangers and steam injection (Metcalf & Eddy, 2004).

These are of three types:

- **Anaerobic Fluidized-Bed Reactor:** In anaerobic fluidized-bed reactor systems, the medium adhered with microbes is fluidized within the reactor. This results in the conversion of organic materials to CH_4 and CO_2. Anaerobic microbes grow on the surface of the medium, expanding the apparent volume of the medium and hence, this reactor is also known as an expanded-bed reactor. Media such as sand, quartzite, alumina, anthracite, granular-activated carbon or cristobalite are usually employed with a particle size of approximately 0.5 mm.
- **Upflow Anaerobic Sludge Blanket Reactor:** An upflow anaerobic sludge blanket (UASB) process capable of affording self-granulation (flocculation) of anaerobic microbes was first reported by Lettinga et al. (1980). In this type of reactor, wastewater enters from the bottom of the reactor and passes through a sludge bed and sludge blanket where organic materials are anaerobically decomposed. The gas produced is then separated by a gas–solid separator and the liquid is discharged over a weir, while the granular sludge naturally settles at the bottom. Granules range in size from 0.5

to 2.5 mm and are in different concentrations, that is 50–100 kg $VSS \cdot m^{-3}$ at the bottom to 5–40 $kg \cdot VSS \cdot m^{-3}$ in the upper part of the reactor. Such UASB systems are primarily used in the treatment of wastewater derived from the food-processing industry. Major parameters in the UASB operation are flow diameter, microbial density and the structure of the gas–solid separator, which effectively retains the microbial granules within the reactor. Granule formation in a UASB system is influenced by the growth of rod-type *Methanothrix* spp. which produces spherical granules.

- **Fixed-Film Digester:** In this type of digester, methane-forming microorganisms grow on supporting media such as wood chips or small plastic rings filling a digestion column. These digesters are also called attached growth digesters or anaerobic filters. The slimy growth coating the media is called a biofilm. Hydraulic retention times of such fixed-film reactors can be shorter than 5 days. Usually, effluent is recycled to maintain a constant upward flow. One of the drawbacks of fixed-film digesters is that manure solids can plug the voids between the supporting media. A solid separator is thus needed to remove particles from the manure before feeding it to the digester. Some potential biogas is also lost due to removal of manure solids.

10.3.3 TWO-STAGE DIGESTER

A high-rate digester is coupled with a second tank in two-stage digestion, and sometimes, it is called a secondary digester (Fig. 10.4). It is merely used to store the digested solids and decant the supernatant. It is neither heated nor mixed.

10.3.4 MESOPHILIC AND THERMOPHILIC DIGESTION

Most of the high-rate digesters are operated in the mesophilic range, with a temperature between 30 and 38 °C (Climent et al., 2007). Anaerobic digestion can also take place at higher temperatures, that is in the thermophilic region where digestion occurs at temperatures between 50 and 57 °C suitable for thermophilic bacteria. Thermophilic digestion is faster than mesophilic

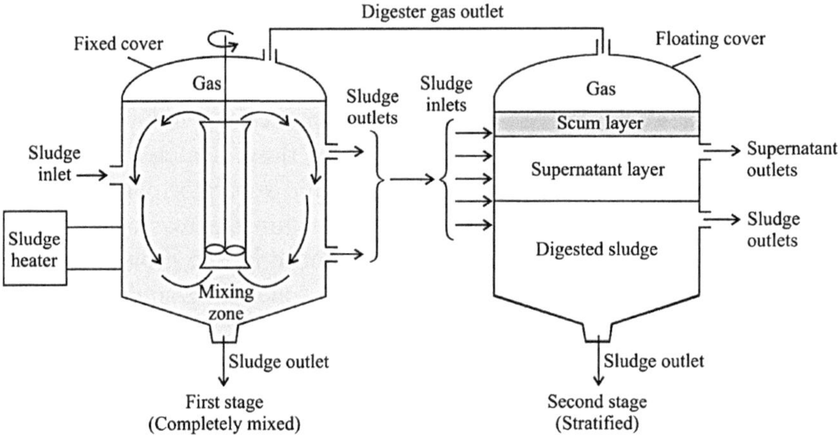

FIGURE 10.4 Two-stage sludge digester.

digestion since the biochemical reaction rates increase with increasing temperature. Thermophilic bacteria are far more sensitive to temperature fluctuations than their mesophilic counterparts (Qasim et al., 1999).

10.4 ULTRASONIC PRETREATMENT

Ultrasound is a sound wave at a frequency above the normal hearing range of human beings (>20 kHz). The rarefactions are the regions of low pressure (excessively large negative pressure), where liquid or slurry is torn apart. Microbubbles are formed in these rarefaction regions, when pressure is reduced. These microbubbles are also known as cavitation bubbles, essentially containing vapourized liquid and gas that was previously dissolved in the liquid. Cavitation occurs above a certain intensity (threshold intensity) when gas bubbles are created. These bubbles increase in size and collapse violently within a few microseconds. The radius of the bubble is inversely proportional to the frequency of the generating US (Tiehm et al., 2001; Young, 1989). The collapse of the bubbles often results in localized temperatures up to 5000 K and pressures up to 180 MPa (Flint and Suslick, 1991).

The sudden and violent collapse of a huge number of these microbubbles generates powerful hydromechanical shear forces in the bulk liquid surrounding the bubbles (Kuttruff, 1991). The collapsing bubbles disrupt the nearby (adjacent) bacterial cells by extreme shear forces, thus rupturing the cell wall and membranes. The localized high temperature and pressure could also assist in the disintegration of sludge at high temperatures. Lipids in the cytoplasmic membrane are also decomposed. It results in holes within the membrane through which intracellular materials are released into the aqueous phase (Wang et al., 2005).

Some additional effects are observed on quality and properties of the sludge when it is treated with US. These are disruption of the extracellular polymeric substances (EPS) and the cell walls of the microorganisms in the sludge (Onyeche et al., 2002), a reduction of the floc size (Dewil et al., 2006a) and a reduced dewaterability (unless extra polyelectrolyte is used), and inconclusive effects on the filamentous WAS components (Neis and Tiehm, 1999).

Sonication was also used as an extraction method to characterize the mineral fraction in extracellular matrix biofilm from activated sludge (Bourven et al., 2011). Sonication disturbs and dissolves the extracellular matrix and increases the accessibility of various inorganic cations, such as Ca^{2+}, Mg^{2+}, Fe^{3+}, Al^{3+}, K^{+} and so forth, trapped in the organic matrix of flocs (Jorand et al., 1995). Li et al. (2010) studied the effect of ultrasonically enhanced two-stage acid-leaching method on the extraction of heavy metals from electroplating sludge. The overall recovery rate of Cu, Ni, Zn, Cr and Fe from electroplating sludge ranged between 97.42 and 100%. Moreover, they were successful in the scale-up of lab-scale results to a pilot-scale industrial test. Deng et al. (2009) showed the mechanism of release of heavy metals from sludge flocs by ultrasonication (Fig. 10.5). Hristozov et al. (2004) compared the two techniques, that is ultrasonication and microwave-assisted methods for the extraction of eight metals, that is Cd, Co, Cr, Cu, Mn, Ni, Pb and Zn from sewage sludge. Both these methods achieved higher rate of metal recovery (75–84%) within similar treatment time. However, ultrasonication method offers several advantages such as it can operate under atmospheric pressure with no acid requirement and cost-effectiveness over the microwave-assisted extraction method.

Sonochemical reactions resulting in the formation of highly reactive radicals (e.g. $^{\bullet}OH$, $HO_2^{\bullet}$, $H^{\bullet}$) and hydrogen peroxide have also been reported to contribute to the ultrasonic disintegration of sludge. This is most likely due to the strong hydrodynamic shear resulting from cavitation effects at low frequency. Wang et al. (2005) explored the mechanism of ultrasonic disintegration of WAS at a frequency of 20 kHz. They added $NaHCO_3$ as a masking agent to eliminate the oxidizing effect of hydroxyl radicals during ultrasonic treatment.

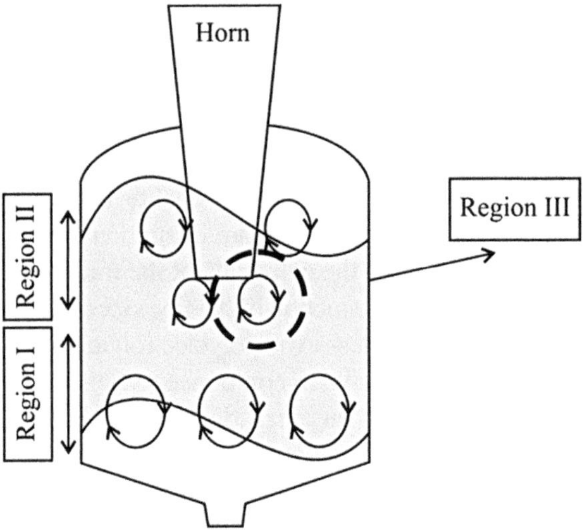

FIGURE 10.5 Regions of acoustic streaming.

Another mechanism operative during sonication of sludge is acoustic streaming. Acoustic streaming has been studied since the early 1930s (Faraday, 1831). It occurs at the solid/liquid (sludge) interface. The major benefit of streaming in sludge processing is mixing. This mixing facilitates uniform distribution of US energy within the sludge mass, convection of the liquid and distribution of heating. Acoustic streaming has three regions (Fig. 10.5). These regions are:

- **Eckart streaming (Region I):** This is the largest region and furthest from the vibrating tool. It has circulating currents that are defined by the shape of the container and the wavelength of the acoustic wave in the liquid.
- **Rayleigh streaming (Region II):** The region near the tooling is called region II. It is located around the horn and has circulating currents. Its size and shape are primarily defined by acoustic tooling. These circulations are typical with much longer wavelengths than that of the acoustic wave in liquid.
- **Schlichting streaming (Region III):** This region is closest to the vibrating tool. It is adjacent to the fluid acoustic boundary layer. In this region, tangential fluid velocity is near the velocity of the horn face. This layer is relatively thin, for example, the acoustic

boundary layer for water at 20°C is <4 μm at 20 kHz (Graff, 1988). All these three regions play a critical role in mixing of the fluid.

10.5 MECHANISMS OF ANAEROBIC DIGESTION

The digestion process begins with hydrolysis of bacteria present in input materials so as to break down insoluble organic polymers such as carbohydrates and make them available for other bacteria. Thereafter acidogenic bacteria convert the sugars and amino acids into carbon dioxide, hydrogen, ammonia and organic acids and it is followed by conversion of these resulting organic acids into acetic acid by acetogenic bacteria, along with additional ammonia, hydrogen and carbon dioxide. Finally, methanogens convert these products to methane and carbon dioxide (Metcalf & Eddy, 2004).

Formation of methane in anaerobic digestion proceeds in four different steps and these are hydrolysis, acidogenesis, acetogenesis and methanogenesis.

Most of the researchers have reported that the rate-limiting step for complex organic substrate is hydrolysis (Fernandes et al., 2009; Heo et al., 2003; Izumi et al., 2010; Duong et al., 2011; Miah et al., 2005; Rafique et al., 2010; Valo et al., 2004) due to the formation of toxic by-products (here, complex heterocyclic compounds) or nondesirable volatile fatty acids (VFAs) formed (Lu et al., 2008; Nevers et al., 2006), whereas methanogenesis is the rate-limiting step for easy biodegradable substrates (Gavala et al., 2003; Rozzi and Remigi, 2004; Skiadas et al., 2005). The anaerobic digestion process can be divided into two phases (Fig. 10.6).

This process involves four basic stages. These are hydrolysis, acidogenesis, acetogenesis and methanogenesis stages.

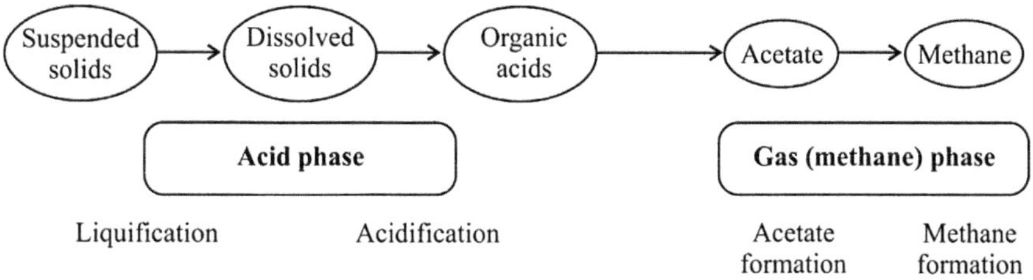

FIGURE 10.6 Phase separation of the anaerobic digestion system.

10.5.1 HYDROLYSIS STAGE

Most of the waste compounds have nonbiodegradable nature and, therefore, it is not possible to treat them directly by microorganisms. Due to this fact, hydrolysis of these complex and insoluble organics is very important so that these can be used by bacteria as an energy and nutrient source. Stabilization of the organic material is not possible during the process of hydrolysis. In this stage, only transformation of the organic materials to another structure is there and therefore, these can be used by microorganisms. This hydrolysis stage is carried out by enzymes produced and given to the environment by bacteria (Filibeli et al., 2000). Hydrolysed complex organic materials, carbohydrates, fats and proteins are fermented to fatty acids, alcohol, carbon dioxide, ammonium, formic acid and hydrogen.

10.5.2 ACIDOGENESIS

The monomers produced in the hydrolytic phase are then taken up by different facultative and obligatory an aerobic bacteria and are further degraded into short-chain organic acids such as acetic acid, butyric acid, propanoic acid, alcohols, hydrogen and carbon dioxide. In general, simple sugars, fatty acids and amino acids are converted into organic acids and alcohols during this phase (Gerardi, 2003).

10.5.3 ACETOGENESIS

The products produced in the acidogenic phase are then consumed as substrates by the other microorganisms which are active in the third phase, which is also called the acetogenic phase. Anaerobic oxidation occurs in this third phase (Aslanzadeh, 2014). Products which cannot be directly converted to methane by methanogenic bacteria are first converted into methanogenic substrates, VFAs and alcohols, which in turn are further oxidized into methanogenic substrates such as acetate, hydrogen and carbon dioxide. VFAs with carbon chains longer than one unit are oxidized into acetate and hydrogen(Al Seadi et al., 2008).

10.5.4 METHANOGENESIS

In the methanogenic phase, the production of methane and carbon dioxide from intermediate products by methanogenic bacteria under strict anaerobic conditions occurs. The complete biochemical processes are shown in Figure 10.7.

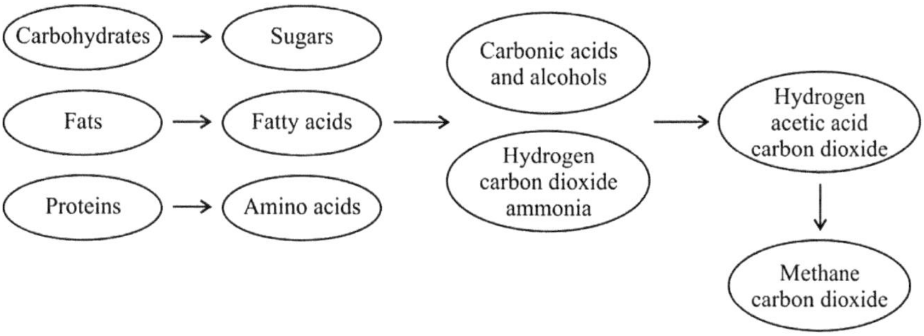

FIGURE 10.7 The key process stages of anaerobic digestion

10.6 FACTORS AFFECTING ULTRASONIC TREATMENT

Sludge pretreatment with US causes a bypass of the rate-limiting hydrolysis step, which leads to either an increased biogas production with a reduced sludge quantity or reduced retention times (Onyeche et al., 2002; Show et al., 2010).There are several factors that influence the effect of US on anaerobic digestion, including organic loading rate (OLR), sludge (solid) characteristics, particle size, sonication time, temperature, pH, specific energy input, operating frequency, sludge retention time, the fraction of the stream and so forth.

10.6.1 ORGANIC LOADING RATE

This factor plays an important role in anaerobic wastewater treatment. Overloading results in biomass washout in the case of nonattached biomass reactors, when the hydraulic retention time is relatively long. This, in turn, leads to process failure. Braguglia et al. (2011) examined the effect of OLR and the degree of disintegration on anaerobic digestion efficiency of ultrasonic-treated sludge. They reported that despite low sonication inputs (5000 $kJ\cdot kg^{-1}$ total solid [TS]), the digestion process (in a semicontinuous reactor) improved in terms of solid degradation (+20.6 %), soluble chemical oxygen demand (SCOD) removal (90 %) and biogas production (+30 %) depending on the soluble organic load.

10.6.2 SLUDGE (SOLID) CHARACTERISTICS

The sludge characteristics such as type of sludge (primary solids, WAS, animal manure, etc.) and TS content, have significant effects on ultrasonic disintegration. The effect of US pretreatment on pulp and paper mill sludge solubilization and subsequent anaerobic digestion was investigated by Saha et al. (2011). Gronroos et al. (2005) reported the maximum SCOD concentration at the highest dry solid (DS) content. However, they did not present the data in terms of $mg\cdot SCOD\cdot g^{-1}$ DS, which made it difficult to understand whether sludge disintegration was efficient at higher solids content. Khanal et al. (2006c) conducted an extensive study to evaluate the effect of TS content on SCOD release at different specific energy inputs. SCOD release showed an increasing trend with increase in both TS content and energy input. However, the release in SCOD slowed down at an energy input of over 35 $kW\cdot s\cdot g^{-1}$ TS for all TS contents. SCOD releases were 1.6, 2.2, 2.5 and 3.2 $mg\cdot kW\cdot s^{-1}$ at TS contents of 1.5, 2.0, 2.5 and 3.0 %, respectively. This corresponds to 38, 59 and 98 % increase in SCOD release at TS contents of 2.0, 2.5 and 3.0 %, respectively, as compared to 1.5 %. Wang et al. (2005) also reported a significant effect of TS content on SCOD release. The SCOD release increased from 3966 to 9019 mgL^{-1}, when the TS content was increased from 0.5 to 1.0 % during 30 min of sonication at an ultrasonic density of 1.44 $W\cdot ml^{-1}$. These findings apparently show that a higher TS content is more energy-efficient for ultrasonic disintegration than the lower one.

The ease of ultrasonic disintegration is also governed by the composition of the sludge matrix. It is believed that nonbiological solids, for example, primary sludge and animal manure, are relatively easy to disintegrate as compared to biological sludge such as WAS. The DS content of the sludge is an important parameter for the disintegration with increased effectiveness at increased concentration of DS in the WAS (Dewil et al., 2006a). More DS creates more nuclei for cavitation and, as a result, more particles were exposed to the resulting shear forces (Lehne et al., 2001).

10.6.3 PARTICLE SIZE

Particle size can affect the rate of anaerobic digestion as it affects the availability of a substrate (i.e. the surface area) to hydrolyzing enzymes, and this is particularly true with plant fibres. Fibre degradation and methane yield improved with decreasing particle size from 100 to 2 mm (Mshandete et al., 2006). Maceration of manure to reduce the size of recalcitrant fibres was found to increase the biogas potential by 16% as compared with a fibre size of 2 mm, and a 20% increase in biogas potential was observed with a fibre size of 0.35 mm; however, no significant difference was found with fibre sizes of 5–20 mm (Angelidaki and Ahring, 2000).

10.6.4 SONICATION TIME

The sonication time has a significant effect on the biogas production and solids reduction. Wang et al. (1999) observed a significant improvement of 11, 20, 38 and 46% in VS removal and 12, 31, 64 and 69% improvement in methane generation during the anaerobic digestion with 11 days solids retention time (SRT) of sludge sonicated (200 W, 9 kHz) for 10, 20, 30 and 40 min, respectively. Tiehm et al. (2001) reported a notable enhancement in VS reduction by 5.6, 27, 46 and 56.7% during the anaerobic digestion of sludge sonicated at 41 kHz for 7.5, 30, 60 and 150 min, respectively.

10.6.5 TEMPERATURE

Anaerobic digestion is strongly influenced by temperature and it can be grouped under one of the following categories (Pol, 1995): psychrophilic or cryophilic (0–20 °C), mesophilic (20–42 °C) and thermophilic (42–75 °C). Although a large section of the work deals with mesophilic operation only. In the mesophilic range, the bacterial activity and growth decrease by one-half for each 108 °C drop below 358 °C. Thus, for a given degree of digestion to be attained, lower is the temperature, the longer will be the digestion time.

10.6.6 pH

Anaerobic reactions are highly pH dependent. The optimal pH range for methane-producing bacteria is 6.8±7.2 while for acid-forming bacteria, a more acidic pH is desirable (Mudrak and Kunst, 1986). The pH of an anaerobic system is typically maintained between methanogenic limits to prevent the predominance of the acid-forming bacteria, which may cause accumulation of VFA. It is essential that the reactor contents provide enough buffer capacity to neutralize any eventual VFA accumulation, thus preventing build-up of localized acid zones in the digester. In general, sodium bicarbonate is used for supplementing the alkalinity, since it is the only chemical, which gently shifts the equilibrium to the desired value without disturbing the physical and chemical balance of the fragile microbial population.

The SCOD release was found to increase when the sludge was sonicated at a higher pH (Wang et al., 2005). It is likely that addition of alkali to raise the pH may have weakened the bacterial cell wall that facilitated better destruction during ultrasonic treatment. Therefore, alkaline treatment of sludge followed by ultrasonic application could lower the energy cost of ultrasonic systems to achieve a desired degree of sludge disintegration.

10.6.7 SPECIFIC ENERGY INPUT

The effect of specific energy input on the sludge solubilization and subsequent anaerobic digestion was also examined by several researchers. Bougrier et al. (2005) observed that biogas generation was 1.48, 1.75, 1.88 and 1.84 times higher during anaerobic digestion (16 days SRT) of the sludge (2% TS) sonicated (225 W, 20 kHz) at specific energy of 1355, 2707, 6951 and 14,547 $kJ \cdot kg^{-1}$ TS, respectively. Donoso-Bravo et al. (2010) reported a significant improvement of 40% in biogas generation at specific energy input of 12,400 $kJ \cdot kg^{-1}$ TS. However, the specific energy of up to 2754 $kJ \cdot kg^{-1}$ did not cause any noteworthy improvement in sludge disintegration and subsequent biogas production during anaerobic digestion. Khanal et al. (2006a) observed that 11.3% more SCOD was removed during anaerobic digestion as compared to the untreated sludge. The ultrasonically pretreated and digested sludge was more stable biologically. Salsabil et al. (2010) observed a 20% improvement in total suspended solids (TSS) removal efficiency for sludge sonicated with a specific energy input of 200,000 $kJ \cdot kg^{-1}$ TS. The TSS removal increased with increasing sludge solubilization.

10.6.8 OPERATING FREQUENCY

Operating frequency is another important parameter that controls the efficacy of ultrasonic systems. This is because operating frequency governs the critical size of cavitational bubbles (Hua and Hoffmann, 2004). The cavitational effect is greatly reduced at high frequency range, whereas lower frequency range generates extremely violent cavitation. Tiehm et al. (2001) reported that hydrodynamic shear forces produced by ultrasonic cavitation are primarily responsible for particle disruption. They found the degree of disintegration based on COD (DDCOD) to be 13.9, 3.6, 3.1 and 1.0%, respectively, at frequencies of 41, 207, 360 and 1068 kHz and concluded that a frequency lower than 41 kHz would yield better sludge disintegration. Sonochemical reactions are particularly predominant at a higher ultrasonic frequency of 200–1000 kHz (Mark et al., 1998).Thus, nearly all sludge disintegration tests are conducted at the lower frequency range of 20 kHz (Khanal et al., 2006a, 2006b).

Several researchers investigated and compared the anaerobic digestion efficiency of the sonicated sludge and control (untreated) sludge at different SRT. Neis (2000) observed a 30% higher VS removal in shorter digestion period of 4 days as compared to the control reactor. The effect of part stream and full stream sonication on the methane generation and COD removal efficiency of anaerobic digester was examined by Perez-Elvira et al. (2010). Pretreating the full stream with US resulted in a higher biogas generation (+41%) in comparison to pretreatment of a part of the stream. On the other hand, Kim and Lee (2012) observed a maximum methane yield, when only 30 and 16% of the sludge was disintegrated. They suggested that excess ultrasonic irradiation transforms VSs into inert or inhibitory compounds, which hamper the methanogenic activity during anaerobic digestion.

The ultrasonic treatment has been successfully applied to enhance the anaerobic digestibility of sludge. Laboratory- and pilot-scale studies showed that the application of sonication with low frequency (20 kHz waves) and limited specific energy (1000–3000 $kJ \cdot kg^{-1}$ TS equal to approximately 20–60 $kJ \cdot L^{-1}$ of sludge) led to a significant increase in biogas generation (Dewil et al., 2006b). Thus, higher sludge solubilization by US pretreatment led to a significant improvement in biogas generation and solids reduction during the biological digestion of sonicated sludge. A remarkable improvement in biogas generation (from 24 to 84%), solids reduction (from 21 to 57% as VS) and specific methane generation (32–104%) was observed in batch, continuous and semicontinuous systems. Moreover, higher F/M (food to microorganisms) ratio (10) and higher percentage of sonicated sludge (full stream sonication) have a positive effect on biodegradability, biogas generation and solid-removal efficiency of biological digesters.

10.7 APPLICATIONS OF ULTRASOUND

Various applications of US are discussed in the following sections.

10.7.1 TREATMENT OF SEWAGE SLUDGE

Tiehm et al. (1997) reported that US treatment is a suitable method to disintegrate sewage sludge. Due to sludge disintegration, organic compounds are transferred from the sludge solids into the aqueous phase resulting in an enhanced biodegradability. In semicontinuously operated fermenters with identical residence times of 22 days, a reduction of VSs (45.8% for untreated sludge and 50.3% for disintegrated sludge) was observed. Reduction of VSs was about 44.3% in the fermenter operated with disintegrated sludge and 8 days residence time. In this fermenter, the production of biogas was significantly enhanced due to the increased throughput. According to the concentration of fatty acids, the biogas production and its composition in the anaerobic systems receiving US treated raw sludge were stable even at reduced digestion times. Therefore, US pretreatment of sewage sludge is a promising method to enhance fermentation rates and to reduce the volume of sludge digesters.

The effect of low-power ultrasonic radiation on anaerobic biodegradability of sewage sludge was investigated by Liu et al. (2009). Soluble substances and variation of microbial system of sewage sludge were subjected to low-power ultrasonic radiation. The well-known hydromechanical shear forces and heating effect of low-frequency US play a major role in the sludge pretreatment process. It was found that the increase insoluble substance may partly result from the destruction of microbial cell by excess ultrasonic pretreatment, but it will inhibit the anaerobic process. By orthogonal tests, the optimal parameters were found as exposure time 15 min, ultrasonic intensity 0.35 $W \cdot cm^{-2}$ and ultrasonic power density of 0.25 $W \cdot ml^{-1}$. Under these conditions, anaerobic biodegradability of sewage sludge was increased by 67.6% and, therefore, it was concluded that low-power ultrasonic pretreatment is a valid method for improving anaerobic biodegradability of sewage sludge.

Alkaline and ultrasonic sludge disintegration can be used as the pretreatment of WAS to promote the subsequent anaerobic or aerobic digestion. Different combinations of these two methods were investigated by Jin et al. (2009). The evaluation was done on the basis of the quantity of SCOD in the pretreated sludge as well as the degradation of organic matter in the subsequent aerobic digestion. The released COD levels were higher than those with ultrasonic or alkaline pretreatment alone in case of WAS samples. NaOH treatment was more efficient than $Ca(OH)_2$ with the ultrasonic treatment for WAS solubilization. The COD levels released in various sequential options of combined NaOH and ultrasonic treatments were in the following order:

Simultaneous treatment >NaOH treatment > ultrasonic treatment >ultrasonic treatment followed by NaOH treatment.

Low NaOH dosage (100 $g \cdot kg^{-1}$ DS) for a short duration (30 min) of NaOH treatment and low ultrasonic specific energy (7500 $kJ \cdot kg^{-1}$ DS) were found suitable in simultaneous treatment for sludge disintegration. The degradation efficiency of organic matter was increased from 38.0 to 50.7% with combined NaOH and ultrasonic pretreatment with optimal parameters, which is much higher than that of ultrasonic (42.5%) or that of NaOH pretreatment (43.5%) in the subsequent aerobic digestion at the same retention time.

Ultrasonication encompasses a comprehensive range of applicability in sludge treatment via enhancement in sludge solubilization and subsequent biodegradability, effective degradation of hazardous compounds such as endocrine-disrupting compounds (EDCs), polycyclic aromatic hydrocarbons (PAHs), heavy metals, apart from resource recovery (bio-fuels, enzymes, metals) and stimulating enzymatic activity, thus enhancing the biodigestion efficiency, sludge minimization by ultrasonication-assisted lysis–cryptic growth and the extraction of chemical compounds, that is, organotin compounds (OTC), enzymes, deoxyribonucleic acid (DNA), PAHs, EDCs, linear alkylbenzenesulphonates (LAS) and EPSs. The higher specific energy input (>5000 $kJ \cdot kg^{-1}$ TS), high-power density/intensity and longer sonication time were found suitable for sludge disintegration as evident from particle size reduction, COD solubilization, supernatant turbidity and morphological changes in the sludge. The quantitative and qualitative enhancements in products, saving in time, space and equipment are the paramount reasons for preferring ultrasonication technique over other conventional methods for sludge disintegration (Tyagi et al., 2014).

Slow degradation of WAS is a disadvantage of anaerobic digestion leading to high sludge retention time in conventional digesters and, therefore, Wonglertarak and Wichitsathian (2014) have investigated the effect of pretreatment on the performance of anaerobic digestion. Treatment efficiency was evaluated under both ambient and thermophilic conditions. They concluded that the biodegradability of WAS was increased by alkaline pretreatment. The SCOD and biochemical oxygen demand (BOD) fractions also increased when pH values were increased.

Two cases of anaerobic digestion of sludge were assessed, that is (i) with pretreatment and (ii) without pretreatment, using mass–energy balance and the corresponding greenhouse gas (GHG) emissions. VS degradation of the control sludge and the ultrasonicated secondary sludge were 51.4 and 60.1 %, respectively, for a digestion period of 30 days (Pilli et al., 2016). It was observed that the quantity of digestate required for dewatering, transport and land application was the lowest (20.2×10^6g dry sludge day^{-1}) for ultrasonicated secondary sludge at 31.4 g $TS \cdot L^{-1}$ and the maximum net energy (energy output − energy input) of total dry solids was 7.89×10^{-6} $kW \cdot h \cdot g^{-1}$ with energy ratio (output/input) 1.0. As the sludge solids concentration was increased, GHG emissions were reduced. Ultrasonication pretreatment has proved to be efficient and beneficial for enhancing anaerobic digestion efficiency of the secondary sludge as compared to the primary and mixed sludge.

The effect of combined microwave-ultrasonic pretreatment on the anaerobic biodegradability of primary, excess activated and mixed sludge was observed (Yeneneh et al., 2013). All sludge samples were subjected to microwave treatment at 2450 MHz, 800 W for 3 min followed by ultrasonic treatment at a density of 0.4 $W \cdot ml^{-1}$, amplitude 90%, intensity 150 W and pulse 55/5 for 6 min. Methane production in microwave-ultrasonic pretreated primary sludge was found to be significantly greater (11.9 $ml \cdot g^{-1}$ TCOD) than the methane yield of the untreated primary sludge (7.9 $ml \cdot g^{-1}$ TCOD).

Tian et al. (2015) observed the sequential combination of ultrasonication and ozonation as sewage sludge treatment prior to anaerobic digestion. They observed synergistic volatile suspended solids (VSS) solubilization when low-energy ultrasonication ($\leq$12 $kJ \cdot g^{-1}$ TS) was followed by ozonation. Nearly 0.048 g O_3g^{-1} TS ozonation induced the maximum VSS solubilization of 41.3%, when the sludge was preultrasonicated at 9 $kJ \cdot g^{-1}$ TS, but the application of same ozone dosage without prior ultrasonication induced only 21.1% VSS solubilization. High molecular weight (MW) components (MW>500 kDa) were found to be the main solubilization products when sludge was only ozonated. However, solubilization products by ozone were mainly in the form of low-MW components (MW<27 kDa) on preultrasonication of sludge. The high-MW products generated by US were effectively degraded in the subsequent ozonation. Anaerobic biodegradability was found to increase by 34.7% when ultrasonication (9 $kJ \cdot g^{-1}$ TS) and ozonation (0.036 g O_3 g^{-1} TS) were combined sequentially. The maximum methane production rate increased from 3.53 to 4.32, 4.21 and 4.54 ml after ultrasonication, ozonation and ultrasonication–ozonation pretreatments, respectively.

An enhancement in the degradation of micropollutants (pesticides) absorbed in agro-food industrial sewage sludge was reported during its disintegration with US treatment (Ibáñez et al., 2015). The US treatment resulted in a significant reduction in the sludge pesticide content, which means that about 90% of the total pesticide mass was removed and, therefore, pollutant accumulation in the sludge was prevented. Sono-degradation of three characteristic pesticides (thiabendazole, acetamiprid and imazalil) revealed the formation of transformation products (TPs). Such TPs were already reported in earlier studies on the degradation of these compounds by advanced oxidation processes, which confirmed the participation of hydroxyl radical reactions during ultrasonication. Thus, this US treatment for the excess sludge reduction is efficient for pesticides and the removal of their TPs.

10.7.2 INDUSTRIAL WASTEWATER

Rajeshwari et al. (2000) reviewed anaerobic digestion for industrial wastewater treatment (effluents from sugar and distillery, pulp and paper, slaughterhouse and dairy units) and the suitability and the status of development of anaerobic reactors for the digestion of selected organic effluents. The presence of biodegradable components in the effluents coupled with the advantages of anaerobic process over other treatment methods makes it an attractive option. In the case of slaughterhouse wastewater, an anaerobic contact reactor can be used without pretreatment whereas for the usage of high-rate digester such as UASB, a pretreatment step for the removal of the suspended solids (SSs) and fats is essential just before anaerobic treatment. Two-phase digestion with pH and temperature control results in a higher biogas production rate with cheese whey waste water digestion. UASB and fixed-film reactors are more commonly used for distillery effluent due to their ability to withstand high OLR. An aerobic posttreatment is necessary to attain the permissible COD and BOD level before discharge. Due to the generation of waste water from various sections of pulp and paper industry, there are variations in the composition and the treatability of effluents. Hence, it is preferable to treat the effluents from each section separately depending on their biodegradability and suitability to the digestion process rather than treating the combined effluent.

Pulp and paper mill effluent (PPME) is a rich cellulosic material which was found to have great potential for biohydrogen production through a photo fermentation process. However, pretreatments were needed for degrading the complex structure of PPME before biohydrogen production. The effect of ultrasonication process on PPME was studied as a pretreatment method and on photo fermentative biohydrogen production using *Rhodobacter sphaeroides* NCIMB (Hay et al., 2015). They demonstrated the potential of using ultrasonication as a pretreatment for PPME as the yield and rate of biohydrogen production were highly enhanced compared to the raw PPME.

Oz and Uzun (2015) investigated the applicability of low-frequency US technology to olive mill wastewaters (OMWs) as a pretreatment step prior to anaerobic batch reactors to improve biogas production and methane yield. OMWs originating from three phase processes were characterized to have high organic content and complex in nature. Thus, the treatment of the wastewater is problematic and alternative treatment options should be searched. For this, first OMW samples were subjected to US at a frequency of 20 kHz with applied powers (50–100 W) under temperature-controlled conditions for different time periods to find out the most effective sonication conditions. The level of organic matter solubilization at US experiments was calculated as the ratio of SCOD and TCOD.

They revealed that the optimum ultrasonic condition for diluted OMW is 20 kHz, 0.4 $W \cdot m \cdot L^{-1}$ for 10 min and its application increased SCOD/TCOD ratio from 0.59 to 0.79. The US was significantly effective on diluted OMW ($p<0.05$) in terms of SCOD parameters, but not for raw OMW ($p>0.05$). This increase has been found to be limited due to the high concentration of SSs for raw OMW. Biogas and methane production rates of anaerobic batch reactor fed with the US pretreated OMW samples were compared with the rates obtained with control reactor fed in case of untreated OMW. The application of low-frequency US to OMW significantly improved both biogas and methane production in anaerobic batch reactor fed with the wastewater ($p<0.05$). Anaerobic batch reactor fed with US pretreated diluted OMW produced approximately 20% more biogas and methane as compared with the untreated one (control reactor). They indicated that low-frequency US pretreatment increased soluble COD in OMW and subsequently biogas production.

10.7.3 ANAEROBIC DIGESTION OF ORGANIC SOLID WASTES

Nickel (2002) reported that biosolids from agriculture, food production and wastewater treatment show a great potential as a source of renewable energy to be transformed by anaerobic processes. Anaerobic degradation is a slow process and therefore, large fermenters are necessary. The rate-limiting hydrolysis of VSs is substituted by ultrasonic disintegration so that a considerable intensification of the anaerobic biogas process takes place. In case of sewage sludge, the anaerobic digestion time could be reduced to one-fourth without any noticeable losses in the yield of biogas. Fermentation times are reduced under US irradiation on one hand and the anaerobic degradation was enhanced on the other. Thus, US disintegration has a strong potential to reduce fermentation times and as a result the volume of anaerobic digesters is minimized.

Fruit and vegetable wastes have low TSs and high VSs and are easily degraded in an anaerobic digester. The rapid hydrolysis of these feed stocks leads to acidification of a digester and the consequent inhibition of methanogenesis. Two-stage reactors effectively use the first stage as a buffer against the high OLR, which offers some protection to the methanogens. Separation of the acidification process from methanogenesis by the use of sequencing batch reactors has been shown to give higher stability, a significant increase in biogas production and an improvement in the effluent quality, when used with fruit and vegetable waste (Bouallagui et al., 2004).

Biomass is a promising feedstock for anaerobic digestion. Grasses, including straws from wheat, rice and sorghum, may provide an adequate supply of biomass, much of which is a waste product of food production. Harvesting time can also significantly affect the biogas yield of plants, as demonstrated by Amon et al. (2007a, 2007b). Maize crops were harvested after 97 days of vegetation at milk ripeness, 122 days of vegetation at wax ripeness and 151 days of vegetation at full ripeness. The maize varieties produced 9-37 % more methane at 97 days of vegetation as compared to 151 days.

Bohdziewicz et al.(2005) reported the influence of ultrasonic field on biodegradation of refractory compounds in leachate and enhancement of treatment efficiency during anaerobic digestion process. They found that in the case of leachate ultrasonication for 300 s at the amplitude of 14 m, the COD removal efficiency was 7% higher as compared to that in fermentation of nonconditioned wastewater. An increase in a biogas production was also noted, where a son the 8th day of the process, a specific methane yield was 22% higher in comparison to that of nonconditioned leachate.

To enhance the anaerobic digestion of municipal WAS, US, thermal and US + thermal (combined) pretreatments were conducted using three US-specific energy inputs (1000, 5000 and 10,000 $kJ \cdot kg^{-1}$ TSS) and three thermal pretreatment temperatures (50, 70 and 90 °C) (Dhar et al., 2012). Combined pretreatments significantly improved VSS reduction by 29–38%. The largest increase in methane production (30%) was observed after 30 min of 90 °C pretreatment followed by 10,000 $kJ \cdot kg^{-1}$ TSSUS pretreatment. Combined pretreatments improved the dimethyl sulphide removal efficiency by 42–72 %. Such a treatment did not show any improvement in hydrogen sulphide removal, when compared with US and thermal pretreatments alone. It was also reported that combined pretreatments can reduce operating costs as compared to conventional anaerobic digestion without pretreatments.

Although anaerobic fermentation is the most applied process for the stabilization of sewage sludge because of its ability in mass reduction, methane production and improved dewatering properties of the fermented sludge, a disadvantage associated with it is the limitation in hydrolysis phase. As a consequence, it is a slow process. This reduced efficiency of anaerobic digestion can be significantly enhanced by using US. US pretreatment aims to convert recalcitrant particulate organics into a soluble, more biodegradable form maximizing anaerobic stabilization and biogas recovery. Some advantages of US pretreatment of sludge can be summarized as below:

- Low cost and efficient operation compared to other pretreatment processes
- Complete process automation
- Potential to control filamentous bulking and foaming in the digester
- Better digester stability
- Improved VS destruction
- Better sludge dewatering ability and improved sludge quality

By applying pretreatments, it is possible to accelerate the hydrolysis of organic matter, thus increasing SCOD and, in many cases, also increasing methane yield. Thus, the use of ultrasonic process for pollutant degradation from wastewater and WAS is a promising method to enhance fermentation rates and to reduce the volume of sludge digesters in order to protect the environment from pollutants as a key technology for future in the world.

REFERENCES

Al Seadi, T.; Ruiz, D.; Prassl, H.; Kottner, M.; Finsterwaldes, T.; Volke, S.; Janssers, R. *Handbook of Biogas;* University of Southern Denmark: Esbjerg, 2008.

Amon, T.; Amon, B.; Kryvoruchko, V.; Machmuller, A.; Hopfner-Sixt, K.; Bodiroza, V.; et al. *Bioresour. Technol.* **2007a,** *98,* 3204–3212.

Amon, T.; Amon, B.; Kryvoruchko, V.; Zollitsch, W.; Mayer, K.; Gruber, L. *Agric. Ecosyst. Environ.* **2007b,** *118,* 173–182.

Angelidaki, I.; Ahring, B. K. *Water Sci. Technol.* **2000,** *41,* 189–194.

Aslanzadeh, S. Pretreatment of Cellulosic Waste and High Rate Biogas Production. Doctoral Thesis on Resource Recovery, University of Borås, Borås, 2014, pp 1–50.

Bohdziewicz, J.; Kwarciak, A.; Neczaj, E. *Environ. Prot. Eng.* **2005,** *31,* 3–4.

Bouallagui, H.; Torrijos, A.; Godon, J. J.; Moletta, R.; Ben Cheikh, R.; Touhami, Y.; Delgenes, J. P.; Di, A. H. *Chem. Eng. J.* **2004,** *21,* 193–197.

Bougrier, C.; Carrere, H.; Delgenes, J. P. *Chem. Eng. J.* **2005,** *106,* 163–169.

Bourven, I.; Joussein, E.; Guibaud, G. *Bioresour. Technol.* **2011,** *102,* 7124–7130.

Braguglia, C. M.; Gianico, A.; Mininni, G. *Bioresour. Technol.* **2011,** *102,* 7567–7573.

Climent, M.; Ferrer, I.; del Mar Baeza, M.; Artola, A.; Vázquez F.;Font, X. *Chem. Eng. J.***2007,***133*(1–3), 335–342.

De Baere, L. *Water Sci. Technol.***2000,***41,* 283–290.

Deng, J.;Feng, X.;Qiu, X. *Chem. Eng. J.* **2009,***152*(1), 177–182.

Dewil, R.;Baeyens, J.;Goutvrind, R. *Environ.Prog.***2006a,***25*(2), 121–128.

Dewil, R.;Baeyens, J.;Goutvrind, R. *Chin. J. Chem. Eng.***2006b,***14,* 105–113.

Dewil, R.;Appels, L.;Baeyens, J.; Degrève, J. *J. Hazard.Mater.***2007,***146*(3), 577–581.

Dhar, B. R.;Nakhla, G.;Ray, M. B. *Waste Manage.***2012,***32*(3),542–549.

Donoso-Bravo, A.; Perez-Elvira, S. I.;Polanco, F. *Chem. Eng. J.***2010,***160,* 607–614.

Duong, T. H. M.; Smits, M.;Vestraete, W.;Carballa, M. *Bioresour. Technol.***2011,***102,* 592–599.

Eastman, J. A.;Ferguson, J. F. (1981). Journal Water Pollution Control Federation, 53 (3), 352–366.

Faraday, M. *Philos. Trans.R. Soc. London***1831,***121,* 299–340.

Fernandes, T. V.;KlaasseBos, G. J.; Zeeman, G.; Sander, J. P. M.;Lier, J. B. *Bioresour. Technol.***2009,***100,* 2575–2579.

Filibeli, A.;Büyükkamacı, N.;Ayol, A. AnaerobikArıtma.DokuzEylülÜniversitesiYayınları No: 280, 2000.

Flint, E. B.;Suslick, K. S. *Science***1991,***253,* 1397–1399.

Gavala, H. N.;Yenal, U.;Skiadas, I. V.;Westermann, P.;Ahring, B. K. *Water Res.***2003,***37,* 4561–4572.

Gerardi, M. H. *The Microbiology of Anaerobic Digesters;*Wiley:Hoboken, 2003; pp. 89–92.

Graff, K. Lecture Notes. WE 795, Independent Study on High Power Ultrasonic, The Ohio State University, Columbus, Ohio, 1988.

Gronroos, A.;Kyllonen, H.;Korpijarvi, K.;Pirkonen, P.;Paavola, T.;Jokela, J.;Rintala, J. *Ultrason.Sonochem.***2005,***12,* 115–120.

Hay, J. X. W.;Wu, T. Y.;Juan, J. C.;Jahim, J. Md. *Energy Convers. Manage.* **2015,***106,* 576–583.

Heo, N. H.; Park, S. C.; Lee, J. S.;Kang, H. *Water Sci. Technol.***2003,***48,* 211–219.

Hristozov, D.; Domini, A. C.; Kmetov, V.; Stefanova, V.; Georgieva, D.; Canals, A. *Anal. Chimica Acta* **2004,** *516,* 187–196.

Hua, I.; Hoffmann, M. R. *Environ. Sci. Technol.* **2004,** *31,* 2237–2243.

Ibáñez, G. R.; Esteban, B.; Ponce-Robles, L.; Casas López, J. L.; Agüera, A.; Sánchez Pérez, J. A. *Chem. Eng. J.* **2015,** *280,* 575–587.

Izumi, K.; Okishio, Y. K.; Niwa, C.; Yamamoto, S.; Toda, T. *Int. Biodeterior. Biodegrad.* **2010,** *64*, 601–608.

Jin, Y.; Li, H.; Mahar, R. B.; Wang, Z.; Nie, Y. *J. Environ. Sci.* **2009,** *21*(3), 279–284.

Jorand, F.; Zartarian, F.; Thomas, F.; Block, J. C.; Bottero, J. Y.; Villemin, G.; Urbain, V.; Manem, J. *Water Res.* **1995,** *29*, 1639–1647.

Khanal, S. K.; Isik, H.; Sung, S.; van Leeuwen, J. Effects of Ultrasound Pretreatment on Aerobic Digestibility of Thickened Waste Activated Sludge. Paper presented at the 7th Specialized Conference on Small Water and Wastewater Systems, Mexico City, Mexico, March 7–10, 2006a.

Khanal, S. K.; Isik, H.; Sung, S.; van Leeuwen, J. Ultrasound Pretreatment of Waste Activated Sludge: Evaluation of Sludge Disintegration and Aerobic Digestibility. I Proceedings of IWA World Water Congress and Exhibition, Beijing, China, September 10–14, 2006b.

Khanal, S. K.; Isik, H.; Sung, S.; van Leeuwen, J. Ultrasonic Conditioning of Waste Activated Sludge for Enhanced Aerobic Digestion. In Proceedings of IWA Specialized Conference—Sustainable Sludge Management: State of the Art, Challenges and Perspectives, Moscow, Russia, May 29–31, 2006c.

Kim, D. J.; Lee, J. *Bioprocess Biosyst. Eng.* **2010,** *35*, 289–296.

Kuttruff, H. *Ultrasonics Fundamentals and Applications;* Elsevier Science: Essex, England, 1991.

Lehne, G.; Mülmler, A.; Schwedes, J. *Water Sci. Technol.* **2001,** *43*(1), 19–26.

Lettinga, G.; van Velsen, A. F. M.; Hobma, S. W.; de Zeeuw, W.; Klapwijk, A. *Biotechnol. Bioeng.* **1980,** *22*(4), 699–734.

Li, C.; Xie, F.; Ma, Y.; Cai, T.; Li, H.; Huang, Z.; Yuan, G. *J. Hazard. Mater.* **2010,** *178*, 823–833.

Liu, C.; Xiao, B.; Dauta A.; Peng, G.; Liu, S.; Hu, Z. *Bioresour. Technol.* **2009,** *100*(24), 6217–6222.

Lu, J.; Gavala, H. N.; Skiadas, I. V.; Mladenovska, Z.; Ahrin, B. K. *J. Environ. Manage.* **2008,** *88*, 881–889.

Malina, J. F.; Pohland, F. G. *Design of Anaerobic Processes for the Treatment of Industrial and Municipal Wastes;* Water Quality Management Library; Technomic Publishing Company: Lancaster, USA, 1992; Vol. 7.

Mark, G.; Tauber, A.; Laupert, R.; Schuchmann, H. P.; Schulz, D.; Mues, A.; von Sonntag, C. *Ultrason. Sonochem.* **1998,** 5, 41–52.

Metcalf & Eddy. *Wastewater Engineering: Treatment and Reuse*, 4th ed; Revised by Tchobanoglous, G., Burton, F. L., Stensel, H. D.; McGraw-Hill: New York, 2004.

Miah, M. S.; Tada, C.; Yang, Y. *J. Mater. Cycles Waste Manage.* **2005,** *7*, 48–54.

Mshandete, A.; Bjornsson, L.; Kivaisi, A. K.; Rubindamayugi, M. S. T.; Mattiasson, B. *Renewable Energy* **2006,** *31*, 2385–2392.

Mudrak K.; Kunst, S. *Biology of Sewage Treatment and Water Pollution Control*; Ellis Horwood Ltd.: England, 1986; 193 p.

Neis, U. *Sewage Treat.* **2000,** *21*, 36–39.

Neis, U.; Tiehm, A. *Ultrasound in Wastewater and Sludge Treatment.* Reports on Sanitary Engineering N 25, Technical University Hamburg: Harburg, 1999.

Nevers, L.; Ribeiro, R.; Oliveira, R.; Alves, M. M. *Biomass Bioenergy* **2006,** *30*, 599–560.

Nickel, K. *Ultrasonic Disintegration of Biosolids—Benefits, Consequences and New Strategies*; Harburg Reports on Sanitary Engineering, Technical University of Hamburg: Harburg, 35, 189-200, 2002.

Onyeche, T. I.; Schlafer, O.; Bormann, H.; Schröder, C.; Sievers, M. *Ultrasonics* **2002,** *40*(1-8), 31-35.

Oz, N. A.; Uzun, A. C. *Ultrason. Sonochem.* **2015,** *22*, 565-572.

Perez-Elvira, S. I.; Ferreira, L. C.; Donoso-Bravo, A.; Fernandez-Polanco, M.; Fernandez-Polanco, F. *Water Sci. Technol.* **2010,** *61*, 1363-1372.

Pilli, S.;Yan, S.;Tyagi, R. D.;Surampalli, R. Y. *J. Environ. Manage.* **2016,** *166*, 374-386.

Pol, H. Waste Characteristics and Factors Affecting Reactor Performance. In *International Course on Anaerobic Wastewater Treatment*; Wageningen Agriculture University, The Delft, Netherlands, 1995.

Qasim, S. R. *Wastewater Treatment Plants: Planning. Design and operation*, 2nd ed.; CRC Press: Boca Raton, 1999.

Rafique, R.; Poulsen, T. G.; Nizami, A. S.; Asamzz, Z. U. Z.; Murphy, J. D.; Kiely, G. *Energy* **2010,** *35*, 4556-4561.

Rajeshwari, K. V.; Balakrishnan, M.; Kansal, A.; Lata, K.; Kishore, V. V. N. *Renewable Sustainable Energy Rev.* **2000,** *4*, 135-156.

Rozzi, A.; Remigi, E. *Rev. Environ. Sci. Biotechnol.* **2004,** *3*, 93-115.

Saha, M.; Eskicioglu, C.; Marin, J. *Bioresour. Technol.* **2011,** *102*, 7815-7826.

Salsabil, M. R.; Laurent, J.; Casellas, M.; Dagot, C. *J. Hazard. Mater.* **2010,** *174*, 323-333.

Show, K. Y.; Tay. J. H.; Hung. Y. T. Ultrasound Pretreatment of Sludge for Anaerobic Digestion. In *Handbook of Environmental Engineering, Environmental Bioengineering*; Wang, L. K., Tay, J. H., Tay, S. T. L., Hung, Y. T. Eds.; Springer/Humana Press: USA, 2010; pp 53-73.

Skiadas, I. V.; Gavala, H. N.; Lu, J.; Ahring, B. K. *Water Sci. Technol.* **2005,** *52*, 161-166.

Tian, X.; Wang, C.; Trzcinski, A. P.; Lin, L.; Ng, W. J. *Chemosphere* **2015,** *140*, 63-71.

Tiehm, A.; Nickel, K.; Neis, U. *Water Sci. Technol.* **1997,** *36*(11), 121-128.

Tiehm, A.; Nickel, K.; Zellhorn, M.; Neis, U. *Water Res.* **2001,** *35*(8), 2003-2009.

Turovskiy, I. S.;Mathai, P. K. *Wastewater Sludge Processing;* Wiley: New York, 2006.

Tyagi, V. K.; Lo, S. L. Appels, L.;Dewil, R. *Crit. Rev. Environ. Sci. Technol.* **2014,** *44*(11), 1220-1288.

Valo, A.; Carrere, H.; Delgenes, J. P. *J. Chem. Technol. Biotechnol.* **2004,** *79*, 1197-1203.

Wang, Q.; Kuninobo, M.; Kakimoto, K.; Ogawa, H. I.; Kato, Y. *Bioresour. Technol.* **1999,** *68*, 309-313.

Wang, F.; Wang, Y.; Ji, M. *J. Hazard. Mater*. **2005,** *B123*, 145-150.

Wonglertarak, W.; Wichitsathian, B. *J. Clean Energy Technol.* **2014,** *2*, 118-121.

Yeneneh, A. M.; Sen, T. K.; Chong, S.; Ang, H. M.; Kayaalp, A. *Comput. Water, Energy, Environ. Eng*. **2013,** *2*, 7-11.

Young, F. R. *Cavitation;* McGraw-Hill: London, 1989.

CHAPTER 11

MEDICAL APPLICATIONS

DIPTI SONI[1], SURBHI BENJAMIN[2]

[1]*Department of Chemistry, PAHER University, Udaipur, India*
E-mail: soni_mbm@rediffmail.com

[2]*Department of Chemistry, PAHER University, Udaipur, India*
E-mail: surbhi.singh1@yahoo.com

CONTENTS

11.1 INTRODUCTION

One of the most significant features of ultrasound is its ability to penetrate the human body. Ultrasound finds therapeutic uses, when its high frequency (~5 MHz) is properly employed at low powers.

Basically, therapeutic ultrasound has two major applications; first one is sonodynamic therapy (SDT). In this medical treatment, a chemotherapy drug is given to a patient, which first gets accumulated in the cancerous tissue and is then targeted by the focused ultrasound so that it is activated inside the affected cells only. In this procedure, sonochemical formation of activated radicals occurs, which play an important role in initiating chain peroxidation of lipids present in cell membranes. This ultimately weakens the cell membrane and helps in cancer treatment.

The second application of therapeutic ultrasound is the high-intensity-focussed ultrasound (HIFU) which generates a very high concentration of energy that can be used to kill the tumour cells with sound. This treatment is mainly used in curing patients suffering from prostrate and soft-tissue cancers.

Another very common and widely used application of ultrasound is the ultrasonography (ultrasonic imaging). For this, frequency of 2 MHz and higher is used. This frequency range with shorter wavelength permits easy resolution of details of internal structures and cells. The power is kept less than 1 W cm^{-2} to prevent cavitation effects and heating in the target object. This technique of ultrasonic imaging is effective in non-destructive testing. It is a medical imaging technique employed to study the size and structure of internal organs, and to obtain tomographic images of the same. The medical sonography using ultrasound has been widely used as a diagnostic tool as it is not very expensive and is very easy in application. While using this technique, care should be taken to use lower powers and to avoid ionizing radiations so as to protect the tissue against any adverse heating or pressure effects.

Ultrasound diagnosis is also observed in veterinary medicine. The ultrasound radiations with desirable frequency are externally and internally used for evaluation of soft tissue and pregnancy detection in animals. This technology also finds use in improving animal health, breeding, and to study the characteristics of unborn calves.

Ultrasound can be used for detection and treatment of pelvic abnormalities, muscle and ligament sprains, joint inflammation, rheumatoid arthritis, osteoarthritis trauma and many more applications.

11.2 APPLICATIONS

Ferrara et al. (2007) reviewed the medical microbubbles and their use in therapeutic delivery and control. The ultrasonic pulse oscillates these minute gas bubbles with a wall velocity of the order of tens to hundreds of metres per second and these can be deflected back to the container wall or broken into nanoparticles. In the same way, targeted ultrasound can be used to disrupt blood vessel walls, thus helping in focussed delivery of drug or gene.

Ishtiaq et al. (2009) reviewed different applications of ultrasound in pharmaceutics, such as drug dispensation, formulation, drug delivery, and so forth and its use. They have also discussed the process such as sonophoresis, ultrasonic microfeeding and pharmaceutical dosing.

Microbubbles have great potential as carriers for different drugs, tiny molecules, nucleic acids and polypeptides. Tinkov et al. (2009) reviewed new frontiers of the pharmaceutical science using ultrasound.. The approach of ultrasound-targeted microbubble destruction for drug delivery was discussed. Manufacturing and drug-loading microbubbles were also presented.

Fyfe and Bullock (1985) reviewed the scientific approach for the therapeutic use of ultrasound in soft-tissue lesions. Yu et al. (2004) reviewed the effects of low-power ultrasound on the biological changes induced in the morphology and functioning of tissues.

Hitchcock and Holland (2010) reviewed ultrasound-activated thrombolysis and its use in the treatment of ischemic stroke. In this review, stable cavitation and clot lysis were analysed. It was concluded that ultrasound-enhanced stable cavitation helped in recombinant tissue plasminogen activator thrombolysis. This effect had the potential to lower the morbidity and death rate of victims of ischemic stroke.

Kremkau (1979) gave a historical review of cancer therapy using ultrasound. He presented that there can be three approaches to this aspect and these are:

- Only ultrasound
- Ultrasound integrated combination with radiotherapy
- Ultrasound in addition to chemotherapy

The mechanism in most cases is absorption heating. The potential of ultrasound to treat tumours has helped to better understand its actions on different kinds of malignancies.

Kennedy et al. (2003) reviewed the recent developments made in the field of HIFU applications. They have discussed the potential of HIFU as a noninvasive surgical technique in the treatment of tumours of various internal organs such as liver, kidney, breast, bone, uterus and pancreas and different other ailments. They suggested that high intensity focussed ultrasound (HIFU) is likely to play a major role as surgery of future.

Rosenthal et al. (2004) reviewed sonodynamic therapy (SDT) which deals with the enhancement of cytotoxic activities of sonosensitizers in the studies of cells in tumour-bearing animals, as the ultrasound energy can be directly focussed on malignancy sites present deep in tissues and can be used to activate a preloaded sonosensitizer. Probable mechanism of SDT include the formation of radicals derived by sonosensitizer, making the cell more susceptible to ultrasound-assisted drug transport across the cell membrane. This is also termed as sonoporation.

Bai et al. (2012) also reviewed the induction of apoptosis of cancer cell using SDT. They discussed that ultrasound could prevent the proliferation or initiate the apoptosis of the malignant cells in vitro or in vivo. It was concluded that low-frequency and low-intensity ultrasound induced apoptosis.

Naor et al. (2016) reviewed scientific aspects of ultrasonic fields, with respect to both space and time, and their impact on neuronal activity. SDT can be defined as the experimental cancer therapy which employs ultrasound to increase the cytotoxic effects of drugs known as sonosensitizers.

Haar (1995) reviewed the ultrasound focal beam surgery. In this surgery, high-intensity beams of ultrasound are focussed at depth within the body, thereby producing selective damage within the focal volume, with no harm to overlying or surrounding tissues. This surgery is used in ophthalmology, urology and oncology.

Harris (2005) discussed instrumentation and techniques, process of calibration, standardization, rate and the effects of nonlinear propagation related to use of ultrasound in medicine and biology.

Mason (2011) has reviewed the use of ultrasound in physiotherapy, surgical instruments, chemotherapy and drug delivery. Certain historical perspectives of clinical use of ultrasound have been presented in this chapter.

Pahk et al. (2015) used ultrasonic histotripsy for tissue therapy. Histotripsy is an extracorporeal noninvasive technique. HIFU has been used for inducing tissue fractionation without causing coagulative necrosis. The proposed mechanisms for this fractionation are combined effect of nonlinear wave propagation, explosive bubble generation and ultrasonic atomization. This work focusses on different types of cavitation activity for both thermally and mechanically induced lesions. Numerical studies were also carried out on the bubble dynamics; both ex and in vivo liver experiments were performed with histological analysis (haematoxylin and eosin stain). It was concluded that the acoustic emissions produced during the thermal ablation and the histotripsy exposure could be distinguished easily and the suggested cell therapy was a potential replacement for disordered hepatocytes.

Ultrasonic waves have been found to be effective in focussing into milimetre-scale regions across the human body and brain, and therefore, have a potential in producing controlled artificial modulation of neuronal activity.

Ultrasound plays an important role in the delivery of genetic, proteinaceous and chemotherapeutic drugs. Cavitating microbubbles act as the mediators by which the energy of pressure waves makes cell membranes permeable to drug carriers and disrupts the vesicles that carry drugs. The presence of microbubbles enhances ultrasonic delivery of smaller chemical agents. Attaching the DNA to the microbubbles or to liposomes increases the gene uptake. Ultrasonic gene delivery has been investigated in different tissues such as cardiac, vascular, skeletal muscle, tumour and fetal tissue. Ultrasonic-enhanced delivery of proteins and hormones helps in transdermal transport of insulin (Pitt et al., 2004).

Mitragotri (2005) observed that ultrasound facilitates the delivery of drugs across the skin, enhances gene therapy to focussed tissues, transports chemotherapeutic drugs into malignant cells and transfers thrombolytic drugs into blood clots in the healing of wounds and fractured bones.

Chemotherapy plays a crucial role in cancer treatment. The main drawback of chemotherapy is the adverse side effects of the anticancer agents and the development of chemoresistance. Yu et al. (2006) discussed that ultrasound assists chemotherapy by overcoming drug resistance. The efficiency of various anticancer agents could be enhanced by ultrasonic exposure either in vitro or in vivo. Ultrasound-assisted chemosensitization could be attained by increasing intracellular drug accumulation. Ultrasound also helps in targeted chemotherapy by transporting anticancer chemicals directly into the lesions.

Taradaj et al. (2007) observed the effect of the sonotherapy and compression therapy on the enhancement of healing ulcers after surgical treatment. They concluded that both these therapies helped in decontamination of the affected area.

Regar et al. (2003) investigated the intracoronary sonotherapy (IST) and its effect on the coronary vessel. After undergoing angioplasty, IST was performed using a 5-French catheter with three serial ultrasound transducers working at 1 MHz. IST was successfully carried on in 36 lesions with success rate of 90 %. IST exposure time per lesion was keptas 718 ± 127 s. It was concluded that IST could be used safely and with acute procedural success.

The major limitation of stent implantation is intimal hyperplasia and in-stent restenosis. Fitzgerald et al. (2001) studied the efficiency of intravascular sonotherapy treatment on intimal hyperplasia in a swine stent model. A total of 48 stented sites underwent sonotherapy using a custom-built, 8-French catheter intravascular sonotherapy system. Ultrasound energy of 700 kHz was used for the treatment group for up to 5 min. It was concluded that intravascular sonotherapy decreased cellular proliferation and reduced in-stent hyperplasia, thus giving intravascular sonotherapy a potential to reduce in-stent restenosis.

Sonoelastography is an imaging modality that has been used to recognize tumours of the breast, thyroid and prostate. Both axial strain elastography and axial shear strain elastography (ASSE) have shown significant potentials to differentiate between benign and malignant tumours.

Johnson et al. (2016) evaluated quantitative ultrasound elastography of benign and malignant musculoskeletal soft-tissue masses. Quantitative evaluation data showed that shear-wave velocity measurement was reproducible and that malignant masses had slower longitudinal shear-wave velocities than benign masses. It was concluded that the sample size of this pilot study formed the basis for larger study designs.

Chaudhry et al. (2016) studied the effect of interstitial fluid pressure (IFP) on ultrasound axial strain and ASSE. IFP developed a novel contrast mechanism in both the axial strain and ASSE images. This in formation would be important for a better elucidation of elastographic images of tumours.

Smith (1995) had discussed several applications of ultrasound for medical treatment, some of which are uterine fibroid ablation, cataract treatment, surgical removal of tissue and haemostasis, transdermal drug delivery, healing of bone fracture, physiotherapeutic uses, sonophoresis, sonoporation, gene therapy, bacterial growth control, dental treatment, thrombolysis and so forth. All these applications work under specific dosage or exposure of the ultrasound waves. Low doses can be used for beneficial effects, whereas high doses cause cell death.

Cancer therapy using HIFU is also very common. Generally, lower frequency sound waves are used for this purpose. Higher frequency sound waves were earlier used for cancer treatment but due to their adverse effects they are not used nowadays. Ultrasound is widely used in the diagnosis of certain health conditions, for example elastographyis beneficial in differentiating healthy tissues from unhealthy tissues. Ultrasound waves are also used for the detection of abnormalities in different body parts.

Misaridis et al. (2005) presented a coded excitation ultrasound imaging system based on a commercial scanner. It had a coded sequence of 4-MHz probe employed for the clinical treatment of the coded (excitation and compression) scheme. The clinical images demonstrated an improvement in contrast and penetration depth. Both axial and lateral resolutions were preserved. An improvement of more than 10 dB in the signal-to-noise ratio of the images was observed with the highest acquisition depth of 15 cm.

Therapeutic ultrasound is different from diagnostic ultrasound as it is used for the treatment of affected organs. Many developments have been made for safe application of ultrasound such as advances in transducer design, determination of acoustic power and so forth. Chemotherapy plays a very significant role in cancer therapy but there are still some disadvantages mainly due to the adverse effects of the anticancer drugs and growing resistance to drugs. Therefore, if ultrasound is employed in conjunction with anticancer agents, it can enhance their efficiency and, thereby, lower the dose amount required and lead to lesser side effects. It can also assist in dealing with drug resistance. Yu et al. (2004) and Rosenthal et al. (2004) also demonstrated that SDT can be given using either conventional low-range ultrasound or by HIFU. Pitt et al. (2004) found that ultrasound also enhances intracellular drug accumulation by the process of sonoporation. As ultrasound has the ability of tissue penetration, this can be used for the treatment of solid tumours present deep inside the body.

The development of excessive skin is a major side effect of acute weight loss. The main evidence-based treatment for this problem of excessive skin is by plastic surgery. Therefore, other treatments need to be developed. Bjerså et al. (2015) investigated the effects of ultrasound therapy on excess skin present on the upper arms after underground bariatric surgery. Fourteen patients were treated for 30 min with active ultrasound by Cellsonic™ on one arm. The other arm was considered as control. The effect was studied by determining the arm volume, circumference and ptosis. Majority of patients reported positive experiences and no side effects were observed. In addition, relief of symptoms related to skin such as pain, looseness and reduced skin burst were also reported.

Ultrasound is widely used for treating most soft-tissue problems, mainly lesions of tendon, ligament and bursa. Clinical evidences of ultrasound in soft-tissue injuries mainly deal with pain and swelling treatment but do not have much effect on wound healing. Ultrasound was also used in the treatment of musculoskeletal pain (Beckermann et al., 1993).

Veronick et al. (2016) studied the effect of low-intensity pulsed ultrasound (LIPUS) to help in bone fracture repair and to treat non-union defects. Ultrasound was also evaluated as a means of generating a transdermal physical force that could initiate osteoblasts that had been encapsulated within collagen hydrogels and led to bony defects. It was also shown that ultrasound generates a physical force and when applied to hydrogel, results in their deformation. The effect became intense as ultrasound intensity was increased and hydrogel stiffness was reduced.

Albornoz et al. (2011) carried out in vitro animal studies using LIPUS leading to osteoinductive effects and observed that it enhanced the healing process and accelerated the bone strength. LIPUS was found to be more effective in smokers and diabetic patients.

Voigt et al. (2011) studied the use of low-frequency ultrasound having range 20–40 kHz as adjunctive therapy for acute wound healing. The main objective of this study was to find out the effect of low-frequency ultrasound employed as an adjunctive therapy in complete treatment and decrease of size of chronic lower limb wounds. They showed that there was a favourable influence of early healing in patients suffering from venous stasis and diabetic foot ulcers by the use of both high- and low-intensity ultrasound.

Bommannan et al. (1992) mainly focussed on:

- Mechanism of sonophoresis
- Pathway of drug permeation in the presence of ultrasound
- Any side effects on skin features and morphology

Electron microscopy was used to follow the permeation of an electron-dense, colloidal tracer, that is lanthanum hydroxide (LH). Experiments have been carried out on using the hairless guinea pig. Colloidal LH suspensions were put on different skin locations, which were then immediately exposed to ultrasound having range at 10–16 MHz, for duration of 5–20 min. It was observed that LH penetrates the skin through the stratum corneum (SC) under ultrasound exposure and LH passed through the epidermis to the upper dermis, even after only 5 min of ultrasound exposure.

Mitragotri et al. (1995) studied the effect of different ultrasound-related phenomena, such as cavitation, thermal effects, production of convective velocities and mechanical effects, in the sonophoresis. It was found that in all the ultrasound-related phenomena, cavitation has the most important role in transdermal drug delivery under the frequency range 1–3 MHz and intensity range 0–2 W cm^{-2}. Confocal microscopy demonstrated that cavitation results in the keratinocytes of the SC upon ultrasound irradiation. It was considered that oscillations of the cavitation bubbles cause disorder in the SC lipid bilayers, thereby, increasing transdermal transport. This hypothesis was supported by using skin electrical resistance measurements. It was concluded that sonophoretic efficiency depends directly on the passive permeant diffusion coefficient, and not on the permeability coefficient through the skin. Specifically, permeants passively diffuse through the skin. Seven permeants were used for the study and these were estradiol, testosterone, progesterone, corticosterone, benzene, butanol and caffeine.

Transdermal drug delivery has many advantages as compared to oral drug delivery such as relief from pain and persistent drug release. However, the use of this technique is restricted to low-skin permeability due to the presence of SC, which is the outermost layer of the skin. Here, sonophoresis comes into scene as it temporarily enhances skin permeability so that various drugs can be transmitted noninvasively. Different studies of sonophoresis in transdermal drug delivery have been carried out focusing on parameter optimization, delivery process, transport route or delivery of drug including hydrophilic and high-molecular weight compounds. Depending on different studies, various possible mechanisms of sonophoresis have been proposed. For example, cavitation is considered to be the most significant mechanism for drug delivery in sonophoresis.

Alam et al. (2010) studied the effect of ultrasound as a safe and an effective technique for facial and neck skin tightening. It was observed that a single ultrasound treatment of the forehead led to brow height elevation of slightly <2 mm. Mild erythema and oedema were observed in most patients.

Anand and Kaczkowski (2009) studied the noninvasive method of estimating the local in situ acoustic heating rate using the heat transfer equation. The heating rate is determined by experimentally calculating the time needed to increase the temperature from a baseline temperature to boiling temperature. The heat transfer equation is solved to obtain the heating rate that is a consequence of boiling. The beginning of boiling is noninvasively determined by observing the time instant of onset of acoustic emissions in the ultrasonic range due to violent collapse of bubbles. This study has potential for its applicability in therapy planning and monitoring for appropriate therapeutic dose optimization.

Burgess et al. (1986) discussed the role of ultrasound in the treatment of glaucoma using HIFU. They summarized results for 170 patients suffering with refractory glaucoma and the treatment with HIFU. The results were studied in terms of the effectiveness of different treatments resulting in complications and category of the patients based on factors such as age and etiology.

Transdermal drug delivery has proved to be an efficient alternative to the conventional drug delivery techniques such as intake and injection. The SC limits the penetration of substances through the skin and acts as a barrier, but when ultrasound is applied to the skin the permeability of skin is improved, which assists in better drug delivery through the skin (Lavon and Kost, 2004).

Hyperthermia (HT) is widely used in the treatment of cancer and other benign diseases. Ultrasound technology proves to be advantageous as it allows efficient control of spatial and dynamic heating compared to other commonly used heating methods. Some of the advantages of ultrasound in this aspect are appropriate range of energy penetration in soft tissue and the capacity to shape the energy deposition methods (Diederich and Hynynen, 1999).

Farinha et al. (2006) reported ultrasonic skin permeation employing Sontramedical corporation, Franklin, MA device (SonoPrep) and used it as a method to reduce skin impedance for electrophysiology analysis. It was concluded that SonoPrep ultrasonic skin permeation decreased skin impedance to lower levels.

HIFU can be used for nonsurgical treatment for body contouring by the process known as liposonix. It destroys adipocytes percutaneously. Fatemi (2009) studied its efficiency and effectiveness. This technique is used to transmit energy through the skin surface at a very low intensity, but it concentrates this energy to the targeted focus in the subcutaneous fat. (Most important part is that at the epidermis, the intensity of the ultrasonic energy is so low that it causes no damage.) The targeting of the ultrasound rays at fixed depth under the epidermis, results in adipose tissue destruction. As soon as adipocytes are disrupted, chemotadic signals initiate body's inflammatory response mechanisms. Macrophage cells reach the area to digest the lipids and cell debris. This ultimately leads to reduction in the overall volume of adipocytes. Liposonix has proved to be a safe and beneficial technique for nonsurgical body shaping by decreasing the fat content.

The influence of exposures varied in terms of intensity and duration of HIFU on nerve conduction were studied in rats by Foley et al. (2008). HIFU has proved to be an alternative for other clinical method for initiating nerve conduction block.

Fry et al. (1954) studied production of focal destructive lesions in the central nervous system with ultrasound.

Gebauer et al. (2005) studied the efficiency of ultrasonic waves in 85 cases with a minimum fracture age of 8 months. The results showed that the LIPUS can influence healing rates similar to those obtained by surgical methods with not much risks and complications.

Gelet et al. (2000) discussed the criteria for measuring response of transrectal HIFU ablation of prostate cancer using progression-free probability calculations. The results showed that 62% of the patients did not report disease progression even after 60 months of transrectal HIFU ablation. Therefore, it was concluded that transrectal HIFU prostate ablation can be a potential alternative for patients suffering from localized prostatic adenocarcinoma.

Hundt et al. (2007) evaluated the effect of HIFU on inducing cell disruption in the case of squamous cell carcinoma. Continuous HIFU and short-pulse HIFU modes were applied to tumour cells. Major changes were exhibited in the MR images of the continuous HIFU mode as compared to the short-pulse HIFU mode. It was observed that tumours exposed to the continuous HIFU mode showed areas of necrosis, whereas in the short-pulse HIFU mode, areas of coagulation necrosis were observed. Gene expression analysis was also carried out. In the continuous HIFU mode, 23 genes were shown to be up regulated and 5 genes were down regulated where as in the short-pulse HIFU mode, 32 genes were up regulated and 16 genes were found to be down regulated. This study offered a potential to obtain molecular targets for imaging and therapeutic treatment.

Magnetic resonance imaging (MRI)-guided focussed ultrasound (MRgFUS) surgery is widely used as a noninvasive thermal ablation technique that makes use of MRI for target definition, treatment methodology and closed-loop monitor of energy storage. Altogether, FUS and MRI as a therapy delivery system help to focus the target and control in real time. This combination helps in treating targeted tissue without damaging other healthy tissues. This gives MRgFUS a potential as an alternative to surgical resection or radiation therapy of cancerous cells in the treatment of uterine fibroids, breast, liver, prostate cancer, brain cancer and many more. FUS, with or without the use of microbubbles, can alter cell membrane permeability temporarily and transmit various compounds for targeted drug delivery (Jolesz, 2009).

Jung et al. (2011) evaluated the complications arising due to HIFU in patients suffering from hepatic and pancreatic cancer. It was observed that all patients experienced skin redness, oedema and pain in the exposed areas. Major complications observed include biliary obstruction, symptomatic pleural effusion, pneumothorax and fistula formation between an abdominal wall abscess and the ablated hepatic tumour. The complications were mainly shown around the targeted lesions or along the ultrasound beam pathway.

Kieran et al. (2007) showed the use of high-intensity pulse-focused ultrasound for tissue ablation using cavitational mechanisms and that too without significant thermal effective creation of nonthermal lesions. It was concluded that further research is required to regulate the parameters for in vivo cavitational tissue ablation, incorporating the effect of tissue perfusion.

Kim et al. (2007a) compared the Cyberwand, which is a new intracorporeal lithotrite that employs coaxial ultrasonic elements working at two frequencies with LithoClast Ultra(R), which was considered as the most efficient commercially available intracorporeal lithotripsy device. An in vitro model system was used to study the efficiency of stone penetration for both the devices. It was observed that the mean ± SD penetration time for the Cyberwand was much shorter than for the LithoClast Ultra(R). Both the devices had no difficulties with high temperature or occlusion. It was concluded that the Cyberwand is more efficient lithotrite than LithoClast Ultra(R).

Kinsey et al. (2008) explored the feasibility and efficiency of a multisectoral tubular array transurethral ultrasound applicator. It was used for prostate thermal therapy of prostrate with potential to facilitate angular and length control of heating under magnetic resonance guidance without mechanical moving of the applicator. This multisectoral tubular array transurethral ultrasound technique showed potential for fast and good targeting of prostate ailments.

Klingler et al. (2008) showed that HIFU allows targeted homogeneous ablation of tissue. The feasibility of HIFU ablation of small renal tumours under laparoscopic monitoring was determined. The histological study showed irreversible and homogeneous thermal damage within the exposed site. A few tumours which were treated showed complete ablation of the entire tumour and a few tumours had a 1- to 3-mm border of viable tissue just near to the HIFU probe. One tumour also revealed a central area having 20% vital tissue. No intra- or postoperative complications were observed.

Lowe and Knudsen (2009) compared ultrasonic, pneumatic and integrated intracorporeal lithotripsy for percutaneous nephrolithotomy. It was observed that integrated lithotripters showed superior efficiency compared to individual ultrasonic and pneumatic lithotripters.

Ultrasound also finds use in liposuction technology (Mann et al., 2008).

HIFU can be used in pulmonary vein isolation (PVI), but certain safety concerns are needed. Therefore, an esophageal temperature (ET)-guided safety algorithm was used with HIFU. In this treatment, HIFU was repeatedly used until PVI was completed. A touch-up radiofrequency ablation was applied, when PVI failed. Periprocedural ET control monitoring and endoscopy were carried out 2 days after ablation. It was concluded that the safety algorithm failed to prevent lethal complications. HIFU did not meet the safety standards needed for treatment of atrial fibrillation (Neven et al., 2010).

Thuroff et al. (2003) studied the use of HIFU for treatment of localized prostate cancer. A phase II/III prospective multicentric clinical trial was applied to study the safety and efficacy of HIFU. They concluded that HIFU can be used for the primary treatment of localized prostate cancer.

Tempany et al. (2003) analysed MRI-guided focused ultrasound surgery for the case of uterine leiomyomas. Periodic sonications were given to nine targets. Temperature-sensitive phase-difference MRI controlled the target area and measured rise in tissue temperature. MR images and hysterectomy specimens were studied. MR thermometry was successful in all the sonication cases. Focal necrotic lesions were observed in all cases at MR.

Drug-delivery vehicles that integrate ultrasonic and molecular targeting are known to locally concentrate a drug at a specific location. The drug-delivery vehicles known as acoustically active lipospheres (AALs) are microbubbles encapsulated by a shell of oil and lipid. Toxicity analysis demonstrated that paclitaxel-containing AALs showed more antiproliferative effect after insonation than paclitaxel-free AALsat the same concentration. Ultrasound and molecular targeting are combined to transmit the drug to the endothelium and interstitium of chorioallantoic membrane vasculature in vivo (Tartis et al. 2006).

White et al. (2007) studied the transcutaneous delivery of intense ultrasound (IUS) energy to focus on the facial superficial musculoaponeurotic system (SMAS), to develop discrete thermal injury zones (TIZs) in the SMAS, and showed the relative sparing of nearby nontargeted layers superficial and deep inside the SMAS layer. The results showed that reproducible TIZs were generated selectively in the SMAS at 7.8 mm depths and the surrounding tissues and epidermis were spared. Higher-energy settings and high-density exposures developed a higher degree of tissue shrinkage. It was concluded that in human cadaveric facial tissue, IUS can noninvasively produce TIZs of reproducible area, size and dimension in the SMAS layer.

Mechanical tissue fractionation was achieved using high-intensity ultrasound pulses in a process known as histotripsy. Histotripsy finds great potential for noninvasive tissue removal. The primary mechanism for histotripsy includes cavitation. Xu et al. (2008) studied the evolution of a cavitating bubble cloud formed by a histotripsy pulse with 10 and 14 cycles at peak negative pressures exceeding 21 MPa. Bubble clouds are produced inside a gelatin phantom and at a tissue–water interface. The dynamics of bubble clouds produced by histotripsy was observed.

Kim et al. (2007 b) studied the biomedical applications of super paramagnetic iron oxide (SPIO) nanoparticles encapsulated within chitosan using sonochemical method. Microspheres made up of SPIO polyglucosamine (chitosan) were prepared as a novel MRI-detectable material. The synthesis followed a sonochemical method. The ferrofluid, solution containing SPIO-embedded chitosan, was sprayed on the surface of an alkaline solution to prepare SPIO-chitosan microspheres. About 100–150 μm microspheres were injected into the blood vessel, which were further transmitted to the kidney. These microspheres were detected in MR images of the kidney.

A one-step ultrasound initiated process starting with the bovine serum albumin (BSA) and tetracycline was used to encapsulate the drug, usually an antibiotic, in small spherical particles of BSA. The tetracycline loading analysis exhibited that the highest tetracycline loading capacity was noted to be 65 %. The antimicrobial activity of tetracycline encapsulated in BSA microspheres was shown on two bacterial strains that are reactive to tetracycline (Avivi et al., 2003).

Medical ultrasound is a diagnostic imaging technique used to see internal body parts such as joints, vessels, muscles, tendons and so forth. This technique also plays a significant role in transmitting drugs to the targeted body parts without affecting the surrounding cells. Thus, sonochemistry is a branch of chemistry, which has a great potential in the medical field. Hence, it can be concluded that sonochemistry has replaced traditional methods of treating different kinds of diseases and it is the widely used technique nowadays. If ultrasound is used by keeping in mind its harmful effects and taking the measures to avoid them, then it is one of the best techniques for certain medical applications.

REFERENCES

Alam, M.; White, L. E.; Martin, N.; Witherspoon, J.; Yoo, S.; West, D. P. *J. Am. Acad. Dermatol.* **2010,** *62*(2), 262–269.

Albornoz, P. M. D.; Khanna, A.; Longo, U. G.; Forriol, F.; Maffulli, N. *Brit. Med. Bull.* **2011,** *100*(1), 39–57.

Anand, A.; Kaczkowski, P. *J. Ultrasound Med. Biol.* **2009,** *35*(10), 1662–1671.

Avivi, S. (Lev i), Nitzan, Y.; Dror, R.; Gedanken, A. *J. Am. Chem. Soc.* **2003,** *125*(51), 15712–15713.

Bai, W. K.; Shen, E.; Hu, B. *Chin. J. Cancer Res.* **2012,** *24*(4), 368–373.

Beckerman, H.; Bouter, L. M.; van der Heijden, G. J.; de Bie, R. A.; Koes, B. W. *Brit. J. Gen. Pract.* **1993,** *43,* 73–77.

Bjerså, K.; Biörserud, C.; Olsén, M. F. *J. Plast. Surg. Hand Surg.* **2015,** *6,* 1–5.

Bommannan, D.; Menon G. K.; Okuyama, H.; Elias, P. M.; Guy, R. H. *Pharm. Res.* **1992,** *9*(8), 1043–1047.

Burgess, S. E.; Silverman, R. H.; Coleman, D. J.; Yablonski, M. E.; Lizzi, F. L.; Driller, J.; Rosado, A.; Dennis, P. H. Jr. *Ophthalmology* **1986,** *93*(6), 831–838.

Chaudhry, A.; Kim, N.; Unnikrishnan, G.; Nair, S.; Reddy, J. N.; Righetti, R. *Ultrason. Imaging* **2016**. DOI: 10.1177/0161734616671713.

Diederich, C. J.; Hynynen, K. *Ultrasound Med. Biol.* **1999,** *25*(6), 871–887.

Farinha, A.; Kellogg, S.; Dickinson, K.; Davison, T. *Biomed. Instrum. Technol.* **2006,** *40*(1), 72–77.

Fatemi, A. *Semin. Cutaneous Med. Surg.* **2009,** *28*(4), 257–262.

Ferrara, K.; Pollard, R.; Borden, M. *Annu. Rev. Biomed. Eng.* **2007,** *9,* 415–447.

Fitzgerald, P. J.; Takagi, A.; Moore, M. P.; Hayase, M.; Kolodgie, F. D.; Corl, D.; et al. *Circulation* **2001,** *103*(14), 1828–1831.

Foley, J. L.; Little, J. W.; Vaezy, S. *Muscle Nerve* **2008,** *37*(2), 241–250.

Fry, W. J.; Mosberg, W. H. Jr; Barnard, J. W.; Fry, F. J. *J. Neurosurg.* **1954,** *11*(5), 471–478.

Fyfe, M. C.; Bullock, M. I. *Aust. J. Physiother.* **1985,** *31*(6), 220–224.

Gebauer, D.; Mayr, E.; Orthner, E.; Ryaby, J. P. *Ultrasound Med. Biol.* **2005,** *31*(10), 1391–1402.

Gelet, A, Chapelon, J. Y.; Bouvier, R.; Rouvière, O.; Lasne, Y.; Lyonnet, D.; Dubernard, J. M. *J. Endourol.* **2000,** *14*(6), 519–528.

Haar, G. T. *Ultrasound Med. Biol.* **1995,** *21*(9), 1089–1100.

Harris, G. R. *IEEE Trans. Ultrason. Ferroelectrics Freq. Control* **2005,** *52*(5), 717–736.

Hitchcock, K. E.; Holland, C. K. *Stroke* **2010,** *41*(10 Suppl), S50–53. DOI: 10.1161/STROKEAHA.110.595348.

Hundt, W.; Yuh, E. L.; Bednarski, M. D.; Guccione, S. *Am. J. Roentgenol.* **2007,** *189*(3), 726–736.

Ishtiaq, F.; Farooq, R.; Farooq, U.; Farooq, A.; Siddique, M.; Shah, H.; Hassan, M.-U.; Ashraf, M. *World Appl. Sci. J.* **2009,** *6*(7), 886–893.

Johnson, M.; Hensor, E. M. A.; Gupta, H.; Robinson, P. *J. Ultrasound Med.* **2016,** *35*, 2209–2216.

Jolesz, F. A. *Annu. Rev. Med.* **2009,** *60*, 417–430.

Jung, S. E.; Cho, S. H.; Jang, J. H.; Han, J. Y. *Abdom. Imaging* **2011,** *36*(2), 185–195.

Kennedy, J. E.; TerHaar, G. R.; Cranston, D. *Brit. J. Radiol.* **2003,** *76*(909), 590–599.

Kieran, K.; Hall, T. L.; Parsons, J. E.; Wolf, J. S. Jr; Fowlkes, J. B.; Cain, C. A.; Roberts, W. W. *J. Urol.* **2007,** *178*(2), 672–676.

Kim, E. H.; Ahn, Y.; Lee, H. S. *J. Alloys Compd.* **2007a,** 434–435, 633–636.

Kim, S. C.; Matlaga, B. R.; Tinmouth, W. W.; Kuo, R. L.; Evan, A. P.; McAteer, J. A.; Williams, J. C. Jr; Lingeman, J. E. *J. Urol.* **2007b,** *177*(4), 1363–1365.

Kinsey, A. M.; Diederich, C. J.; Rieke, V.; Nau, W. H.; Pauly, K. B.; Bouley, D.; Sommer, G. *J. Med. Phys.* **2008,** *35*(5), 2081–2093.

Klingler, H. C.; Susani, M.; Seip, R.; Mauermann, J.; Sanghvi, N.; Marberger, M. J. *Eur. Urol.* **2008,** *53*(4), 810–816.

Kremkau, F. W. *J. Clin. Ultrasound* **1979,** *7*(4), 287–300.

Lavon, I.; Kost, J. *Drug Discovery Today* **2004,** *9*(15), 670–676.

Lowe, G.; Knudsen, B. E. *J. Endourol.* **2009,** *23*(10), 1663–1668.

Mann, M. W.; Palm, M. D.; Sengelmann, R. D. *Semin. Cutaneous Med. Surg.* **2008,** *27*(1), 72–82.

Mason, T. J. *Ultrason. Sonochem.* **2011,** *18*(4), 847–852.

Misaridis, T.; Jensen, J. A. *IEEE Trans. Ultrason. Ferroelectrics Freq. Control* **2005,** *52*(2), 192–207.

Mitragotri, S. *Nat. Rev. Drug Discovery* **2005,** *4*, 255–260.

Mitragotri, S.; Edwards, D.; Blankschtein, D.; Langer, R. *J. Pharm. Sci.* **1995,** *84*, 697–706.

Naor, O.; Krupa, S.; Shoham, S. *J. Neural Eng.* **2016,** *13*(3), 031003. DOI:10.1088/1741-2560/13/3/031003.

Neven, K.; Schmidt, B.; Metzner, A.; Otomo, K.; Nuyens, D.; De Potter, T.; Chun, K. R.; Ouyang, F.; Kuck, K. H. *Circ. Arrhythmia Electrophysiol.* **2010,** *3*(3), 260–265.

Pahk, K. J.; Dhar, D. K.; Malago, M.; Saffari, N. *J. Phys.: Conf. Ser.* **2015,** *581*(1). DOI:10.1088/1742-6596/581/1/012001.

Pitt, W. G.; Husseini, G. A.; Staples, B. J. *Expert Opin. Drug Delivery* **2004,** *1*, 37–56.

Regar, E.; Thury, A.; van der Giessen, W. J.; Sianos, G.; Vos, J.; Smits, P. C.; Carlier, S. G.; deFeyter, P.; Foley, D. P.; Serruys, P. W. *Catheterization Cardiovasc. Interventions* **2003,** *60*(1), 9–17.

Rosenthal, I.; Sostaric, J. Z.; Riesz, P. *Ultrason. Sonochem.* **2004,** *11*(6), 349–363.

Smith, Y. *Ultrasound Med. Biol.* **1995,** *21*(9), 1089–1100.

Taradaj, J.; Franek, A.; Dolibog, P.; Cierpka, L.; Błaszczak, E. *Polski Merkuriusz Lekarski* **2007,** *23*(138),426–429.

Tartis, M. S.; McCallan, J.; Lum, A. F.; LaBell, R.; Stieger, S. M.; Matsunaga, T. O.; Ferrara, K. W. *Ultrasound Med. Biol.* **2006,** *32*(11), 1771–1780.

Tempany, C. M.; Stewart, E. A.; McDannold, N.; Quade, B. J.; Jolesz, F. A.; Hynynen, K. *Radiology* **2003,** *226*(3), 897–905.

Thüroff, S.; Chaussy, C.; Vallancien, G.; Wieland, W.; Kiel, H. J.; Le, Duc, A.; Desgrandchamps, F.; De La Rosette, J. J.; Gelet. A. *J. Endourol.* **2003,** *17*(8), 673–677.

Tinkov, S.; Bekeredjian, R.; Winter, G.; Coester, C. *J. Pharm. Sci.* **2009,** *98*(6), 1935–1961.

Veronick, J.; Assanah, F.; Nair, L. S.; Vyas V.; Huey, B.; Khan, Y. *Exp. Biol. Med.* **2016,** *241*(10), 1149–1156.

Voigt, J.; Wendelken, M.; Driver V.; Alvarez, O. M. *Int. J. Lower Extremity Wounds* **2011,** *10*(4), 190–199.

White, W. M.; Makin, I. R.; Barthe, P. G.; Slayton, M. H.; Gliklich, R. E. *Archines Facial Plast. Surg.* **2007,** *9*(1), 22–29.

Xu, Z.; Raghavan, M.; Hall, T. L.; Mycek, M. A.; Fowlkes, J. B. *IEEE Trans. Ultrason. Ferroelectrics Freq. Control* **2008,** *55*(5), 1122–1132.

Yu, T.; Wang, Z.; Mason, T. J. *Ultrason. Sonochem.* **2004,** *11*(2), 95–103.

Yu, T.; Li, S.; Zhao, J.; Timothy, B. S.; Mason, J. *Technol. Cancer Res. Treat.* **2006,** *5*, 51–60.

CHAPTER 12

INDUSTRIAL APPLICATIONS

ANIL KUMAR CHOHADIA[1], YASMIN[2], NEELAM KUNWAR[3]

[1]*Department of Chemistry, M. P. Govt PG College, Chitttorgrh, India*
E-mail: anilchohadia@yahoo.co.in

[2]*Department of Chemistry, Techno India NJR Institute of Technology, Udaipur, India, E-mail: ali_yasmin2002@yahoo.com*

[3]*Department of Chemistry, PAHER University, Udaipur, India*
E-mail: neelamkunwar13@yahoo.com

CONTENTS

12.1 INTRODUCTION

In recent years, power ultrasound has been used in various industries such as dyeing, pharma, sugar, leather, extraction and so forth for various purposes such as desizing, washing, cleaning, bleaching, finishing, scouring, descaling, drug delivery, welding and so on. It is also used in the enzymatic scouring process, since it increases the mass transfer effect. Ultrasonic waves travelling in a liquid result in cavitation and thus, produce bubbles. The bubbles exhibit rich dynamic behaviours such as translation, oscillation, growth and collapse in response to the varying acoustic pressure (Choi et al., 2016). Due to the high temperature and pressure inside the bubbles created by strong ultrasonic waves during strong collapse, water vapour inside the bubbles is dissociated and chemical products such as $^{\cdot}OH$, $^{\cdot}O$ and $^{\cdot}H$ radicals, as well as hydrogen peroxide and ozone molecules are created inside the bubbles (Yasui et al., 2005). Energy caused by ultrasound enhances the dispersibility of dyestuffs and auxiliaries and stimulates the emulsifying ability and solubility of dyestuffs leading to higher homogenization of the auxiliaries and dyestuffs. Such types of studies on the uses of ultrasound in the textile industry are quite new. Nowadays, several studies are available on the use of ultrasound in textile industry (El-Shishtawy et al., 2003; Merdan et al., 2004; Akalin et al., 2004; Vouters et al., 2004).

Main environmental benefits of using ultrasound are:

- Lower process temperatures (Energy savings)
- Shorter running times (Energy savings)
- Minimization of auxiliary consumption

It has been observed that ultrasonic technology enhances mass transfer during some textile-processing steps such as desizing, scouring, bleaching, mercerizing and dyeing of natural fabrics apart from cleaning of materials. The cleaning action of ultrasonic energy is due to cavitations. There is an implosion of vapour bubbles inside the cleaning auxiliaries and near the surface to be cleaned. It produces such a stress on the surface to be cleaned that the contaminants are destroyed and thus, impurities are removed.

Stable cavitations may also cause the dispersion of the particles of contaminant removed from the surface. Such an ultrasonic system for the continuous washing of textiles in liquid layers has been designed. High-power operation without the interaction of perturbing undesired vibration modes has shown very good washing behaviour. Attempts have been made to use acoustic cavitation for washing textiles and ultrasonic cleaning has been widely used to remove very small contaminant particles adhering to solid substrates.

Soltani et al. (2015) used a porous clay-like support for immobilizing ZnO nanostructures (as a sonocatalyst) for the decolourization of methylene blue dye in the aqueous phase. They concluded that the ZnO–biosilica nanocomposite can be a suitable sonocatalyst for the sonocatalytic decolourization of coloured solutions with high reusability potential and cost efficiency.

12.2 TEXTILE INDUSTRIES

Textile applications of ultrasound include a number of processes such as bleaching open-width fabric, scouring, bast fibre separation, removing weaving sizes, nonwoven fabric consolidation, preparation for dyeing, chemical coating, heat treatment and so forth.

Some fundamental processes are affected by ultrasound, which leads to a better design for the dyeing process in a pilot plant as well as commercial scale-up applications. The use of ultrasound in dyeing industries may replace expensive thermal energy and chemicals required for the treatment of wastewater. It is also likely to improve the quality of dyed fibre; thus, the competitiveness of textile industry is favourably enhanced by the use of ultrasound.

Washing time to wash reactive dyes off of a stationary fabric can also be reduced by over 60 % by using ultrasound. This saving may be reflected in increase in the productivity, decrease in water consumption and also in steam consumption, if the washing is to be done with hot water.

Ultrasound increased diffusion coefficients by 30 % and permeability coefficients by more than 300 %, thereby enhancing the dye penetration. The apparent activation energy of dye diffusion was also lowered approximately to its one-fourth, which shows a potential for reducing dyeing temperatures. Its use also increased the reactivity of fibre-reactive dyes. Some promising results were also obtained in the dyeing of nylon with acid dyes, polyester dyed with disperse dyes, and so forth.

Wet processing of textiles requires large quantities of water, and electrical and thermal energy. Majority of these wet processes involve the use of different chemicals for assisting, accelerating or retarding their rates, and therefore, these processes must be carried out at higher temperatures to transfer mass from processing liquid medium to the surface of the textile material and that too in a reasonable time. Similar to all the other chemical processes, these transport processes are time- as well as temperature-dependent. However, compromising with these conditions could adversely affect product quality. Ultrasonics may be utilized to reduce processing time and energy consumption, and also for maintaining or improving product quality. An improvement and acceleration of textile processes using ultrasound has been reported.

Intensity of the waves is found to be greatly affected by the temperature of the medium. As the temperature of the liquid was kept higher, the effects of intensity were reduced. the effect of cavitation reaches its maximum at about 50 °C in water. The effect of cavitation is several hundred times greater in a heterogeneous system as compared to homogeneous systems.

Applications of ultrasound are related to preparation of auxiliary baths for processes such as preparation of sizes, emulsions, dye dispersions and thickeners. Starch solution can be prepared more rapidly and at a lower temperature in the presence of ultrasound. It was observed that starch-sizing agents prepared with ultrasound were superior to conventionally prepared starch solution. Ultrasonic waves are also used to prepare solutions and emulsions of lubricants for fibre that help in reducing interfibre friction and static electricity.

Oil–water emulsions were prepared using an ultrasound device, which remained stable for more than 200 h, whereas conventionally prepared emulsion separated into phases after 12 h only. Ultrasound has also been used in the preparation of paraffin/styrene emulsions to obtain a homogeneous emulsion. It was observed that reduced particle size of 1 μ was obtained as compared to 3 μ obtained by the conventional method. All such applications make use of the dispersion effect of ultrasound.

The dispersing action of ultrasound is commonly used for cleaning purpose on commercial scale in the cleaning of machine parts. Other examples include needles in knitting machines, spinnerets forming the chemical fibres, open-end yarn-spinning rotors, and so forth.

12.2.1 DYEING

Reactive dyes are mostly used in dyeing and printing of cotton fibres. Reactive dyes have a specific reactive nature due to the presence of some active groups. These groups form covalent bonds with –OH groups of cotton either through substitution and/or addition mechanism. Some conventional eco-friendly methods for dyeing cotton with reactive dyes are available which do not require thermal energy and result in high dye fixation, but at least 12 h or more batching time is required for dye fixation. Khatri et al. (2011) proposed cold pad-batch dyeing method for dyeing of cotton fabric with reactive dyes and reduced this time using ultrasonic energy. The dyeing of cotton fibre was carried out with reactive red-195 and Reactive Black-5 under ultrasonic radiation. They also showed that such use of ultrasonic energy significantly reduces not only the batching time and the concentrations of alkalis but also the colour strength, relative straightening of fibre and dye fixation was also enhanced with no adverse effect on colour fastness of the dyed fabric.

Both low- and high-frequency ultrasonic waves have been used to study their effects on the quality of dispersion of dyes, dye absorption by textile materials and the change in solubility of water-soluble dyes. This dispersion had a longer life as compared to conventional stirring. Ultrasound also affects the quality of dispersions and the particle size of the dispersion.

Kubidus (1962) also reported a change in the solubility of direct dyes in cold water. An increase in the solubility of two direct dyes, blue-M and brown-mX, was also observed, on using low-frequency ultrasound, but there was no change observed in the solubility of direct sky blue dye. The effectiveness of the treatment was found to depend upon the nature of dyes and their physicochemical properties, particularly their solubility in water.

Alexander and Meek (1953a, 1953b, 1953c) dyed cotton with direct dyes, wool with acid dyes, and nylon and acetate with disperse dyes in the presence of ultrasound at 17.3 kHz. It was concluded that it is more beneficial on hydrophobic fibres dyed with water-insoluble dyes. Ultrasound was found to be less effective for water-soluble dyes. Quality dyeing was also obtained with water-soluble systems on a larger scale with ultrasound. The influence of ultrasound on the dyeing system has been proposed to show the following three effects:

- **Dispersion:** This process involves breaking up of micelles and high-molecular weight aggregates into uniform dispersions in the dye bath.
- **Degassing:** This process involves expulsion of gas (dissolved or entrapped) or air molecules from fibre capillaries and interstices at the crossover points of fabric into liquid and removal by cavitation, thus facilitating dye-fibre contact.
- **Diffusion:** This process involves accelerating the rate of diffusion of dye inside the fibre by piercing the insulating layer covering the fibre and accelerating the interaction or chemical reaction, if any, between dye and fibre.

Ultrasound accelerates the rate of dyeing, increases the colour yield and also improves quality of the fabric by reducing the warp streaks. Dyeing of polyester and acetate fibres showed only a marginal increase in colour yield but substantial increase in rate of dyeing.

Shimizu et al. (1989) dyed amorphous and unoriented nylon 6 films at 20, 40 and 60 °C temperatures in an ultrasonic field at a frequency of 27 kHz. The absorption of all the dyes was found to increase while activation energy was decreased with ultrasound. The amount of decrease was the largest for disperse dyes whereas it was the lowest for fibre-reactive dyes. It was reported that high temperatures can be avoided during dyeing synthetic fibres using ultrasound.

Kamel et al. (2009) also reported dyeing of cationized cotton fabric with Solfix-E using colouring matter extracted from Cochineal dye using ultrasonic techniques. Babar et al. (2017) presented the dyeing results of lyocell fabrics dyed with conventional pad-batch and pad-ultrasonic-batch processes. The dyeing of lyocell fabrics was carried out with two commercial dyes namely Drimarine Blue CL-BR and Remazol Blue RGB. The dyeing of acrylic fabrics was studied with both conventional and ultrasonic techniques using Astrazon Basic Red-5BL by Kamel et al. (2010). Bamboo is a regenerated cellulose fibre and it is usually dyed with reactive dyes such as Reactive Black-5 and Reactive Red-147. Batch-wise dyeing of bamboo cellulose fabric with reactive dyes has been carried out using ultrasonic energy by Larik et al. (2015).

Ultrasonic preparation of cationic cotton and its application in ultrasonic natural dyeing has been reported by Guesmi et al. (2013). Cationization of cotton fabric was carried out under sonication of cellulose with bromoacetyl bromide, followed by the substitution of the terminal bromo groups by triethylamine. Ultrasonic energy has also been used in dyeing polyamide (microfibre)/Lycra fabrics with reactive dyes. The polyamide (microfibre)/Lycra blends were dyed using conventional and ultrasonic dyeing techniques with three reactive dyes containing different chromophore and reactive groups by Merdan et al. (2004). Ferrero and Periolatto (2012) proposed that the ultrasound can be used for dyeing of wool with acid dye at low temperature. Ultrasound energy was found to accelerate dye uptake and interaction of fibre with reactive dye on knitted cotton fabric at low temperatures (Tissera et al., 2016).

Hao et al. (2012) prepared a stable anionic nanoscale pigment suspension using a polymeric dispersant to colour cationize cotton with the exhaust method. It was found that nanoscale pigment has higher adsorption rate on using ultrasonic method, as it promotes the diffusion of pigment through the fibre–liquid boundary layer. The colour difference reveals the fact that nanoscale pigment can be deposited on cotton surface more uniformly under ultrasonic conditions with improved product quality. Sun et al. (2010) investigated the acceleration efficiency of ultrasound during dyeing process in an ultrasound cleaner. They showed that the ultrasound pretreatment could slightly improve the dye exhaustion and fixation. Ben et al. (2016) also improved the dyeability of modified cotton fabrics by the natural aqueous extract from red cabbage using ultrasonic energy.

Textile dyeing assisted by ultrasonic energy has gained a significant interest. Ultrasonic-assisted dyeing of cellulosic fibres has already proved to be a better choice among conventional dyeing processes.

12.2.2 SCOURING AND BLEACHING

Conventional hydrogen peroxide bleaching process is an important and a specific step for wet processors in textile industries, but it has certain problems such as long time and high-energy consumption. On the other hand, ultrasonic energy can be used as an alternative method in bleaching (Mistik et al., 2005). Fibre damage was also relatively less when ultrasound was used as compared to conventionally scoured wool fibres. An increase in the bleaching rate and reduction in required time was there in peroxide bleaching under ultrasonication. The whiteness of the fabric was also found to be improved over conventionally bleached fabrics. Whiteness was improved over conventionally scoured and bleached flax fibres during processing of flax fibres—an application of ultrasound.

The scouring of raw wool fibres is a textile chemical processing that needs a lot of detergents and water due to the presence of high amount of impurities (Simpson and Crawshaw, 2002). Effluents produced by this treatment are extremely polluted with chemicals and impurities washed out from the fibres. Ultrasound washing can effectively remove different substances from the textile surfaces without using surfactants because of cavitations occurring at certain parameters of the ultrasound field.

This energy could destroy and remove the pollutants from textile wastewater on textile surfaces even if additional auxiliaries are not used (Makino et al., 1983). The synergistic effect between power ultrasound and enzymes in an enzymatic scouring process has been reported by Agrawal et al. (2010). They used scouring enzymes such as *Fusarium solani pisi* cutinase and pectate lyase. It was observed that ultrasound shortens the enzymatic scouring process time dramatically. Rositza et al. (2011) worked on enzyme-assisted ultrasound scouring of raw wool fibres. Kadam et al. (2013) also analysed the wool scouring using ultrasound irradiation at intermediate stages and compared it with the wool scoured without ultrasonic energy. No cuticle damage was detected on exposing wool fibre to ultrasound as evident from scanning electron microscopy. Fibre diameter, single fibre strength and moisture content also did not show any significant change after ultrasound irradiation.

12.2.3 WASHING AND CLEANING

Washing of textiles using ultrasound is another process which is well studied.

The use of ultrasonics for washing fabrics after dyeing and printing has also been reported. An ultrasonic washing machine can be used to wash sheets or strands of textiles. It has been reported that the washing of flax can be improved by ultrasonic vibration. This method also removed noncellulosic material more effectively than mechanical agitation and thus, improved whiteness of the flax fibre.

Vigorous physical effects including microjet and micro-streaming can be induced in heterogeneous systems by acoustic cavitation. Such a process is useful for the removal of pollutants from contaminated soil particles. Son et al. (2012) compared diesel removal efficiencies in ultrasonic, mechanical and combined soil-washing processes considering the electrical energy consumptions for these processes. This type of combined process showed synergistic effects for both removal efficiency and effective volume, and it also has the advantage of a short operation time as compared to the sequential processes. Hence, the ultrasonic soil-washing process with mechanical mixing can be considered a promising technology for industrial use.

Ultrasound can be applied to washing of textile as a mechanical action for soil removal. Polyester fabric was soiled with carbon black or oleic acid as a model contaminant. It was washed with the original fabric in aqueous solutions with and without alkali or surfactant by applying ultrasound, shaking or stirring action. The detergency and soil redeposition were evaluated from the change in the surface reflectance due to washing of these artificially soiled fabrics. Ultrasound was found to remove the particulate and oily soils quiet effectively in a relatively shorter time and at low bath ratio as compared to shaking and stirring actions with increasing ultrasound power (Gotoh and Harayama, 2013). It was observed that detergency of both soils increased. It was also shown that ultrasound washing caused little mechanical damage to the fabric using three standard fabrics.

Wet textile washing processes were set up for wool and cotton fabrics to evaluate the potential of ultrasound transducers in improving dirt removal. Textile samples were contaminated with an emulsion of carbon soot in vegetable oil and aged for 3 h in fan oven. These fabrics were soaked for 3 min in a standard detergent solution before washing and then washed in a water bath. The removal of dirt was evaluated through colorimetric measurements. The total colour differences of the samples were measured with respect to an uncontaminated fabric before and after each washing cycle. Its percentage variation was calculated and correlated to the dirt removal (Peila et al., 2015). They showed that the use of ultrasound transducers enhanced the dirt removal and the temperature is an important parameter influencing the efficiency of the cleaning process. Much better results were obtained at a lower process temperature.

The cleaning of wool is both energy- and time-consuming process. It produces large amounts of waste water with high chemical and biological oxygen demand. Cleaning occurs in a number of ways and that too at different times during the production and use of a wool textile. Most cleaning processes affect the quality of wool fabric by causing felting shrinkage, particularly in case of fine wool. Wool cleaning using ultrasound provides a method of reducing the ecological and quality impacts. It is an efficient mass-transfer mechanism that can help us in improving the removal of contaminant and also in reducing the aggressive movement of liquor in the cleaning bath; thus entanglement of wool fibres is checked to a greater extent.

The use of ultrasonic irradiation within a bath containing water, wool and detergent improves cleaning of wool and that too at lower cleaning temperatures and with less detergent. The effect of ultrasonic irradiation on the structure and properties of wool fibres and fabrics has also been observed.

No adverse effects on colour or fabric tensile strength were reported on the repeated use of ultrasound. A wool fabric having stains of coffee, tea, red wine, orange carbonated sugar drink and engine oil can be cleaned using ultrasound. Colour measurement of the stain before and after cleaning revealed that dirt can be better released using ultrasound. Both alkaline and acidic detergents can be used for removing stain of all these types.

A significant benefit to the environment can also be achieved during wool cleaning by ultrasonic irradiation as it reduces energy inputs and the amount of detergent required for cleaning. As a result, the level of detergent can be reduced in waste water.

A wool textile needs cleaning at multiple stages during its life cycle . Raw wool is contaminated by dust, dirt, sweat, food, grease, oils, smells, body fluids and so forth. These contaminants should be removed before yarn production.

12.2.4 FINISHING

Finishing of textiles involving ultrasonics has been studied by several researchers from time to time (Antonescu and Grunichevici, 1979; Kunert, 1979; Dupont and Co, 1984). Scinkivich et al. (1975a, 1975b) studied the formaldehyde resin treatment of cotton fabric using ultrasound at 8 and 18 kHz frequency and measured the change in physical properties before and after 60 washing cycles. The crease recovery angle was much higher even after 60 washings with ultrasound than without ultrasound, but this treatment resulted in a small decrease in tensile strength. A military fabric was also treated with a liquid repellent fluorochemical finish in the presence of high-frequency ultrasonic waves (Last and McAndless, 1981). This method resulted in an increase in finish add-on.

Dyeing kinetics of nylon-6 fibre with different reactive dyes was compared using conventional and ultrasonic conditions (El-Shishtawy et al., 2003) The time/dye uptake isotherms revealed the fact that enhanced dye uptake takes place in the second phase of dyeing (diffusion phase). The results fitted well with the integrated form of the first-order rate equation. Ultrasonic efficiency in accelerating the dyeing rate relative to conventional heating was examined for all dyes and it was found that ultrasonics are most effective for a dyeing system particularly when it is difficult to achieve high dye uptake. Time-dependent ultrasonic pretreatment of nylon-6 fibres was observed to clarify the role of fibre fine structure. It was observed that the percentage of nylon-6 fibre crystallinity becomes higher in ultrasonically treated fibre as compared to the conventionally treated one. Increased crystallinity of the fibre due to ultrasonic dyeing process would retard uptake of dye, but the enhanced effect of power ultrasonic may be high enough to overcome this side effect on the fibre.

Akalin et al. (2004) studied the effects of ultrasonic energy on the wash fastness of reactive dyes having three different reactive groups. The samples were applied with three types of washing processes simultaneously after dyeing with the traditional method—conventional, ultrasonic probe and ultrasonic bath. A comparative study was made in these three processes; three different fixing agents were used in the washing processes. Different parameters such as colourfastness, staining fastness, magnitude of total colour difference and lightness difference of the colour values of dyed samples were measured.

Potential applications of ultrasound exist for finishing of textiles. Specific applicators of ultrasonic energy were developed, which could be adapted on jigger, a widespread textile finishing machine (Vouters, et al., 2004). Their effects on fibre structure were observed and validated by trials in dynamic conditions.

A facile one-step sonochemical route has been suggested for uniform deposition of inorganic nanoparticles (NPs) on the surface of solid substrates, including textiles. The antimicrobial finishing is very important for medical textiles as decreasing the risk of hospital-acquired infections are reduced by it. Petkova et al. (2016) reported a simultaneous sonochemical/enzymatic process for coating of cotton with zinc oxide (ZnO) NPs as it produces ready-to-use antibacterial medical textiles in a single step. A multilayer coating of uniformly dispersed NPs was obtained in this process. The pretreatment of cotton fabrics with enzymes causes better adhesion of the ZnO NPs on its surface. It was reported that NPs-coated cotton fabrics inhibited the growth of *Escherichia coli* and *Staphylococcus aureus* by 100 and 67%, respectively.

Processing baths containing textile substrates are also affected by ultrasonic waves. Ultrasonic equipment has been used in preparation processes, for example desizing, scouring, bleaching, dyeing, finishing and washing. The objective of using ultrasound is different in all these processes, that is removing natural material or impurities (soil) from the surface of the fibre in preparation and washing operations, to transport or diffuse dyes or chemicals into the fibre in dyeing and finishing processes as these processes are complex and so forth. In all these processes, the rates were found to increase by the use of ultrasound; however, it is little difficult to conclude how ultrasound affects these widely diverse processes but it is all attributed to cavitation process. The use of ultrasound in textile wet processing has many advantages including energy savings, process enhancement and reduced processing times.

12.3 SUGAR INDUSTRIES

Scaling is one of the major problems in evaporators, boilers and heaters in various industries such as sugar, paper making, chemical fertilizer and so forth. Industries use lot of liquids, which must be evaporated and this causes scale formation. Scaling causes economic loss due to low heat-transfer capacity. These losses were in billions annually in sugar factories all over the globe. Scaling resulted in an increase in energy requirement and decrease of production capacity. There was an estimated British national economic loss of about 0.5 billion pounds in the late 1980s (Qiu and Yao, 1999).

At present, various chemical, mechanical and physical methods are commonly used for cleaning the scales. As this scale sticks to the walls of tubes or vessels very strongly, its removal by an individual method is not very successful and therefore, quite often, it requires synergistic methods. Irrespective of the method used, it is important to stop evaporation process. Evaporators tend to become worn and corroded. Different methods of preventing scale formations have been proposed to solve this problem, such as deploying some antiscalants, ion exchange and electric or magnetic fields. Unfortunately, these methods have not been widely adapted in the sugar industry for various reasons (Qiu and Hu, 2002). Since ultrasound has been reported to have many effects on compounds in solution, its influence on scale in evaporators was studied. It has been reported that ultrasound not only inhibited the formation of scale but also removed the scaling efficiently. A newer technique with equipment of scale control was developed and used in a sugar factory.

The transducers are installed on the outside of the bottom of the rectangular reaction pipe. The main part is a powerful ultrasonic wave generator. Its electro-circuit design is also special. Forty-eight transducers were divided into 4 groups having 12 transducers per group. Each group of these transducers is controlled by its own integrated circuit with a single switch. Any group out of these four groups can be chosen to run so that ultrasonic power can be changed according to the requirement of operation parameters. All transducers can tolerate high temperatures (95–105 °C) and protect themselves automatically.

Ultrasonic frequency can be automatically monitored and in that condition, ultrasonic generator and transducer will run in a good condition for a long time without much care. The operation of such equipment system is also easy.

The ultrasonic equipment is normally placed between the juice sedimentation vessel and the evaporator. Scale formation on the heating area of heat-transfer equipment conforms to the mechanism of mass crystallization. Heterogeneous nucleation provides more nucleation spots on a clean heating surface because an uneven surface can decrease the surface energy, which is necessary for nucleation. Adsorption action causes scale nuclei to form on the uneven surface in the initial stage. The induction period of nucleation of various scale-forming materials is shortened due to ultrasonic cavitation; thus, crystal scale nuclei are produced in a relatively shorter period of time. A large proportion of the nonsugar material (it may be inorganic or organic impurities) can be deposited onto the nuclei, in place of the surface of the heating tubes. Such deposits remain suspended in the sugar solution and flow out of the evaporators with the syrup, where these can be separated from syrups. Thus, the amount of precipitate deposited onto the surface of evaporator tube scan can be greatly reduced (Senapati, 1991; Chepurnoi, 1990; Qiu, 1994; Yao and Qiu, 1999).

Such an investigation was made on scale control by ultrasound in a sugar factory using five evaporation systems. It was observed that heat transfer coefficients (HTCs) of the evaporators were low, and deposition of scale was a very serious problem. Therefore, lots of chemical reagents were used to clean it, and two evaporators had to be cleaned in turn daily.

Seven parameters were investigated to know more about the performance of the ultrasonic technique in the removal of scale. These are:

- HTC of evaporators
- Time to clean scale
- Viscosity of juice/syrup
- Efficiency of removing scale (ERS)
- Evaporation intensity (EI)
- Usage of chemical detergents
- Effects of ultrasound on white sugar

It was found that the EI, HTC and ERS of syrup were improved. Interestingly, the scale with ultrasound was found to be loose, soft and white in colour (on drying), which could be easily removed by just tapping. On the other hand, the scale without using ultrasound was dense, hard and yellowish in colour after drying. This scale could not be removed completely even using a steel brush. Therefore, the scaling could be reduced to a significant extent using ultrasound and its physical character was also changed.

Viscosity is a force between two layers of liquids, and therefore, it is related to the force among molecular forces of liquid. Like any other sound wave, ultrasound is transmitted via waves, which alternately compress and stretch the molecular structure of the syrup while passing through it. As a result, the average distance between the molecules in the syrup will vary as the molecules oscillate about their mean position. The molecular distance will increase, when the acoustic pressure is the pressure on rarefaction. Otherwise, it decreases on the application of a sufficiently large negative pressure on the syrup and the distance between the molecules exceeds the critical molecular distance necessary to hold the liquid intact. As a result, the syrup liquid breaks down; the inner friction force of molecules in syrup decreases, resulting into a decrease the viscosity of syrup. A series of physical and chemical changes are responsible for affecting the viscosity of syrup during the evaporation process. The syrup was found to be less viscous, when ultrasound was used.

The products received after the treatment of ultrasound were all within the desired specifications although the colour, haze and impurity with ultrasound were a little higher than that without ultrasound.

12.4 PHARMA INDUSTRIES

Chemical effects on the reaction system can be induced by ultrasonic waves, such as the generation of free radicals, which increase the rate of reaction. Besides this, ultrasound may have other mechanical effects on the reaction system, such as increasing the surface area between the reactants and accelerating dissolution rate. Sonochemistry has several applications in the pharmaceutical industry including sonocrystallization, sonophoresis, lowering extraction time, solution atomization, melt sonocrystallization, crystallization by sonication and particle rounding technology.

Numerous other domains of ultrasound in the pharmaceutical industries have been explored for future development, for example formation of aerosols, enhancing inhalation drug delivery, sonochemical preparation of biomaterials, enhanced drug delivery, drying conducive in polymerization and depolymerization, extraction, chemotherapy, filtration, cell therapy, homogenization and synthesis. It also has some significance in transdermal drug delivery, haemotherapy and cell therapy.

The drug engineering has been aimed to alter and improve the primary and physical properties of drugs such as particle size, shape, crystal habit, crystal form, density, porosity and so forth, as well as secondary properties such as flowability, compressibility, compactibility, consolidation, dust generation and air entrapment during processing. In the pharmaceutical industry, efficient production of small and/or substantially uniform particles is desired not only due to its impact on the performance during processing and storage but also due to its impact after consumption by patients. Micronization of drugs is routinely carried out for increasing dissolution rate. For inhalation drug-delivery system, the particle size, shape and surface properties determine its interaction with the container and amount of dose deposited in lungs, which is generally the desired site of action (Paradkar and Dhumal, 2011). Application of ultrasound measurements as process analytical technology (PAT) tools for industrial crystallization process development of pharmaceutical compounds has been reported by Helmdach et al. (2015).

Ultrasound is a simple and effective technique for producing drug nanocrystals. With the use of Barbell Horn Ultrasonic Technology (BHUT), the process is directly scalable and can be used in the commercial production of high-quality nanocrystals, nanoemulsions and liposomes for the pharmaceutical industry. This technology is also making it possible to implement laboratory accomplishments in an industrial production environment, guaranteeing reproducible and predictable results at any scale.

12.4.1 SONOCRYSTALLIZATION

Crystallization is an omnipresent operation for the manufacture of fine chemicals, pharmaceuticals or intermediates, whether derived from chemical or biochemical processing. Nucleation and crystallization can be mediated by acoustic cavitation and streaming. Sonocrystallization is mediated by the bubbles caused by such acoustic effects. These bubbles are transient microreactors that facilitate faster chemical reaction and crystallization. Significant energy is transferred to molecules suspended or in solution over a very short time. Sonocrystallization is involved in one or all of the crucial steps in the nucleation and crystallization process. The sonocrystallization process involves:

- Improved mass transport, which improves clustering and templating
- Rapid cooling after cavitation collapse
- Transient high supersaturation close to the collapsing bubble
- Increase in pressure reducing the temperature for crystallization
- Shock waves to assist in nucleation
- Overcoming energy barriers for nucleation

The widely used explanation is the so-called hot-spot theory, which attributes nucleation to local hot spots, created by the concentration of kinetic energy in the collapsing cavity or due to rapid cooling afterwards. Local temperatures in excess of 5000 K in the gas phase and 2000 K in the liquid phase have been reported due to these hot spots. Heat dissipation happens within 2 μm, and hence, cooling rates are of the order of 109 $K \cdot s^{-1}$. Another popular mechanism is based on the fact that the pressure shock wave caused by cavity collapse creates high pressures locally. There are substances for which the solubility reduces with pressure, increasing local supersaturation, which could induce nucleation. A hypothesis related to the shock wave effect states that nucleation is initiated due to segregation of the solute and solvent near the bubble wall. This is caused by high pressures occurring in the ultimate phase of bubble collapse. There is also an opinion that nucleation occurs during bubble expansion. Solvent evaporating into the bubble or cooling of the liquid interface layer increases local supersaturation, which could lead to nucleation around the cavity. Electrical theory is also of interest which proposes that the consequences of cavitation are caused by electrical charges on the cavity interface layer (Fig. 12.1).

Sonocrystallization can be used mundanely for polymorph control, improving crystal size distribution and morphology, reducing impurities and superior solid–liquid separation. It can also help in augmenting secondary nucleation by disrupting crystals or agglomerates. Significantly, sonocrystallization is a viable manufacturing option, whether it

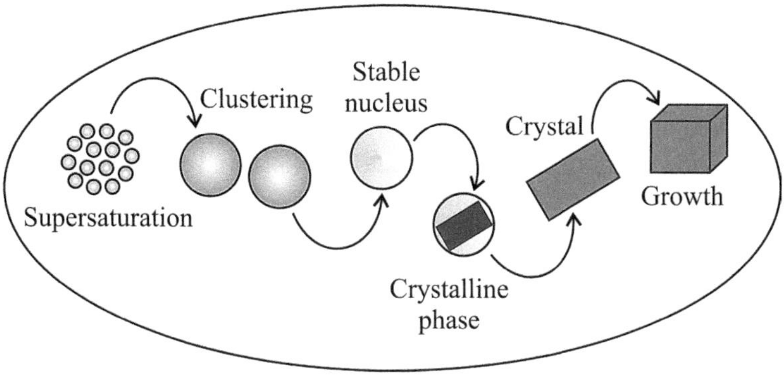

FIGURE 12.1 Crystal growth.

is continuous flow mode, batch mode or for in situ generation of seed crystals. Crystallization of drug actives and intermediates is a ubiquitous process for the removal of impurities and procuring a suitable solid state form in readiness for formulation and milling. Generally, it is straightforward but quite often it can be a troubling process step. Crystallization processes include cooling, evaporation, anti-solvent and salt formation variants and so forth. A key challenge in pharmaceutical active crystallization is manufacturing the desired solid form with the desired chemical and physical properties.

Ruecroft and Collier (2005) studied the sonocrystallization particle engineering for inhalation and improved respiratory medicines. Ultrasound-mediated amorphous to crystalline transition (UMAX) and dispersive crystallization with ultrasound (DISCUS) are being developed for the manufacture of respiratory particles for asthma and chronic obstructive pulmonary disease (COPD) drug products. The particles have optimal size, shape, surface rugosity, surface free energy, crystallinity and stability. UMAX can be used to manufacture combination particles, where two or more mutually synergistic pharmaceutical ingredients can be processed into highly crystalline particles. Compounds processed by these methodologies include corticosteroids, long-acting beta agonists (LABA), long-acting muscarinic agents (LAMA) and combinations.

Dhumal et al. (2009) produced fine elongated crystals of salbutamol sulphate by sonocrystallization for pulmonary delivery and compared with micronized and spray-dried salbutamol sulphate for in vitro aerosolization behaviour. Application of ultrasound during anti-solvent crystallization resulted in fine elongated crystals compared to aggregates of large irregular crystals obtained without sonication. The major drawback of simvastatin to formulate a dosage form is its poor aqueous solubility. Therefore, Tripathy et al. (2016) used a solvent and carrier-less technique (called melt sonocrystallization) and utilized it to form tiny crystals of simvastatin with enhanced solubility in distilled water. This technique has also been reported to improve solubility, micromeritic properties and rheological properties of drugs such as piroxicam (Gupta et al. 2013). Attempts have been made to control the release of simvastatin by forming hydrogels (Park et al., 2013), porous polymer scaffolds (Gentile et al., 2016) and by the formation of matrix tablets by using gums (Mantry et al., 2013).

Melt sonocrystallized curcumin was developed by Khan et al. (2015), and its therapeutic potential was validated by in vitro cytotoxicity studies against human oral cancer cell line KB. Manish et al. (2005) also used melt sonocrystallization technique for ibuprofen agglomerates and characterization of their physicochemical, micromeritic and compressional properties. By using melt sonocrystallization technique, an improvement in compressional properties and reduction in sticking was also observed due to thc change in crystal habit.

Integrated drug substance and drug product design for an active pharmaceutical ingredient using particle engineering has been suggested by Kougoulos et al. (2011). Particle engineering techniques such as sonocrystallization, high-shear wet milling and dry impact milling were used to manufacture samples of an active pharmaceutical ingredient with diverse particle size and size distributions.

The lack of systematic knowledge of physics underlying the ultrasound-assisted compaction of pharmaceutical powders prevented optimizing the processing steps for industrial pharmacy utilization. Many groups are involved in the application of ultrasound in compaction with the aim to design a controlled release of a variety of active agents from prepared systems taking the advantage of the physical changes occurring during the ultrasound compaction (Motta, 1994; Rodriguez et al., 1998; Fini et al., 1997; Sancin et al., 1999; Fini et al., 2002a, 2002b; Cavallari et al., 2005) and the possibility to obtain a direct formation of tablets even with powders of poor compactibility. Levina et al. (2000a, 2000b, 2002) found that coherent ibuprofen or paracetamol tablets could be prepared by ultrasound-assisted compaction at pressures as low as 20–30 MPa. Effect of ultrasound on the compaction of ibuprofen/isomalt systems has also been reported by Fini et al. (2009).

12.5 EXTRACTION

Extraction techniques are commonly required before any analytical determination in samples.

One of the important requirements of most of the extraction techniques is that solvents at high temperature or pressure must be used, but such operations can be performed with ultrasonic processors at ambient temperature and normal pressure, and relatively mild chemical conditions.

Sonication is normally recommended for pretreatment of solid environmental samples such as soils, sludges and wastes for the extraction of nonvolatile and semi-volatile organic compounds. Sonication is considered as an effective method in comparison with different methods available for analyte extraction as not much sophisticated instrumentation is required and solid liquid separations can be easily performed in a short time using diluted reagents and that too at low temperatures. Most of the applications of ultrasonic extraction have been carried out for organic compounds, but ultrasound has also been used for element extraction. Some examples of solid liquid extraction of some elements using ultrasound are given in Table 12.1.

TABLE 12.1 Percentages of Metals Extracted by Sonication.

Sample	Element (%)	References
Cabbage leave	Cd (89), Pb (1)	Dobrowolski et al. (1993)
Carbon	Cu (69), Cr (2)	Miller-Ihli (1993)
Lemon leaves	Cd (67), Cu (88), Mn (98)	Minami et al. (1996)
Orchard leaves	Cd (100), Cu (88), Pb (98)	Minami et al. (1996)
Prawns	Se (88)	Mierzwa et al. (1997)
Spinach	Cu (98), Cr (74)	Miller-Ihli (1993)
Talc	As (59), Cr (61), Ni (74)	Mierzwa and Dhindsa (1988)
Tomato leaves	Mn (70), Fe (70), Cr (51)	Miller-Ihli (1990)
Wheat flour	Mn (97), Fe (88)	Miller-Ihli (1990)

It was observed that parameters, such as sonication time, vibrational amplitude of the probe, acid concentration, particle size and solid concentration in the liquid, affect ultrasound-assisted extraction processes.

However, strongly bound analytes should be more difficult to extract, and hence, require more stringent extraction conditions.

It was observed that extraction efficiency could be increased with ultrasound by the addition of glass beads, which promote particle disruption by focusing the energy released by cavitation and by physical crushing. The use of a bubbling gas during sonication gives rise to formation of more H_2O_2 and hydroxyl radicals ($^{\cdot}OH$) and thus, aids to analyte extraction from oxidizable materials.

Near-field cavitation breaks down cell walls, so that cell contents are released into the surrounding liquid. Such a method is used to extract active antigens for making vaccines. Ultrasonic extraction is done with cooling so that it may cause minimal degradation of active contents.

Perfumes from flowers, essential oils from hops, juices from fruits and chemicals from plants can be extracted sonochemically. The demand for natural products from agricultural and marine bioresources is growing rapidly worldwide. The quality and yield of the extracts of natural products is also improved by ultrasound-assisted extractions. One of the best uses of ultrasound is in preparing tinctures, which are normally mixtures of water and ethanol-containing extracts from different medicinal herbs. The time taken by conventional methods in obtaining a tincture ranges from 1 to 2 weeks, depending on a particular herb. The same or sometimes even better-quality tincture can be obtained by ultrasonically assisted extraction in hours instead of days (Vinatoru et al., 1997; Toma et al., 2001; Vinatoru, 2001). It has also been proved to be successful on industrial scale in preparing water–alcohol extracts of medicinal herbs. The cavitation bubbles collapse near herb cell walls creating shock waves and liquid jets in the presence of ultrasound. This causes the cells walls to break and release their contents into the solvent. This speeds up and improves the diffusion process, which is very slow under normal solvent extraction conditions.

The usefulness of ultrasound was also demonstrated in mayonnaise and ketchup manufacturing (Povey and Mason, 1998). The power of ultrasound was fully employed to create emulsions in food-processing processes. Recently, the ultrasonic power has been used to emulsify normally immiscible liquids and produced a biodiesel from oil seed oils

(Stavarache et al., 2003, 2004; Maeda et al., 2003, 2004). The advantages of using ultrasound to perform the transesterification of vegetable oils with methanol or ethanol to give biodiesel are shorter reaction time, reduced catalyst requirement, lowering of reaction temperatures and so forth.

The extraction conditions for *Memecylon edule* shoots in order to achieve the highest polyphenols levels and antioxidant activities were based on the suitable solvent and duration of sonication (Falleh et al., 2012). Zou et al. (2011) showed the optimization of ultrasound-assisted extraction of anthocyanins from Mulberry. Wang et al. (2011) made use of ultrasound-assisted extraction to extract three dibenzylbutyrolactone lignans, including tracheloside, hemislienoside and arctiin from *Hemistepta lyrata.*

12.6 CLEANING

Ultrasonic cleaning is an oldest industrial application of power ultrasound. Here, high-frequency sound waves (above the upper range of human hearing or about 18 kHz) are utilized to remove a variety of contaminants from materials immersed in aqueous media. These contaminants can be dirt, oil, grease, buffing/polishing compounds, mold release agents and so forth. Materials that can be cleaned by this technique include metals, glass, ceramics, and so forth. It is powerful enough to remove contaminants without damaging the substrate. It provides excellent penetration and cleaning in the smallest cracks, chasm, cranny and between tightly spaced parts in a cleaning tank.

Micro-sized bubbles originate and grow in process of cavitation due to alternating positive and negative pressure waves generated in the solution. These bubbles subjected to alternating pressure waves continue to grow till they attain a resonant size., There is a tremendous amount of energy stored inside the bubble just before its implosion.

Temperature and pressures inside a cavitating bubble are extremely high and, as a result, implosion occurs near a hard surface, which changes the bubble in the form of a jet about one-tenth of the bubble size, and it travels at very high speed up to 400 $km \cdot h^{-1}$ towards the hard surface. With the combination of these factors, that is pressure, temperature and velocity, the jet detaches any contaminants from the substrate. Due to this inherently small size of the jet and relatively large energy, ultrasonic cleaning process has the capability to reach into small cranny cracks and chasm and remove any entrapped material (dirt) very efficiently.

Ultrasonic transducers used in the cleaning industry have a frequency range of 20–80 kHz. Low-frequency transducers create larger bubbles with more energy and tend to form larger dents, whereas cleaners with higher frequency will form much smaller dents.

A mechanical vibrating device is required so that positive and negative pressure waves are produced in the aqueous medium. Manufacturers of ultrasonic instruments used a diaphragm attached to high-frequency transducers. These transducers are vibrating at their resonant frequency because of a high-frequency electronic generator source and induced amplified vibration of the diaphragm, which is the source of positive and negative pressure waves propagating through the solution transmitted through water. Oscillation of stable cavitational bubbles and the resultant micro-streaming also contribute to cleaning as these pressure waves create the cavitation processes.

Ultrasonic transducers are mainly of two types:

- Piezoelectric
- Magnctostrictive

Both these types of transducers have same function, but they are dramatically different in their performance. A ceramic (usually lead zirconate) crystal is sandwiched between two strips of tin and it creates a displacement in the crystal on applying the voltage across the strips. This is called piezoelectric effect. When such transducers are mounted to a diaphragm on surface such as walls or bottom of a tank, the displacement in the crystal results in a movement of the diaphragm. Thus, a pressure wave is generated and it is transmitted through the medium. However, piezoelectric transducers have several disadvantages as far as industrial cleaning is concerned.

On the other hand, magnetostrictive transducers are established for their ruggedness and durability and, as such, these can be used in industrial applications. Zero-space magnetostrictive transducers consist of nickel laminations which are attached tightly together with an electrical coil placed over the nickel stack. A magnetic field is created when current flows through this coil. This is similar to the deformation of a piezoelectric crystal, when a voltage is applied. When an alternating current is passed through the magnetostrictive coil, the nickel stack vibrates at the frequency of the current.

There are some advantages of zero-space magnetostrictive transducers and these are:

- They are silver brazed for permanent bonding with no damping effect.
- A consistent performance is provided throughout the life of this unit and there is no degradation of transducers.
- The high mass of this type of transducers results in high energy, but less load sensitivity.
- Its thick diaphragm prevents its erosion and wear through.

It is very important to select a solution (or solvent) to be used in ultrasonic cleaning. Water is considered as an excellent solvent as it is nontoxic, nonflammable and also environmental friendly. Ultrasonic cleaning is better used for relatively hard materials such as metals, ceramics, glass, plastics, and so forth as these materials reflect the sound rather than absorbing it. Thus various things such as ball bearings, carburetor parts and vessels having complex internal cavities can be effectively cleaned. The layers of maximum cavitational intensity repeat approximately every 1.3–3.8 in. producing quite uniform cleaning throughout. However, the parts may be moved during exposure for getting more uniformity.

Power density of ultrasonic cleaners is relatively low (usually $<10\ W \cdot in.^{-2}$ of the driving area). If an attempt is made to overdrive, these may be loss of the far field effect, and it causes pronounced cavitation resulting in wear at the driving surface.

The choice of cleaner frequency is normally decided on the basis of its application. As cavitational shock intensity is higher at lower frequencies (2.5 kHz), such cleaner will have harsher cleaning ability than with a higher-frequency cleaner (40 kHz). However, such lower frequencies are likely to damage some delicate parts, and therefore, 40 kHz cleaners may be preferred for cleaning of semiconductors and in that case, cleaning is also simpler and silent.

Standard industrial cleaners typically have power ranging from one hundred to a couple of thousand watts , depending on capacities of tank. Although higher kilowatt systems with larger tank capacities are available for the past few decades, low-power and low-cost cleaners have become available in the past few years making ultrasonic cleaning readily accessible to restaurants, shops and laboratories.

At present, there seems to be no dramatic breakthroughs in the process of ultrasonic cleaning; however, tank materials and design can definitely be further improved to extend the life of ultrasound cleaners and enhance the cleaning process. Since the process of cavitational choice of optimum cleaning parameters can be tricky, behaviour depends on nature of solvents and temperature and thus, further advancement in this area is possible in years to come.

12.7 WELDING

Some of the welding techniques are discussed in detail in the following sections.

12.7.1 PLASTIC WELDING

People may come in contact daily with some or the other ultrasonically welded plastic parts. The process of plastic welding was developed in the past few years and it was quickly accepted for assembling toys and different thermoplastic appliances. A big breakthrough came with discovery of far field welding, so as to make welding of rigid thermoplastics possible and it was later extended beyond welding of plastic films known at that time.

Ultrasonic welding is a fast and clean process, as it does not require consumables. It is quite extensively used in the automobile industry for the assembly of various parts such as taillights, dashboards, heater ducts, in which plastics have almost replaced the traditional use of glass and metal.

High-frequency vibration produces large amount of heat which melts the plastic. Ultrasonically induced heat is generated precisely and selectively at the interface of the parts to be joined without indiscriminate heating of the surrounding material. Therefore, less energy is required for welding resulting in little distortion and degradation of material. Since the heat is generated within the plastic and it is not conducted in surrounding materials, such as tools, ultrasound welding can be accomplished in completely inaccessible places.

Most of the thermoplastics have the desired characteristics suitable for ultrasonic welding. This is all due to their ability to transmit and to absorb vibration, as well as their low thermal conductivity which facilitate local build-up of heat.

Heating in plastic is a function of ultrasonic stress which varies as the square of amplitude of this stress. The contact area between the parts to be joined is reduced, just to maximize the effect stress in the welded region. The parts are clamped together during welding and are kept there for a fraction of a second after ultrasonic exposure. This will allow the plastic to solidify and then they are unclamped. Normally ultrasonic welding takes less than a second for its completion.

Ultrasonic plastic welding (UPW) requires much higher power densities (about hundreds of watts per square inch), as compared to ultrasound cleaning. Plastic welding horns operate at amplitudes at about 10 times higher than cleaning for such power densities. Very sharp mechanical resonances are exhibited due to high energy storage. They must accommodate a wide variety of loads depending on the application, and mechanical loading may vary during the welding. In this case, a new class of equipment had to be developed to satisfy these requirements taking ultrasonic power technology to a new level.

Modern ultrasonic plastic welders normally operate around 20 kHz at power outputs below 1000 W. They lock automatically on the horn resonance, and hold vibrational amplitude constant for different mechanical loads. Another improvement is using mechanical amplitude transformers which facilitate matching of equipment to the load. Ultrasonic exposure is controlled by accurate electronic timers.

Progress has been made in the past few years in the size of plastic parts that can be welded ultrasonically, particularly using higher-power equipment and improvement in ultrasonic horn. Such improvements in joint design have extended the area of ultrasonic welding to more difficult plastics and shapes. The development in ultrasonic welding process has been mainly affected by ultrasonic staking, spot welding and inserting of metal parts into plastics.

Ultrasonic welding of woven and nonwoven fibres deserve special consideration. Thermoplastic textiles can be sewn ultrasonically with up to 35 % natural fibre content. The advantages include absence of thread and its colour-matching problems, execution of several stitches at the same time and numerous other variations of concurrent cut and seal operations.

UPW is a welding technique, where two phases are present:

- Vibration phase
- Consolidation phase

In the vibration phase, an ultrasonic vibration perpendicular to the joining interface is applied to heat the thermoplastic by frictional and visco-elastic heating above its melting temperature. After intermolecular diffusion, it is again cooled down to room temperature under applied pressure during consolidation phase. This process is completed within 10 s or so.

This heating process of UPW is a combination of interfacial friction heating and viscoelastic heating (Tolunay et al., 1983; Zhang et al., 2009; Levy et al., 2012). The frictional heating process for amorphous and semi-crystalline thermoplastics is through surface asperities which results in a large proportion of the heating process in the beginning, whereas above the glass transition temperature, major role is played by viscoelastic heating (Zhang et al., 2009; Benatar et al., 1989).

12.7.2 METAL WELDING

Commercial equipment for ultrasonic metal welding was available in the late 1950s. The process of metal welding, known as micro-bonding, was accepted in the semiconductor industry for welding of miniature conductors. Development in the design of equipment and the requirement for better ways of welding of high-conductivity metals has created interest in the use of ultrasonics for this purpose. Now, equipment is available to weld even thicker and larger parts, depending on the material and configuration of the part.

Interestingly, ultrasonic metal welding is a relatively cold process. Although some heating occurs, but this type of metal welding depends more on cleaning than on melting of the material. Ultrasonic shear causes abrasion of the surfaces to be joined, breaking contamination including dispersing oxides. The exposed, plasticized, metal surfaces are brought together under pressure and solid-state bonding takes place. Ultrasonic welding resembles spin welding or pressure welding with a major difference that there is no gross movement of parts or large displacement of material.

Ultrasonic metal welds require low heat and welding temperatures are below the melting temperatures of the metals and relative distortion is also low. It avoids embrittlement and the formation of high-resistance intermetallic compounds, when welding dissimilar metals.

As no role is played by electrical conductivity in this process, those applications that are difficult or almost impossible with resistance welding can also be successfully done ultrasonically. Worth mentioning examples are welding of high-conductivity metals (electric grade aluminium and copper) and also metals of different resistivities (copper and steel). Similarly, welding of parts that are quite different in heat capacities (foil to thick sections) is much difficult with heat dependent methods but it can be easily achieved using ultrasound. Another important use is in sealing of liquid filled containers and packing heat-damageable contents and explosives.

Ultrasonic metal welding has the desirable characteristics of UPW but it also has more competition from other metal joining methods. Apart from microbonding, it is mainly found useful in electric and electronic industries particularly in the assembly of electric motors, transformers, switches and relays. Replacement of copper by aluminium is associated by ultrasonic welding, as there are not many reliable alternate methods for joining aluminium conductors.

Far field welding is not practical as metal welding requires shear ultrasonic motion parallel to the plane of the weld. Therefore, the method is more suitable for producing spot welds and line welds. Here, continuous seaming of metal foil and sheet is also possible.

Ultrasonic power densities at the contact with the welding tip are relatively very high and it is in the order of 10,000 $W \cdot in^{-2}$. High power density causes tip wear and makes ultrasonic welding almost impractical for welding of hard metals, which due to requirement for the mutual abrading ability must not be too different in hardness and therefore, compatibility of materials also pose a limitation on its use.

Ultrasound finds great use in different industrial processes. Various industries such as textile, dyeing, leather, food, pharma, and so forth are employing this technique on a wide scale. Different industrial processes such as drying, freezing, extraction, and so forth, also use sound waves. The basic reason for the wide range use of ultrasound in industrial processes is because of their ability to enhance rate of reaction, product yield, easiness and feasibility of operation. The use of sonochemistry has proved that it has a great potential in various industries.

REFERENCES

Agrawal, P. B.; Nierstrasz, V. A.; Warmoeskerken, M. M. C. G. *Biocatal. Biotransform.* **2010,** *28*(5–6), 320–328.

Akalin, M.; Merdan, N.; Kocak, D.; Usta, I. *Ultrasonics* **2004,** *42*, 161–164.

Alexander, P.; Meek, G. A. *Melliand Textilber.* **1953a,** *34*, 57–59.

Alexander, P.; Meek, G. A. *Melliand Textilber.* **1953b,** *34*, 133–139.

Alexander, P.; Meek, G. A. *Melliand Textilber.* **1953c,** *34*, 214–216.

Antonescu, I.; Grunichevici, E. *Ind. Usoara* **1979,** *30*(6), 254.

Babar, A. A.; Peerzada, M. H.; Jhatial, A. K.; Bughio, N. U. *Ultrason. Sonochem.* **2017,** *34*, 993–999.

Ben Ticha, M.; Haddar, W.; Meksi, N.; Guesmi, A.; Mhenni, M. F. *Carbohydr. Polym.* **2016,** *154*, 287–295.

Benatar A.; Eswaran, R. V.; Nayar, S. K. *Polym. Eng. Sci.* **1989,** *29*(23), 1689–1698.

Cavallari, C.; Albertini, B.; Rodriguez, L.; Rabasco, A. M.; Fini, A. *J. Controlled Release* **2005,** *102*, 39–47.

Chepurnoi, M. N. *Pishch. Tekhnol.* **1990,** *4*, 68–70.

Choi, J.; Kim T.-H.; Kim, H.-Y.; Kim, W. *Ultrason. Sonochem.* **2016,** *29*, 563–567.

Dhumal, R. S.; Biradar, S. V.; Paradkar, A. R.; York, P. *Int. J. Pharm.* **2009,** *368*(1–2), 129–137.

E. I. Du Pont De Nemours & Company; Strohmair, A. J. U. S. Patent. 4,431,684, 1984.

El-Shishtawy, R. M.; Kamel, M. M.; Hanna, H. L.; Ahmed, N. S. E. *Polym. Int.* **2003,** *52*(3), 381–388.

Falleh, H.; Ksouri, R.; Lucchessi, M.-E.; Abdelly, C.; Magne, C. *Trop. J. Pharm. Res.* **2012,** *11*(2), 243–249.

Ferrero, F.; Periolatto, M. *Ultrason. Sonochem.* **2012,** *19*(3), 601–606.

Fini, A.; Fernàndez-Hervàs, M. J.; Holgado, M. A.; Rodriguez, L.; Cavallari, C.; Passerini, N.; Caputo, O. *J. Pharm. Sci.* **1997,** *86*, 1303–1309.

Fini, A.; Holgado, M. A.; Rodriguez, L.; Cavallari, C. *J. Pharm. Sci.* **2002a,** *91*, 1880–1890.

Fini, A.; Rodriguez, L.; Cavallari, C.; Albertini, B.; Passerini, N. *Int. J. Pharm.* **2002b,** *247*, 11–22.

Fini, A.; Cavallari, C.; Ospitali, F. *Pharmaceutics* **2009,** *1*, 3–19.

Gentile, P.; Nandagiri, V. K.; Daly, J.; Chiono, V.; Mattu, C.; Tonda-Turo, C. et al. *Mater. Sci. Eng.: C* **2016,** *59*, 249–257.

Gotoh, K.; Harayama, K. *Ultrason. Sonochem.* **2013,** *20*(2), 747–753.

Guesmi, A.; Ladhari, N.; Sakli, F. *Ultrason. Sonochem.* **2013,** *20*(1), 571–579.

Gupta P. S.; Sharma V.; Pathak K. *Expert opin. Drug Delivery* **2013,** *10*(17–32), 1744–7593.

Hao, L.; Wang, R.; Liu, J.; Liu, R. *Carbohydr. Polym.* **2012,** *90*(4), 1420–1427.

Helmdach, L.; Feth, M. P.; Ulrich, J. *Org. Process Res. Dev.* **2015,** *19*(1), 110–121.

Kadam, V. V.; Goud, V.; Shakyawar, D. B. *Indian J. Fibre Text. Res.* **2013,** *38*, 410–414.

Kamel, M. M.; El Zawahry, Ahmed, N. S.; Abdelghaffar, F. *Ultrason. Sonochem.* **2009,** *16*(2), 243–249.

Kamel, M. M.; Helmy, H. M.; Mashaly, H. M.; Kafafy, H. H. *Ultrason. Sonochem. 2010, 17*(1), 92–97.

Khan, M. A.; Akhtar, N.; Sharma, V.; Pathak, K. *Pharmaceutics* **2015,** *7*, 43–63.

Khatri, Z.; Memon, M. H.; Khatri, A.; Tanwari, A. *Ultrason. Sonochem.* **2011,** *18*(6), 1301–1307.
Kougloulos, E.; Smales, I.; Verrier, H. M. *Off. J. Am. Assoc. Pharm. Sci.* **2011,** *12*(1), 287–294.
Kubidus, Y. Y. *Tekstil Prom* **1962,** *22*(6), 69.
Kunert, K. A. *J. Polym. Sci.: Polym. Lett. Ed.* **1979,** *17*, 363–367.
Larik, S. A.; Khatri, A.; Ali, S.; Kim, S. H. *Ultrason. Sonochem.* **2015,** *24*, 178–183.
Last, A. J.; McAndless, J. N. U.S. Patent 4,302,485, 1981.
Levina, M.; Rubinstein, M. H. *J. Pharm. Sci.* **2000,** *89*(6), 705–723.
Levina, M.; Rubinstein, M. H. *Drug Dev. Ind. Pharm.* **2002,** *28*(5), 495–514.
Levina, M.; Rubinstein, M. H.; Rajabi-Siahboomi, A. R. *Pharm. Res.* **2000,** *17*(3), 257–265.
Maeda, Y.; Vinatoru, M.; Stavarache, C. E.; Iwai, K.; Oshige, H. European Patent Application No 03,023,081.7, 2003
Maeda, Y.; Vinatoru, M.; Stavarache, C. E.; Iwai, K.; Oshige, H. US Patent Application No US2004/0,159,537, 2004.
Makino, K.; Mossoba, M.; Riesz, P. *J. Phys. Chem.* **1983,** *87*(8), 1369–1377.
Manish, M.; Harshal, J.; Anant, P. *Eur. J. Pharm. Sci.* **2005,** *25*(1), 41–48.
Mantry, S.; Reddy, K. V. N.; Sriram, N.; Sahoo, C. S. *Indo Am. J. Pharm. Res.* **2013,** *3*(5), 4031–4041.
Merdan, N.; Akalin, M.; Kocak, D.; Usta, I. *Ultrasonics* **2004,** *42*(1–9), 165–168.
Mistik, S. I.; Yukseloglu, S. M. Hydrogen Peroxide Bleaching of Cotton in Ultrasonic Energy. *Ultrasonics* **2005,** *43*(10), 811–814.
Motta, G. International Patent, WO 94/14421, 1994.
Paradkar, A.; Dhumal, R. Ultrasound-Assisted Particle Engineering. In *Handbook on Applications of Ultrasound: Sonochemistry for Sustainability;* Mudhoo, A, Ed.; CRC Press, 2011.
Park, Y. S.; Davis, A. E.; Park, K. M.; Lin, C.; Than, K. D.; Lee, K. et al. *Am. Assoc. Pharm. Sci.* **2013,** *15*(2), 367–376.
Peila, R.; Actis Grande, G.; Giansetti, M.; Rehman, S.; Sicardi, S.; Rovero, G. *Ultrason. Sonochem.* **2015,** *23*, 324–332.
Petkova, P. Francesko, A.; Perelshtein, I.; Gedanken, A.; Tzanov. T. *Ultrason. Sonochem.* **2016,** *29*, 244–250.
Povey, M. J. W.; Mason, T. J. Eds. *Ultrasound in Food Processing;* Blackie Academic: London, 1998.
Qiu, T.-Q. *Int. Sugar J.* **1994,** *96*, 523–526.
Qiu, T.-Q.; Hu, A.-j. *Appl. Acoust.* **2002,** *21*(2), 8–11.
Qiu, T.-Q.; Yao, C.-C. *Sugar Cane Cane Sugar* **1999,** *4*, 29–34
Rodriguez, L.; Cini, M.; Cavallari, C.; Passerini, N.; Saettone, M. F.; Fini, A.; Caputo, O.*Int. J. Pharm.* **1998,** *170*, 201–208.
Rositza, B.; Dancho, Y.; Lubov, Y. J. *Biomater. Nanobiotechnol.* **2011,** *2*, 65–70.
Ruecroft, G.; Collier, A. *Org. Process Res. Dev.* **2005,** *9*(6), 923–932.
Sancin, P.; Caputo, O.; Cavallari, C.; Passerini, N.; Rodriguez, L.; Cini, M.; Fini, A. *Eur. J. Pharm. Sci.* **1999,** *7*, 207–213.
Scinkivich, N. N.; Pugachevski, G. F.; Fredrman, V. M. *Izv. VysshUcheben, Zaved Tekhmol Legb Prom-Sti.***1975a,** *4*, 104.
Scinkivich, N. N.; Pugachevski, G. F.; Fredrman, V. M. *Tekstil Prom. (Moscow)* **1975b,** *11*, 65.

Senapati, N. *Ultrasound in Chemical Processing;* JAI Press: London, 1991.

Shimizu, Y.; Yamamoto, R.; Simizu, H. *Text. Res. J.* **1989,** *59*, 684–687.

Simkovich, N. N.; Yastrebinski, A. A. *Tekh Tekstil Prom.* **1975,** *106*(3), 15.

Simpson, W. S.; Crawshaw, G. Eds. *Wool: Science and Technology*; Published in Association with the Textile Institute, CRC Press: Cambridge, 2002.

Soltani, R. D. C.; Jorfi, S.; Ramezani, H.; Purfadakari, S. *Ultrason. Sonochem.* **2015,** *28*, 69–78.

Son, Y.; Nam, S.; Ashokkumar, M. Khim, J. *Ultrason. Sonochem.* **2012,** *19*(3), 395–398.

Stavarache, C.; Vinatoru, M.; Yim, B.; Maeda, Y. *Chem. Lett.* **2003,** *32*(8), 716–717.

Stavarache, C.; Vinatoru, M.; Nishimura, R.; Maeda, Y. *Ultrason. Sonochem.* **2004,** *12*, 367–372.

Sun, D.; Guo, Q.; Liu, X. *Ultrasonics* **2010,** *50*(4–5), 441–446.

Tissera, N. D.; Wijesena, R. N.; de Silva, K. M. *Ultrason. Sonochem.* **2016,** *29*, 270–278.

Toma, M.; Vinatoru, M.; Larisa, Paniwnyk, Mason, T. J. *Ultrason. Sonochem.* **2001,** *8*(2), 137–142.

Tripathy, S.; Singh, B. K.; Patel, D. K.; Palei, N. N. *J. Appl. Pharm. Sci.* **2016,** *6*(09), 041–047.

Vinatoru, M. *Ultrason. Sonochem.* **2001,** *8*, 303–313.

Vinatoru, M.; Toma, M.; Radu, O.; Filip, P. I.; Lazurca, D.; Mason, T. J. *Ultrason. Sonochem.* **1997,** *4*, 135–138.

Vouters, M.; Rumeau, P.; Tierce, P.; Costes, S. *Ultrason. Sonochem.* **2004,** *11*(1), 33–38.

Wang, W.; Wu, X.; Han, Y.; Zhang, Y.; Sun, T.; Dong, F. *J. Appl. Pharm. Sci.* **2011,** *1*(9), 24–28.

Yao, C.-C.; Qiu, T.-Q. *Int. Sugar J.* **1999,** *101*, 602–605.

Yasui, K.; Tuziuti, T.; Iida, Y. *Ultrason. Sonochem.* **2005,** *12*, 43–51.

Zhang, Z.; Wang, X.; Luo, Y.; Zhang, Z.; Wang, L. *J. Thermoplast. Compos. Mater.* **2009,** *23*(5), 647–664.

Zou, T.-B.; Wang, M.; Gan, R.-Y.; Ling, W. *Int. J. Mol. Sci.* **2011,** *12*(5), 3006–3017.

CHAPTER 13

SONOCHEMISTRY: A VERSATILE APPROACH

RAKSHIT AMETA[1]

[1]*Department of Chemistry, J.R.N. Rajasthan Vidyapeeth (Deemed to be University), Udaipur, India E-mail: rakshit_ameta@yahoo.in*

Organisms like dolphins and bats use sound waves to develop an unseen picture of things around them to move around safely. This depends on the reflection of ultrasound wave and its detection by these animals. This technique is called echolocation.

Individual particles are held together by some attraction forces, which may be of physical and chemical nature. These forces include surface tension in liquids, van der Waal forces, and so forth. These attraction forces must be increased so as to deagglomerate and disperse fine particles in liquid media. Such a process is used in different products such as shampoo, ink, beverages, paints, varnish and other polishing materials. Ultrasonic cavitation generates higher shear forces that have enough energy to break down some particle agglomerates into single dispersed particles. High-intensity ultrasound is found effective for such dispersion and deagglomerization of powders in liquids.

Emulsions are dispersion of two or more immiscible liquids. Here also, high-intensity ultrasound power is required to disperse a liquid phase (dispersive phase) in the form of very small droplets into another phase (continuous phase). Emulsification is a very important process as it is used in the preparation of wide range of intermediates as well as final products such as cosmetic, skin lotion, varnishes, paints, lubricants, fuels, pharmaceuticals, creams, etc., which are in the form of emulsion either completely or partially. The cavitation bubbles generate intensive shock waves on implosion in the liquid in dispersing zone. It forms liquid jets of high liquid velocity. As a result, microemulsion is obtained with an average droplet size less than 1 μ.

Fibrous cellulosic materials can be broken into fine particles by ultrasonic exposure by processes such as milling and grinding that lead to breakage of the wall of the cell structure and, therefore, the intracellular materials such as starch, sugar are released into the liquids. With ultrasonic treatment, much more intracellular material will be available to the enzymes that will convert starch into sugars. Ultrasonic treatment can also be used for digestion, fermentation and other conversion of organic matter, where surface area exposed to the enzymes is increased resulting in enhancement of the rate of these processes accompanied by higher yields. In this way, the production of ethanol from biomass can be increased. The process of extraction of proteins by enzymes and other potential bioactive compounds within the body of plants and seeds by a solvent can be dramatically improved. Ultrasound can also help us in wet milling and micro grinding of bigger particles, where it has many additional advantages as compared to other traditional milling techniques, such as ball mills, disc mills, bead mills and jet mills. Processing volume can be reduced by ultrasonication, which allows the processing of high concentration. It is an efficient technique for preparing micro-sized and nanosized materials such as metal oxide, calcium carbonate and different ceramics.

Conversion of oil into biodiesel through transesterification is also increased in terms of rate and yield in the presence of ultrasound. Animal fats can also be used for such conversion. In the ultrasonic method production, batch processing is changed to continuous processing making the process cost-effective. Vegetable oils and animal fats are converted to biodiesel through base-catalysed transesterification of fatty acids with methanol or ethanol providing respective methyl and ethyl esters. The time taken in the processing and separation can be reduced significantly along with the higher yield of biodiesel (around 99%) by ultrasonic radiations.

Degassing is also one of the important applications of ultrasonics, where it removes smaller suspended gas bubbles from the liquid and the level of dissolved gas is brought below its natural equilibrium level. It is also used for the detection of leaks in bottles and cans as immediate release of CO_2 confirms leakage in container filled with carbonated beverages. Cavitation generated by ultrasonic power is used as an eco-friendly technique to remove dust, soap, oil, grease and so forth. It can also be used for cleaning of continuous materials such as tubes, wires, cables, tap and others.

Cheese, milk, ice cream, and so on can be made healthier for human consumption by using ultrasound. Small bubbles of water can be injected into the fat so as to reduce the fat content making it better in quality along with keeping intact the properties of the products and their external appearances. The processing time is also reduced from minutes to seconds. Such a method is also used for injecting additives such as menthol, fruits, berries, etc.

If ultrasound is applied in crystallizer in nucleation stage, it provides much clear and higher-quality crystal. In addition, the crystals will be of uniform size. Most probably, convection caused by microjets plays an important role here.

Grignard reagents are quite useful in organic synthesis. During conventional preparation of this reagent, one has to use the pure form of magnesium as well as distilled dry ether; whereas no prior treatment of magnesium is required and commercial sample of ether can also be used, if the reaction is carried out in the presence of ultrasound. A strong base such as lithium diisopropylamide can be obtained using lithium metal by ultrasonic exposure. This type of preparation is not possible otherwise.

If a reaction is to be carried out in two phases and one reagent is soluble in one phase while the other is soluble in other phase. In such cases, phase transfer catalysts are required. These reactions can be carried out even in the absence of phase transfer catalyst with high yield and reduced time, provided these reactions are carried out in the presence of ultrasonic radiations.

Ultrasound radiation creates cavitations or microjets depending on the environment and these not only help in increasing the rate of reaction and yield of the products, but also take care of the environment. In the presence of ultrasound, a new route can be also developed apart from the traditional pathway. It adds a new dimension in the field of sonochemistry known as sonochemical switching. Sonochemistry can be utilized in a number of diverse fields. It is developing as a green chemical technology and is likely to occupy a prominent position as compared to other eco-friendly techniques in the years to come.

INDEX

M

N

O

R

S